Handbook on Sourdough Biotechnology

Marco Gobbetti • Michael Gänzle
Editors

Handbook on Sourdough Biotechnology

Springer

Editors
Marco Gobbetti
Department of Soil, Plant and Food Science
University of Bari Aldo Moro
Bari, Italy

Michael Gänzle
Department of Agricultural, Food, and Nutritional Science
University of Alberta
Edmonton, Canada

ISBN 978-1-4614-5424-3 ISBN 978-1-4614-5425-0 (eBook)
DOI 10.1007/978-1-4614-5425-0
Springer New York Heidelberg Dordrecht London

Library of Congress Control Number: 2012951618

Printed on acid-free paper

Springer is part of Springer Science+Business Media (www.springer.com)

Contents

Chapter 1
History and Social Aspects of Sourdough

Stefan Cappelle, Lacaze Guylaine, M. Gänzle, and M. Gobbetti

1.1 Sourdough: The Ferment of Life

The history of sourdough and related baked goods follows the entire arc of the development of human civilization, from the beginning of agriculture to the present. Sourdough bread and other sourdough baked goods made from cereals are examples of foods that summarize different types of knowledge, from agricultural practices and technological processes through to cultural heritage. Bread is closely linked to human subsistence and intimately connected to tradition, the practices of civil society and religion. Christian prayer says "Give us this day our daily bread" and the Gospels report that Jesus, breaking bread at the Last Supper, gave it to the Apostles to eat, saying, "This is my body given as a sacrifice for you". Language also retains expressions that recall the close bond between life and bread: "to earn his bread" and "remove bread from his mouth" are just some of the most common idioms, not to mention the etymology of words in current use: "companion" is derived from *cum panis*, which means someone with whom you share your bread; "lord", is derived from the Old English vocabulary *hlaford*, which translates as guardian of the bread [1]. The symbolic assimilation between bread and life is not just a template that has its heritage in the collective unconscious, but it is probably a precipitate of the history of culture and traditions. Throughout development of the human civilization, (sourdough) bread was preferred over unleavened cereal products, supporting

S. Cappelle (✉) • L. Guylaine
Puratos Group, Industrialaan 25, Groot-Bijgaarden, Belgium
e-mail: SCapelle@puratos.com

M. Gänzle
Department of Agricultural, Food and Nutritional Science,
University of Alberta, Edmonton, Canada

M. Gobbetti
Department of Soil, Plant and Food Science, University of Bari Aldo Moro, Bari, Italy

M. Gobbetti and M. Gänzle (eds.), *Handbook on Sourdough Biotechnology*,
DOI 10.1007/978-1-4614-5425-0_1, © Springer Science+Business Media New York 2013

the hypothesis of a precise symbolism between the idea of elaborate and stylish, and that of sourdough. Fermentation and leavening makes bread something different from the raw cereals, i.e. an artifact, in the sense of "made art". Besides symbolism, sourdough bread has acquired a central social position over time. Bread, and especially sourdough bread, has become central in the diet of peasant societies. This suggests that the rural population empirically perceived sensory and nutritional transformations, which are also implemented through sourdough fermentation. In other words, the eating of bread, and especially of sourdough bread, was often a choice of civilization.

The oldest leavened and acidified bread is over 5,000 years old and was discovered in an excavation in Switzerland [2]. The first documented production and consumption of sourdough bread can be traced back to the second millennium B.C. [3]. Egyptians discovered that a mixture of flour and water, left for a bit of time to ferment, increased in volume and, after baking along with other fresh dough, it produced soft and light breads. Much later, microscopic observations of yeast as well as measurements of the acidity of bread from early Egypt demonstrate that the fermentation of bread dough involved yeasts and lactic acid bacteria – the leavening of dough with sourdough had been discovered [4]. Eventually, the environmental contamination of dough was deliberately carried out by starting the fermentation with material from the previous fermentation process. Egyptians also made use of the foam of beer for bread making. At the same time, Egyptians also selected the best variety of wheat flour, adopted innovative tools for making bread, and used high-temperature ovens. The Jewish people learned the art of baking in Egypt. As the Bible says, the Jews fleeing Egypt took with them unleavened dough.

In Greece, bread was a food solely for consumption in wealthy homes. Its preparation was reserved for women. Only in a later period, does the literature mention evidence of bakers, perhaps meeting in corporations, which prepared the bread for retail sale. The use of sourdough was adopted from Egypt about 800 B.C. [4]. Greek gastronomy had over 70 varieties of breads, including sweet and savoury types, those made with grains, and different preparation processes. The Greeks used to make votive offerings with flour, cereal grains or toasted breads and cakes mixed with oil and wine. For instance, during the rites dedicated to Dionysus, the god of fertility, but also of euphoria and unbridled passion, the priestesses offered large loaves of bread. The step from the use of sacrificial bread to the use of curative bread was quick. Patients, who visited temples dedicated to Asclepius (the god of medicine and healing), left breads, and, upon leaving the holy place, received a part of the breads back imbued with the healing power attributed to the god [5, 6].

The use of sourdough is also part of the history of North America. The use of sourdough as a leavening agent was essential whenever pioneers or gold prospectors left behind the infrastructure that would provide alternative means of dough leavening. Examples include the Oregon Trail of 1848, the California gold rush of 1849, and the Klondike gold rush in the Yukon Territories, Canada, in 1898. During the 1849 gold rush, San Francisco was invaded by tens of thousands of men and women in the grip of gold fever. Following the gold rush, sourdough bread remained an element that distinguishes the local tradition until today. Some bakeries in San Francisco claim to use sourdough that has been propagated for over 150 years. The predominant yeast in San Francisco sourdoughs is not brewer's yeast but *Kazachstania exigua* (formerly *Saccharomyces exiguus*), which is tolerant to more acidic environments. *Lactobacillus sanfranciscensis* (formerly *Lactobacillus brevis* subsp. *lindneri* and *sanfrancisco*) was first described as a new species in San Francisco sourdough [7]. The use of sourdough during the Klondike gold rush in 1898 resulted in the use of "sourdough" to designate inhabitants of Alaska and the Yukon Territories and is even in use today. The Yukon definition of sourdough is "someone who has seen the Yukon River freeze and thaw", i.e. a long-term resident of the area.

From antiquity to most recent times, the mystery of leavening has also been unveiled from a scientific point of view. The definitive explanation of microbial leavening was given in 1857 by Louis Pasteur. The scientific research also verified an assumption that the Greeks had already advanced: sourdough bread has greater nutritional value. Pliny the elder wrote that it gave strength to the body. The history and social significance of the use of sourdough is further described below for countries such as France, Italy and Germany where this traditional biotechnology is widely used, and where its use is well documented.

1.2 History and Social Aspects of Sourdough in France

The history of sourdough usage in France was linked to socio-cultural and socio-economic factors. There is little information about sourdough usage and bakery industries (it seems to be more appropriated than baking), in general, in France before the eighteenth century. It seems as if sourdough bread was introduced in Gaul by the Greeks living in Marseille in the fourth century B.C. In 200 B.C., the Gauls removed water from the bread recipe and replaced it with *cervoise*, a drink based on fermented cereal comparable to beer. They noticed that the cloudier the *cervoise*, the more the dough leavened. Thus, they started to use the foam of *cervoise* to leaven the bread dough. The bread obtained was particularly light.

During the Middle Ages (400–1400 A.D.), bread making did not progress much and remained a family activity. In the cities, the profession of the baker appeared. The history of bread making in France was mainly linked to Parisian bakers because of the geographic localization of Paris. The regions with the biggest wheat production were near Paris, and Paris had major importance in terms of inhabitants. In that period, the production of bread was exclusively carried out using sourdough fermentation, the only method known at that time. Furthermore, the use of sourdough, thanks to its acidity, permitted baking without salt, an expensive and taxed (*Gabelle*) raw material, and allowed one to produce breads appropriate for eating habits in the Middle Ages [8].

The seventeenth century marked a turning point in the history of French bakery. Until then, sourdough was used alone to ensure fermentation of the dough even if in some French regions wine, vinegar or rennet was added. Toward 1600 A.D., French bakers rediscovered the use of brewer's yeast for bread making. The yeast came from Picardie and Flanders in winter and from Paris breweries in summer. The breads obtained with this technique were named *pain mollet* because of the texture of the dough, which was softer than the bread produced up to that point (*pain brie*). Two French queens, Catherine de Medicis (Henri II's wife) and Marie de Medicis (Henri IV's wife) contributed to the success and development of these yeast-fermented breads. In 1666, the use of brewer's yeast was authorized for bread making but, after a great deal of debate, in 1668, the use of brewers' yeast was prohibited. Following the request of Louis XIV, the Faculty of Medicine of the Paris University studied the consequences of yeast usage on public health. According to the doctors, yeast was harmful to human health, because of its bitterness, coming from barley and rotting water. Despite this negative conclusion by the Faculty, Parliament, in its decision of 21st March 1670, authorized the use of brewer's yeast for bread making in combination with sourdough. Besides the apparition of yeast in bread making, during that period, eating habits evolved towards less acidic foods. Thus, back-slopping techniques were adapted in order to reduce bread acidity [9].

The seventeenth century was also a period of development of the French philosophic and encyclopaedic mind and, fortunately, bread making did not escape this movement. Two books detail the art of bread making and provide information on bread-making techniques and knowledge of that period: "L'Art de la Boulangerie" [10] and "Le Parfait Boulanger" [11]. We have already learned that sourdough was obtained from a part of the leavened dough prepared on the day in question. The volume of this dough piece is progressively increased through addition of flour and water (back slopping) to prepare a sourdough that is ready to be used to ferment the dough. The original piece of dough, called *levain-chef*, must not be too old or too sour. The weight of the *levain-chef* is doubled or tripled by addition of water and flour leading to the *levain de première*. After 6 or 7 hours of fermentation, water and flour are added to give the *levain de seconde*, which is fermented for 4 or 5 hours. Again, water and flour are added. The dough obtained is called *levain tout point* and after 1 or 2 hours of fermentation is added to the bread dough. This technique called *travail sur 3 levains* was recommended by Parmentier [11], who imputed the bad quality of Anjou bread to bread making based only on one sourdough. Bread making based on two or three sourdoughs was predominantly used in that period. In addition, it was understood that outside Paris, bread was mainly produced at home by women. It is interesting to note that Malouin had already made the distinction between sourdough and artificial sourdough in 1779 [10]. Artificial sourdough refers to sourdough obtained from a dough that may contain yeast. This distinction between sourdough and artificial sourdoughs remained in the nineteenth century.

Until 1840, the yeast was always used in association with sourdough to initiate fermentation. On this date, an Austrian baker introduced a bread-making process in France based on yeast fermentation alone. This technique was called *poolish*. The bread obtained, called *pain viennois*, had much success but use of this method remained limited. In the middle of the nineteenth century, bread making based on three sourdoughs progressively disappeared and was replaced by bread making based on two sourdoughs. Indeed, the back slopping, necessary to maintain the fermentative activity of sourdoughs, imposed a hard working rhythm on the bakers. In 1872, the opening of the first factory for the production of yeast from grain fermentation in France by

Fould-Springer facilitated the development of bread making based on yeast to the detriment of sourdough bread making. This yeast was more active, more constant, with a nice flavour and most of all had a longer shelf life than brewer's yeast. As a consequence, from 1885, bread making based on polish fermentation was becoming more wide spread. Sourdough bread was, from that time on, called French bread.

In 1910, a bill that prohibited night work and, in 1920, the reduction of working hours, necessitated modification within fermentation processes. Sourdough bread making regressed to a greater and greater extent in the cities when bread making based on three sourdoughs totally disappeared even though, in 1914, the first *fermentôlevain* appeared. After the First World War, the use of yeast was extended from Paris to the provinces. Indeed, yeast that was produced on molasses from 1922 had a better shelf life and was thus easier to distribute over long distances. However, homemade loaves were still produced, even though they no longer existed in the cities, in the country until 1930 in the form of the *levain chef*, kept in stone jugs, and passed on from one family to another. The return of war in 1939 led to a further reduction in the use of homemade sourdough bread. In 1964, Raymond Calvel [12] wrote that "sourdough bread making does not exist anymore". Indeed, baker's yeast was systematically added to promote dough leavening, which permitted one to obtain lighter breads. In addition, the use of baker's yeast permitted one to better manage bread quality and to reduce quality variations. Two sourdough bread-making methods remained in this period. The first was a method based on two sourdoughs, which was mainly used in West and South Loire, and the second, more commonly used, method was based on one sourdough with a high level of baker's yeast. Between 1957 and 1960, the sensory qualities of bread decreased as a consequence of cost reduction. Fermentation time was reduced to a minimum. Sourdough bread was no longer produced. It was only during the 1980s that sourdough bread making gained popularity again thanks to consumer requests for authentic and tasty breads. Since 1990, the availability of starter cultures facilitated the re-introduction of sourdough in bread-making processes. Indeed, these starters permit one to obtain a *levain tout-point* with a single step and simplify the bread-making process. A regulation issued on 13th September 1993 [13] defined sourdough and sourdough bread. According to Article 4, sourdough is "dough made from wheat or rye, or just one of these, with water added and salt (optional), and which undergoes a naturally acidifying fermentation, whose purpose is to ensure that the dough will rise. The sourdough contains acidifying microbiota made up primarily of lactic bacteria and yeasts. Adding baker's yeast (*Saccharomyces cerevisiae*) is allowed when the dough reaches its last phase of kneading, to a maximum amount of 0.2% relative to the weight of flour used up to this point". This definition allowed one to dehydrate sourdough with the flora remaining active (amounts of bacteria and yeast are indicated). Sourdough can also be obtained by addition of starter to flour and water. Article 3 of the same regulation declares that "Breads sold under the category of *pain au levain* must be made from a starter as defined by Article 4, just have a potential maximum pH of 4.3 and an acetic acid content of at least 900 ppm". The *syndicat national des fabricants de produits intermédiaires pour boulangerie, patisserie et biscuiterie* is working on a new definition of sourdough in order to be closer to the reality of sourdough bread.

1.3 History and Social Aspects of Sourdough in Italy

The people in early Italy mainly cultivated barley, millet, emmer and other grains, which were used for preparation of non-fermented *focacce* and polenta. Emmer was not only used for making foods, but also performed as a vehicle of transmission in sacred rituals. At first, the Romans mainly consumed roasted or boiled cereals, seasoned with olive oil and combined with vegetables. After contact with Greek civilization, the Romans learned the process of baking and the technique of building bread ovens. Numa Pompilius sanctioned this gastronomic revolution with the introduction of celebrations dedicated to Fornace, the ancient divinity who was the guardian for proper functioning of the bread oven. The Romans gave a great boost to improvements in the techniques of kneading and baking of leavened products, and regulated manufacture and distribution by bakers (*pistores*). Cato the Elder described many varieties of bread in *De agri coltura* (160 B.C.), which by then had already spread to Rome: the *libum* or votive bread, the *placenta*, a loaf of wheat flour, barley and honey, the *erneum*, a kind of pandoro, and the *mustaceus*, bread made with grape must. In the first century A.D., Pliny the Elder [14] refers to several alternative methods of dough leavening, including sourdough that was air-dried after 3 days of fermentation, the use of dried grapes as a starter culture, and particularly the use of back-slopping of dough as the most common method to achieve dough leavening. Pliny the Elder specifically refers to sourdough in his indication that "it is an acid substance carrying out the fermentation". According to Pliny the Elder, it was generally acknowledged that "consumption of fermented bread improves health" [14].

After the triumph of classical baking, there were no novel developments in this field throughout the Middle Ages. Finding bread and flour in these centuries was difficult, because of involution of agriculture and the famine and epidemics raging at this time. The bread was divided into two categories: black bread, made from flours of different cereals, of little value and reserved for the most humble people, and white bread, made from refined flour, which was more expensive and present on the tables of the rich. A special bread, whose tradition has been preserved to this day in different national or regional varieties, is the *Brezel*, originating from the South of Germany. It has a characteristic shape of a knotted and dark red crust, which is generated by application of alkali prior to baking, and is sprinkled with coarse salt crystals. According to legend, it was invented by a German court baker in Urach in South West Germany, who, to avoid the loss of his job, was asked by the Duke of Württemberg to develop a bread that allows the sun to shine through three times. This special bread requires 2 days of working: the first to prepare the sourdough with wheat flour, and the second to mix it with water, flour, salt, lard and malt.

During the Renaissance, the practice of holding banquets in the courts of the nobles was a triumph for bread, which was presented in various forms in support of the different dishes. In Venice "*fugassa*" was prepared for the Easter holidays, a sweet bread made with sugar, eggs and butter. In Tuscany, they used to prepare "*pane impepato*", while in Milan it appeared as "*panettone*". Only towards the end of the 1600s was the use of yeast re-introduced for the distribution of luxury bread, which was salty and had added milk. In 1700, a very important innovation in the art of bread making was disseminated: the millstones in mills were replaced with a series of steel rollers. This allowed cheaper refining of flour. Also, pioneering mixers were set up. With the advances brought by the industrial revolution, bread was increasingly emerging as a staple food for workers. Rather than making the bread at home, people preferred to buy it from bakers. This change was criticized as distorting traditional values. At the same time, a health movement that originated in America started a battle against leavened bread, stating it was deleterious to health. Baker's yeast was considered a toxic element, perhaps because it was derived from beer, while the sourdough gave a bad taste to the bread, which was remediated by the addition of potash, equally harmful. When Louis Pasteur discovered that microorganisms caused the fermentation, the concern over the toxicity of biological agents was amplified. Pasteur's discovery eventually benefitted the supporters of the bread, as they stated that the use of selected yeast and related techniques was helpful in the manufacture of bread with a longer shelf life. The education of taste in different food cultures explains, however, the different relationship that has existed between the perception of the quality of bread and its level of acidity.

During the First World War, the so-called "military bread" was used in Europe, which was a loaf of 700 g weight with a hard crust. It was initially distributed to soldiers and then also passed on to the civilian population. In the post-war period, thanks to the much-discussed Battle of Wheat, strongly supported by Mussolini, the production of wheat was plentiful and the bread was brought to the table of the general population. The Second World War again resulted in an insufficient supply of bread. With the arrival of the American allies, the bread of liberation – a square white bread – became disseminated. Today, bread is regaining some importance. With a turnaround in the culinary habits of Westerners, bread made with unrefined flour, so-called black bread, is more widely consumed.

A brief mention should be made, finally, of the various breads that are currently made with modern baking practices. Typical breads, with PDO (Denomination of Protected Origin) or PGI (Protected Geographical Indication) status, are the Altamura bread, the bread of Dittaino, the Coppia Ferrasese, the bread of Genzano and the Cornetto of Matera. The manufacture of these breads is based on new processes, but still at an artisanal level [15].

1.4 History and Social Aspects of Sourdough in Germany

Acidified and leavened bread has been consistently produced in Central Europe (contemporary Austria, Germany, and Switzerland) for over 5,000 years. Leavened and acidified bread dating from 3,600 B.C. was excavated near Bern, Switzerland [2]; comparable findings of bread or acidified flat bread were made in Austria (dating from 1800 B.C.) and Quedlinburg, Germany (dating from 800 B.C.) [16]. It remains unknown whether these breads represent temporary and local traditions or a permanent and widespread production of leavened and acidified bread; however, these archaeological findings indicate that the use of sourdough for production of leavened breads developed independently in Central Europe and the Mediterranean.

Paralleling the use of leavening agents in France, sourdough was used as the sole leavening agent in Germany until the use of brewer's yeast became common in the fifteenth and sixteenth centuries [4, 16]. In many medieval monasteries, brewing and baking were carried out in the same facility to employ the heat of the baking ovens to dry the malt, and to use the spent brewer's yeast to leaven the dough. The close connection between brewing and baking is also documented in the medieval guilds. In Germany, bakers and brewers were often organized in the same guild. In many cities, bakers also enjoyed the right to brew beer [17].

Baker's yeast has been produced for use as a leavening agent in baking since the second half of the nineteenth century [4, 16, 18]. Baker's yeast was initially produced with cereal substrates, but the shortage of grains in Germany in the First World War forced the use of molasses as a substrate for baker's yeast production [4]. Although artisanal bread production relied on the use of sourdough as the main leavening agent until the twentieth century, the use of baker's yeast widely replaced sourdough as the leavening agent. Maurizio indicates in 1917 that baker's yeast was the predominant leavening agent for white wheat bread, whereas whole grain and rye products continued to be leavened with sourdough [19]. In 1954, Neuman and Pelshenke referred to baker's yeast as the main or sole leavening agent for wheat bread and as an alternative leavening agent in rye bread [20]. The industrial production of baker's yeast to achieve leavening in straight dough processes was followed by the commercial production of sourdough starter cultures in Germany from 1910.

The continued use of sourdough in Germany throughout the twentieth century particularly relates to the use of rye flour in bread production. Rye flour requires acidification to achieve optimal bread quality. Acidification inhibits amylase activity and prevents starch degradation during baking. Moreover, the solubilisation of pentosans during sourdough fermentation improves water binding and gas retention in the dough stage. Following the introduction of baker's yeast as a leavening agent, the aim of sourdough fermentation in rye baking shifted from its use as a leavening agent to its use as an acidifying agent [18]. This use of sourdough for acidification of rye dough in Germany is paralleled in other countries where rye bread has a major share of the bread market, including Sweden, Finland, the Baltic countries, and Russia. For example, the industrialization of bread production in the Soviet Union in the 1920s led to the development of fermentation equipment for the large scale and partially automated production of rye sourdough bread [21].

Chemical acidulants for the purpose of dough acidification became commercially available in the twentieth century as alternatives to sourdough fermentation. However, artisanal as well as industrial bakeries continued to use sourdough fermentation owing to the substantial difference in product quality. To differentiate between chemical and the more labour-intensive and expensive biological acidification, German food law provided a definition of sourdough as dough containing viable and metabolically active lactic acid bacteria, and defines sourdough bread as bread where acidity is exclusively derived from biological acidification. Sourdough is thus one of very few intermediates of food production that is regulated by legislation, and recognized by many consumers [4]. The consumer perception as well as the regulatory protection of the term "sourdough" in Germany and other European countries facilitated the recent renaissance of sourdough use in baking. In comparison, the term "sourdough" is not protected in the United States and the widespread labelling of chemically acidified bread as "sourdough bread" resulted in a widespread consumer perception of sourdough bread as highly acidic bread, and the use of alternative terminology to label bread produced with biological acidification.

The commercialization of dried sourdough with high titratable acidity constituted a compromise between economic bread production based on convenient use of baking improvers, and the use of sourdough fermentation for improved bread quality. These products were introduced in the 1970s [18]. Their economic importance rapidly surpassed the importance of sourdough starter cultures. Dried or stabilized sourdoughs produced for acidification provided the conceptual template for the increased use of sourdough products as baking improvers over the last 20 years. Sourdough fermentation was thus no longer confined to small-scale, artisanal fermentation to achieve dough leavening and/or acidification. Sourdough fermentation is also carried out in industrial bakeries at a large scale matching large-scale bread production, and in specialized ingredient companies for production of baking improvers specifically aimed at influencing the storage life as well as the sensory and nutritional quality of bread.

References

1. Mc Gee H (1989) Il cibo e la cucina. Scienza e cultura degli alimenti. Muzzio, Padova
2. Währen M (2000) Gesammelte Aufsätze und Studien zur Brot- und Gebäckkunde und –geschichte. In: Eiselen H (ed) Deutsches Brotmuseum Ulm, Germany
3. Adrrario C (2002) "Ta" Getreide und Brot im alten Ägypten. Deutsches Brotmus eum, Ulm
4. Brandt MJ (2005) Geschichte des Sauerteiges. In: Brandt MJ, Gänzle MG (eds) Handbuch Sauerteig, 6th edn. Behr's Verlag, Hamburg, pp 1–5
5. Moiraghi C (2002) Breve storia del pane. Lions Club Milano Ambrosiano, Milano
6. Guidotti MC (2005) L'alimentazione nell'antico Egitto, in Cibi e sapèori nel Mondo antico. Sillabe, Livorno, pp 18–24
7. Kline L, Sigihara RF (1971) Microorganisms of the San Fransisco sour dough bread process. II. Isolation and characterization of undescribed bacterial species responsible for the souring activity. Appl Microbiol 21:459–465
8. Roussel P, Chiron H (2002) Les pains français: évolution, qualité, production, Sciences et Technologie des Métiers de Bouche. Maé-Erti, Vezoul
9. Dewalque Marc, La lecture du levain au XVIIIième siècle sur http://www.boulangerie.net/forums/bnweb/dt/lecturelevain/lecturelevainacc.php, consultée le 07/06/2012 à 14h42
10. Malouin PJ (1779) L'Art de la boulangerie ou La description de toutes les méthodes de pétrir, pour fabriquer les différentes sortes de pastes et de pains, 2nd edn. Paris
11. Parmentier AA (1778) Le parfait boulanger ou Traité complet sur la fabrication & le commerce du pain. Imprimerie royale, Paris

12. Calvel R (1964) Le pain et la panification. Que sais-je ? Presses universitaires de France, Paris
13. Décret n°93-1074 du 13 septembre 1993 pris pour l'application de la loi du 1er août 1905 en ce qui concerne certaines catégories de pains
14. Pliny the Elder G (1972) Naturalis Historia XVIII, 102–104, edition of Le Biniec H; Pline L'Ancien, Historie Naturelle, Livre XVIII, Societé D'Editions le Belles Lettres, Paris
15. Buonassisi V (1981) Storia del pane e del forno. SIDALM, Milano
16. Spicher G, Stephan H (1982) Handbuch Sauerteig, 1st edn. Behr's Verlag, Hamburg
17. Krauß I (1994) Heute back' ich, morgen brau' ich. Eiselen Stiftung Ulm, Ulm
18. Brandt MJ (2007) Sourdough products for convenient use in baking. Food Microbiol 24:161–164
19. Maurizio A (1917) Die Nahrumgsmittel aus Getreide. Parey, Berlin
20. Neumann MP, Pelshenke PF (1954) Brotgetreide und Brot, 5th edn. Parey, Berlin
21. Böcker G (2006) Grundsätze von Anlagen für Sauerteig. In: Brandt MJ, Gänzle MG (eds) Handbuch sauerteig, 6th edn. Behr's Verlag, Hamburg, pp 329–352

Chapter 2
Chemistry of Cereal Grains

Peter Koehler and Herbert Wieser

2.1 Introductory Remarks

Cereals are the most important staple foods for mankind worldwide and represent the main constituent of animal feed. Most recently, cereals have been additionally used for energy production, for example by fermentation yielding biogas or bioethanol. The major cereals are wheat, corn, rice, barley, sorghum, millet, oats, and rye. They are grown on nearly 60% of the cultivated land in the world. Wheat, corn, and rice take up the greatest part of the land cultivated by cereals and produce the largest quantities of cereal grains (Table 2.1) [1]. Botanically, cereals are grasses and belong to the monocot family *Poaceae*. Wheat, rye, and barley are closely related as members of the subfamily *Pooideae* and the tribus *Triticeae*. Oats are a distant relative of the *Triticeae* within the subfamily *Pooideae*, whereas rice, corn, sorghum, and millet show separate evolutionary lines. Cultivated wheat comprises five species: the hexaploid common (bread) wheat and spelt wheat (genome AABBDD), the tetraploid durum wheat and emmer (AABB), and the diploid einkorn (AA). Triticale is a man-made hybrid of durum wheat and rye (AABBRR). Within each cereal species numerous varieties exist produced by breeding in order to optimize agronomical, technological, and nutritional properties.

The farming of all cereals is, in principle, similar. They are annual plants and consequently, one planting yields one harvest. The demands on climate, however, are different. "Warm-season" cereals (corn, rice, sorghum, millet) are grown in tropical lowlands throughout the year and in temperate climates during the frost-free season. Rice is mainly grown in flooded fields, and sorghum and millet are adapted to arid conditions. "Cool-season" cereals (wheat, rye, barley, and oats) grow best in a moderate climate. Wheat, rye, and barley can be differentiated into

P. Koehler (✉) • H. Wieser
German Research Center for Food Chemistry,
Lise-Meitner-Strasse 34, 85354 Freising, Germany
e-mail: peter.koehler@tum.de

M. Gobbetti and M. Gänzle (eds.), *Handbook on Sourdough Biotechnology*,
DOI 10.1007/978-1-4614-5425-0_2,

Table 2.1 Cereal production in 2010 [1]

Species	Cultivated area (million ha)	Grain production (million tons)
Corn	162	844
Rice	154	672
Wheat	217	651
Barley	48	123
Sorghum + millet	76	85
Oats	9	20
Triticale	4	13
Rye	5	12

winter or spring varieties. The winter type requires vernalization by low temperatures; it is sown in autumn and matures in early summer. Spring cereals are sensitive to frost temperatures and are sown in springtime and mature in midsummer; they require more irrigation and give lower yields than winter cereals.

Cereals produce dry, one-seeded fruits, called the "kernel" or "grain", in the form of a caryopsis, in which the fruit coat (pericarp) is strongly bound to the seed coat (testa). Grain size and weight vary widely from rather big corn grains (~350 mg) to small millet grains (~9 mg). The anatomy of cereal grains is fairly uniform: fruit and seed coats (bran) enclose the germ and the endosperm, the latter consisting of the starchy endosperm and the aleurone layer. In oats, barley, and rice the husk is fused together with the fruit coat and cannot be simply removed by threshing as can be done with common wheat and rye (*naked* cereals).

The chemical composition of cereal grains (moisture 11–14%) is characterized by the high content of carbohydrates (Table 2.2) [2, 3]. Available carbohydrates, mainly starch deposited in the endosperm, amount to 56–74% and fiber, mainly located in the bran, to 2–13%. The second important group of constituents is the proteins which fall within an average range of about 8–11%. With the exception of oats (~7%), cereal lipids belong to the minor constituents (2–4%) along with minerals (1–3%). The relatively high content of B-vitamins is, in particular, of nutritional relevance. With respect to structures and quantities of chemical constituents, notable differences exist between cereals and even between species and varieties within each cereal. These differences strongly affect the quality of products made from cereal grains. Because of the importance of the constituents, in the following we provide an insight into the detailed chemical composition of cereal grains including carbohydrates, proteins, lipids, and the minor components (minerals and vitamins).

2.2 Carbohydrates

Cereal grains contain 66–76% carbohydrates (Table 2.2), thus, this is by far the most abundant group of constituents. The major carbohydrate is starch (55–70%) followed by minor constituents such as arabinoxylans (1.5–8%), β-glucans (0.5–7%), sugars (~3%), cellulose (~2.5%), and glucofructans (~1%).

Table 2.2 Chemical composition of cereal grains (average values) [2, 3]

	Wheat	Rye	Corn	Barley	Oats	Rice	Millet
	(g/100 g)						
Moisture	12.6	13.6	11.3	12.1	13.1	13.0	12.0
Protein (N × 6.25)	11.3	9.4	8.8	11.1	10.8	7.7	10.5
Lipids	1.8	1.7	3.8	2.1	7.2	2.2	3.9
Available carbohydrates	59.4	60.3	65.0	62.7	56.2	73.7	68.2
Fiber	13.2	13.1	9.8	9.7	9.8	2.2	3.8
Minerals	1.7	1.9	1.3	2.3	2.9	1.2	1.6
	(mg/kg)						
Vitamin B_1 (thiamine)	4.6	3.7	3.6	4.3	6.7	4.1	4.3
Vitamin B_2 (riboflavin)	0.9	1.7	2.0	1.8	1.7	0.9	1.1
Nicotinamide	51.0	18.0	15.0	48.0	24.0	52.0	18.0
Panthothenic acid	12.0	15.0	6.5	6.8	7.1	17.0	14.0
Vitamin B_6	2.7	2.3	4.0	5.6	9.6	2.8	5.2
Folic acid	0.9	1.4	0.3	0.7	0.3	0.2	0.4
Total tocopherols	41.0	40.0	66.0	22.0	18.0	19.0	40.0

2.2.1 *Starch*

Starch is the major storage carbohydrate of cereals and an important part of our nutrition. Because of its unique properties starch is important for the textural properties of many foods, in particular bread and other baked goods. Finally, starch is nowadays also an important feedstock for bioethanol or biogas production (for reviews see [4, 5]).

2.2.1.1 Amylose and Amylopectin

Starch occurs only in the endosperm and is present in granular form. It consists of the two water-insoluble homoglucans amylose and amylopectin. Cereal starches are typically composed of 25–28% amylose and 72–75% amylopectin [6]. Mutant genotypes may have an altered amylose/amylopectin ratio. "Waxy" cultivars have a very high amylopectin level (up to 100%), whereas "high amylose" or "amylostarch" cultivars may contain up to 70% amylose. This altered ratio of amylose/amylopectin affects the technological properties of these cultivars [7, 8]. High-amylose wheat has been suggested as a raw material for the production of enzyme-resistant starch [9].

Amylose consists of α-(1,4)-linked D-glucopyranosyl units and is almost linear. Parts of the molecules also have α-(1,6)-linkages providing slightly branched structures [10, 11]. The degree of polymerization ranges from 500 to 6,000 glucose units giving a molecular weight (MW) of 8×10^4 to 10^6. Amylopectin is responsible for the granular nature of starch. It contains 30,000–3,000,000 glucose units and, therefore, it has a considerably higher MW (10^7–10^9) than amylose [12]. Amylopectin is a highly branched polysaccharide consisting of α-(1,4)-linked D-glucopyranosyl chains, which are interconnected via α-(1,6)-glycosidic linkages, also called branch points [13]. The α-(1,4)-linked chains have variable length of 6 to more than 100 glucose units depending on the molecular site at which they are located. The unbranched A- or outer chains can be distinguished from the branched B- or inner chains, which can be subdivided into B1-, B2-, B3-, and B4-chains [14]. The molecules are "terminated" by a single C-chain containing the reducing glucose residue [15]. Amylopectin has a tree-like structure, in which clusters of chains occur at regular intervals along the axis of the molecule [16]. Short A- and B1-chains of 12–15 glucose residues form the clusters which have double-helical structures. The longer, less abundant B2-, B3-, and B4-chains interconnect 2, 3 or 4 clusters, respectively. B2-chains contain approximately 35–40, B3-chains 70–80, and B4-chains up to more than 100 glucose residues [12, 17].

2.2.1.2 Starch Granules

In the endosperm starch is present as intracellular granules of different sizes and shapes, depending on the cereal species. In contrast to most plant starches, wheat, rye, and barley starches usually have two granule populations differing in size. Small spherical B-granules with an average size of 5 μm can be distinguished from large ellipsoid A-granules with mean diameters around 20 μm [18]. In the polarization microscope native starch granules are birefringent indicating that ordered, partially crystalline structures are present in the granule. The degree of crystallinity ranges from 20 to 40% [19] and is primarily caused by the structural features of amylopectin. It is thought that the macromolecules are oriented perpendicularly to the granule surface [12, 16] with the nonreducing ends of the molecules pointing to the surface.

A model of starch granule organization from the microscopic to the nanoscopic level has been suggested [12]. At the microscopic level alternating concentric "growth rings" with periodicities of several hundreds of nanometers can be observed. They reflect alternating semicrystalline and amorphous shells [12]. The latter are less dense, enriched in amylose, and contain noncrystalline amylopectin. They further consist of alternating amorphous and crystalline lamellae of about 9–10 nm [20]. Crystalline regions contain amylopectin double helices of A- and B1-chains oriented in parallel fashion and possibly 18 nm-wide, left-handed superhelices formed from double helices. Amorphous regions represent the amylopectin branching sites, which may also contain a few amylose molecules. The lamellae are organized into larger spherical blocklets, which vary periodically in diameter between 20 and 500 nm [21]. The amylopectin double helices may be packed into different crystal types. The very densely packed A-type is found in most cereal starches, while the more hydrated tube-like B-type is found in some tuber starches, high amylose cereal starches, and retrograded starch [12, 19]. Mixtures of A- and B-types are designated C-type.

2.2.1.3 Changes in Starch Structure During Processing

In many cereal manufacturing processes flour and also starch is usually dispersed in water and finally heated. In particular heating induces a series of structural changes. This process has been termed gelatinization [22]. Depending on water content, water distribution, and intensity of heat treatment the molecular order of the starch granules can be completely transformed from the semicrystalline to an amorphous state.

The mixing of starch and excess water at room temperature leads to a starch suspension. During mixing starch absorbs water up to 50% of its dry weight (1) because of physical immobilization of water in the void space between the granules, and (2) because of water uptake due to swelling. The latter process increases with temperature. If the temperature is below the gelatinization temperature, the described changes are reversible. As the temperature increases, more water permeates into the starch granules and initiates hydration reactions. Firstly, the amorphous regions are hydrated thereby increasing molecular mobility. This also affects the crystalline regions, in which amylopectin double helices dissociate and the crystallites melt [23, 24]. These reactions are endothermic and irreversible. They are accompanied by the loss of birefringence, which can be observed under the polarization microscope. Endothermic melting of crystallites can also be followed by differential scanning calorimetry (DSC). Viscosity measurements, for example in an amylograph or a rapid visco analyzer, also allow one to monitor the gelatinization process. Characteristic points are the onset temperature (*T*o; ca. 45 °C), which reflects the initiation of the process, as well as the peak (*T*p; ca. 60 °C) and conclusion (*T*c; ca. 75 °C) temperatures. These temperatures are subject to change depending on the botanical source of the starch and the water content of the suspension. The loss of molecular order and crystallinity during gelatinization is accompanied by further granule swelling due to increased water uptake and a limited starch solubilization. Mainly amylose is dissolved in water, which strongly increases the viscosity of the starch suspension. This phenomenon has been termed "amylose leaching," and it is caused by a phase separation between amylose and amylopectin, which are immiscible [25]. During further heating beyond the conclusion temperature of gelatinization swelling and leaching continue and a starch paste consisting of solubilized amylose and swollen, amorphous starch granules is formed. The shapes of the starch granules can still be observed unless shear force or higher temperatures are applied [23, 26].

Upon cooling with mixing the viscosity of a starch paste increases, whereas a starch gel is formed on cooling without mixing at concentrations above 6%. The second process is relevant in cereal baked goods. The changes that occur during cooling and storage of a starch paste have been summarized as "retrogradation" [22]. Generally, the amorphous system reassociates to a more ordered, crystalline state. Retrogradation processes can be divided into two subprocesses. The first is related to amylose and occurs in a time range of minutes to hours, the second is caused by amylopectin and takes place within hours or days. Therefore, amylose retrogradation is responsible for the initial hardness of a starch gel or bread, whereas amylopectin retrogradation determines the long-term gel structure, crystallinity, and hardness of a starch-containing food [27].

On cooling granule remnants that are enriched in amorphous amylopectin become incorporated into a continuous amylose matrix. Amylose molecules that are dissolved during gelatinization reassociate to local double helices interconnected by hydrated parts of the molecules, and a continuous network (gel) forms [27]. As amylose retrogradation proceeds, double helix formation increases and, finally, very stable crystalline structures are formed, which cannot be melted again by heating. Amylopectin retrogradation takes several hours or days and occurs in the granule remnants embedded in the initial amylose gel [27]. Crystallization mainly occurs within the short-chain outer A- and B1-chains of the molecules. The amylopectin crystallites melt at ca. 60 °C and, therefore, aged bread can partly be "refreshed" by heating. This so-called "staling endotherm" can be measured by DSC to evaluate amylopectin retrogradation. Amylopectin retrogradation is strongly influenced by a number of conditions and substances, including pH and the presence of low-molecular-weight (LMW) compounds such as salts, sugars, and lipids [26].

2.2.1.4 Interaction with Lipids

Amylose is able to form helical inclusion complexes in particular with polar lipids and this can occur in native (starch lipids; see below) as well as in gelatinized starch [28]. During gelatinization amylose forms a left-handed single helix and the nonpolar moiety of the polar lipid is located in the central cavity [16]. The inclusion complexes give rise to a V-type X-ray diffraction pattern. The presence of polar lipids strongly affects the retrogradation characteristics of the starch, because amylose-lipid complexes do not participate in the recrystallization process [26]. Complex formation is, however, strongly affected by the structure of the polar lipid [29]. For example, monoglycerides are more active than diglycerides and saturated fatty acids more active than unsaturated ones, because inclusion complexes are preferably formed with linear hydrocarbon chains and with compounds having one fatty acid residue. In addition, lipids, in particular lysophospholipids (lysolecithin), are minor constituents of cereal starches in amounts of 0.8–1.2% [30]. As so-called starch lipids they are associated with amylose as well as with the outer branches of amylopectin [28]. These lipid complexes lead to a delay of the onset of gelatinization and affect the properties of the starch especially in baking applications.

2.2.2 Nonstarch Polysaccharides (NSP)

Polysaccharides other than starch are primarily constituents of the cell walls and are much more abundant in the outer than in the inner layers of the grains. Therefore, a higher extraction rate is associated with a higher content of NSP. From a nutritional point of view NSP are dietary fiber, which has been associated with positive health effects. For example, cereal dietary fiber has been related to a reduced risk of chronic life style diseases such as cardiovascular diseases, type II diabetes, and gastrointestinal cancer [31–36]. In addition, technological functionalities have been described for the arabinoxylans (AX) of wheat (reviewed by [4]) and rye.

2.2.2.1 Arabinoxylans

AX are the major fraction (85–90%) of the so-called pentosans. Different cereal species contain different amounts of AX. The highest contents are present in rye (6–8%), whereas wheat contains only 1.5–2% AX. On the basis of solubility AX can be subdivided into a water-extractable (WEAX) and a water-unextractable fraction (WUAX). The former makes up 25–30% of total AX in wheat and 15–25% in rye [37]. In particular WEAX has considerable functionality in breadmaking.

AX consist of linear β-(1,4)-D-xylopyranosyl-chains, which can be substituted at the O-2 and/or O-3-positions with α-L-arabinofuranose [38, 39]. A particular minor component of AX is ferulic acid, which is bound to arabinose as an ester at the O-5 position [40]. AX of different cereals may vary substantially in content, substitutional pattern and molecular weight [41–43]. WEAX mainly consist of two populations of alternating open and highly branched regions, which can be distinguished by their characteristic arabinose/xylose ratios, ranging between 0.3 and 1.1 depending on the specific structural region [44]. WUAX can be solubilized by mild alkaline treatment yielding structures that are comparable to those of WEAX [37, 45–48].

The unique technological properties of AX are attributable to the fact that AX are able to absorb 15–20 times more water than their own weight and, thus, form highly viscous solutions, which may increase gas holding capacity of wheat doughs via stabilization of the gas bubbles [49]. In total, WEAX bind up to 25% of the added water in wheat doughs [50]. Under oxidizing conditions, in particular under acidic pH, the so-called "oxidative gelation" [51] leads to AX gel formation by inducing di- and oligoferulic acid cross-links [52, 53]. This is thought to be one major structure-forming reaction in rye sourdoughs. Because of covalent cross-links to the cell wall structure WUAX do not dissolve in water. Although they have high water-holding capacity and assist in water binding during dough mixing they are considered to have a negative impact on wheat breadmaking as they form physical barriers against the gluten network and, thus, destabilize the gas bubbles. However, the baking performance can be affected by adding endoxylanases, which preferentially hydrolyze WUAX. This produces solubilized WUAX, which have techno-functional effects comparable to WEAX [54, 55].

Beside AX the pentosan fraction contains a small part of a water-soluble, highly branched arabinogalactan peptide [41]. It consists of β-(1,3) and β-(1,6) linked galactopyranose units with α-glycosidically bound arabinofuranose residues. The peptide is attached by 4-*trans*-hydroxyproline. Unlike AX, arabinogalactan peptides have no significant effects in cereal processing.

2.2.2.2 β-Glucans

β-Glucans are also called lichenins and are present particularly in barley (3–7%) and oats (3.5–5%), whereas less than 2% β-glucans are found in other cereals. The chemical structure of these NSP is made up of linear D-glucose chains linked via mixed β-(1,3)- and β-(1,4)-glycosidic linkages. β-Glucans show a higher water solubility than AX (38–69% in barley, 65–90% in oats) and form viscous solutions, which in the case of barley may interfere in wort filtration during the production of beer.

2.3 Proteins

The average protein content of cereal grains covers a relatively narrow range (8–11%, Table 2.2), variations, however, are quite noticeable. Wheat grains, for instance, may vary from less than 6% to more than 20%. The content depends on the genotype (cereal, species, variety) and the growing conditions (soil, climate, fertilization); amount and time of nitrogen fertilization are of particular importance. Proteins are distributed over the whole grain, their concentration within each compartment, however, is remarkably different. The germ and aleurone layer of wheat grains, for instance, contain more than 30% proteins, the starchy endosperm ~13%, and the bran ~7% [3]. Regarding the different proportions of these compartments, most proteins of grains are located in the starchy endosperm, which is the source of white flours obtained by milling the grains and sieving.

White flours are the most important grain products. Therefore, the predominant part of the literature on cereal proteins deals with white flour proteins. The amino acid compositions of flour proteins from various cereals are shown in Table 2.3. Typical of all flours is the fact that glutamic acid almost entirely occurs in its amidated form as glutamine [56]. This amino acid generally predominates (15–31%), followed by proline in the case of wheat, rye, and barley (12–14%). Further major amino acids are leucine (7–14%) and alanine (4–11%). The nutritionally essential amino acids tryptophan (0.2–1.0%), methionine (1.3–2.9%), histidine (1.8–2.2%), and lysine (1.4–3.3%) are present only at very low levels. Through breeding and genetic engineering, attempts are being made to improve the content of essential amino acids. These approaches have been successful in the case of high-lysine barley and corn.

2.3.1 *Osborne Fractions*

Traditionally, cereal flour proteins have been classified into four fractions (albumins, globulins, prolamins, and glutelins) according to their different solubility and based on the fractionation procedure of Osborne [57]. Albumins are soluble in water,

Table 2.3 Amino acid composition (mol-%) of the total proteins of flours from various cereals [56]

Amino acid	Wheat	Rye	Barley	Oats	Rice	Millet	Corn
Asx[a]	4.2	6.9	4.9	8.1	8.8	7.7	5.9
Thr	3.2	4.0	3.8	3.9	4.1	4.5	3.7
Ser	6.6	6.4	6.0	6.6	6.8	6.6	6.4
Glx[a]	31.1	23.6	24.8	19.5	15.4	17.1	17.7
Pro	12.6	12.2	14.3	6.2	5.2	7.5	10.8
Gly	6.1	7.0	6.0	8.2	7.8	5.7	4.9
Ala	4.3	6.0	5.1	6.7	8.1	11.2	11.2
Cys	1.8	1.6	1.5	2.6	1.6	1.2	1.6
Val	4.9	5.5	6.1	6.2	6.7	6.7	5.0
Met	1.4	1.3	1.6	1.7	2.6	2.9	1.8
Ile	3.8	3.6	3.7	4.0	4.2	3.9	3.6
Leu	6.8	6.6	6.8	7.6	8.1	9.6	14.1
Tyr	2.3	2.2	2.7	2.8	3.8	2.7	3.1
Phe	3.8	3.9	4.3	4.4	4.1	4.0	4.0
His	1.8	1.9	1.8	2.0	2.2	2.1	2.2
Lys	1.8	3.1	2.6	3.3	3.3	2.5	1.4
Arg	2.8	3.7	3.3	5.4	6.4	3.1	2.4
Trp	0.7	0.5	0.7	0.8	0.8	1.0	0.2
Amide group	31.0	24.4	26.1	19.2	15.7	22.8	19.8

[a] *Asx* Asp+Asn, *Glx* Glu+Gln

while globulins are insoluble in pure water but soluble in dilute salt solutions. Prolamins are classically defined as cereal proteins soluble in aqueous alcohols, for example 60–70% ethanol. Originally, glutelins were described as proteins that were insoluble in water, salt solution, aqueous alcohols and soluble in dilute acids or bases. Later, it was ascertained that notable portions of glutelins are insoluble in dilute acids such as acetic acid, and that extraction with strong bases destroys the primary structure of proteins. Nowadays, complete solubility of glutelins is achieved by solvents containing a mixture of aqueous alcohols (e.g., 50% propanol), reducing agents (e.g., dithiothreitol), and disaggregating compounds (e.g., urea).

Regarding their functions, most of the albumins and globulins are metabolic proteins, for example enzymes or enzyme inhibitors (see Sect. 2.3.4). Oats are an exception containing considerable amounts of legume-like globulins such as 12S globulin [58]. Albumins and globulins are concentrated in the aleurone layer, bran, and germ, whereas their concentration in the starchy endosperm is relatively low. Predominantly, prolamins and glutelins are the storage proteins of cereal grains (see Sects. 2.3.2 and 2.3.3). Their only biological function is to supply the seedling with nitrogen and amino acids during germination. They are located only in the starchy endosperm; in white flours, their proportions based on total proteins amount to 70–90%. In general, none of the Osborne fractions consists of a single protein, but of a complex mixture of different proteins. A small portion of proteins does not fall into any of the four solubility fractions. Together with starch, they remain in the insoluble residue after Osborne fractionation and mainly belong to the class of lipo (membrane) proteins.

The prolamin fractions of the different cereals have been given trivial names: gliadin (wheat), secalin (rye), hordein (barley), avenin (oats), zein (corn), kafirin (millet, sorghum), and oryzin (rice). The glutelin fraction of wheat has been termed glutenin. Terms for the other glutelin fractions such as secalinin (rye), hordenin (barley), and zeanin (corn) are scarcely used today. Gliadin and glutenin fractions of wheat have been combined in the terms gluten or gluten proteins.

The content of the Osborne fractions varies considerably and depends on genotype and growing conditions. Moreover, the results of Osborne fractionation are strongly influenced by experimental conditions, and the fractions obtained are not clear-cut. Therefore, data from the literature on the qualitative and quantitative composition of Osborne fractions is differing and, in parts, contradictory. On average, the smallest proportion of total protein is present in the globulin fraction, followed by the albumin fraction. An exception is oat globulins amounting to more than 50% of total proteins. In most cereal flours, prolamins are the dominating fractions, oat prolamins, however, are minor protein components and rice flour is almost free of prolamins. Beside quantitative aspects the Osborne procedure is still useful for the preparation and characterization of flour proteins and the enrichment of different protein types.

2.3.2 *Storage Proteins of Wheat Rye, Barley, and Oats*

2.3.2.1 Classification and Primary Structures

Storage proteins (prolamins and glutelins) have been extensively investigated by the analysis of amino acid compositions, amino acid sequences, MW, and intra- and interchain disulfide linkages. The results indicated that, in accordance with phylogeny (see Sect. 2.1), the storage proteins of wheat, rye, and barley are closely related, whereas those of oats, in particular their glutelins, are structurally divergent. According to common structures storage proteins have been classified into three groups by two different principles. Shewry and coworkers [59] defined all storage proteins as prolamins and grouped them into the high-molecular-weight (HMW), sulfur-poor (S-poor) and sulfur-rich (S-rich) prolamins based on differences in MW and sulfur (cysteine, methionine) content. To prevent confusion, however, the term "prolamin" is not used for total storage proteins in the present paper, since classically the term prolamins comprises only the alcohol-soluble portions of storage proteins and does not include glutelins. We classified storage proteins according to related amino acid sequences and molecular masses into the following groups [60, 61]: (1) a HMW group; (2) a medium-molecular-weight (MMW) group; and (3) a LMW group. The proteins of these groups can be divided into different types on the basis of structural homologies (Table 2.4). Each type contains numerous closely related proteins; the small differences in their amino acid sequences can be traced back to substitutions, insertions, and deletions of single amino acids and short peptides.

Table 2.4 Characterization of storage protein types from wheat, rye, barley, and oats [61, 63]

Group/type	Code	Residues	State[a]	Repetitive unit[b,c] (frequency)	Partial amino acid composition (mol-%)[b]					
					Q	P	F+Y	G	L	V
HMW group										
HMW-GS x	Q6R2V1	815	a	QQPGQG (72×)	36	13	5.8	20	4.4	1.7
HMW-GS y	Q52JL3	637	a	QQPGQG (50×)	32	11	5.5	18	3.8	2.3
HMW-secalin x	Q94IK6	760	a	QQPGQG (66×)	34	15	6.7	20	3.7	1.5
HMW-secalin y	Q94IL4	716	a	QQPGQG (60×)	34	12	5.0	18	3.2	1.8
D-hordein	Q40054	686	a	QQPGQG (26×)	26	11	5.5	16	4.1	4.1
MMW group										
ω5-gliadin	Q402I5	420	m	(Q)QQQFP (65×)	53	20	10.0	0.7	3.1	0.2
ω1,2-gliadin	Q6DLC7	373	m	(QP)QQPFP (42×)	42	29	9.9	0.8	4.0	0.5
ω-secalin	O04365	338	m	(Q)QPQQPFP (32×)	40	29	8.6	0.6	4.4	1.8
C-hordein	Q40055	328	m	(Q)QPQQPFP (36×)	37	29	9.4	0.6	8.6	0.3
LMW group										
α/ß-gliadin	Q9M4M5	273	m	QPQPFPPQQPYP (5×)	36	15	7.4	2.6	8.1	5.1
γ-gliadin	Q94G91	308	m	(Q)QPQQPFP (15×)	36	18	5.2	2.9	7.2	4.6
LMW-GS	Q52NZ4	282	a	(Q)QQPPFS (11×)	32	13	5.7	3.2	8.2	5.3
γ-40 k-secalin[d]	–	–	m	QPQQPFP	34	18	5.5	2.4	7.4	4.7
γ-75 k-secalin	Q9FR41	436	a	QQPQQPFP (32×)	38	22	6.1	1.6	4.8	5.3
γ-hordein	P17990	286	m	QPQQPFP (15×)	28	17	7.7	3.1	7.0	7.3
B-hordein	P06470	274	a	QQPFPQ (13×)	30	19	7.3	2.9	8.0	6.2
avenin	Q09072	203	m	PFVQQQQ (3×)	33	11	8.4	2.0	8.9	8.3

[a]*a* aggregative, *m* monomeric
[b]One-letter-code for amino acids
[c]Basic unit frequently modified by substitution, insertion, and deletion of single amino acid residues
[d]Gellrich et al. [65]

The nomenclature of types is rather confusing and inconsequential. On the one hand prolamins have been termed according to their electrophoretic mobility in acid polyacrylamide gel electrophoresis (PAGE) with band regions designated as ω (lowest mobility), γ (medium mobility), and α/β (highest mobility). On the other hand, the nomenclature is based on their apparent sizes (after reduction of disulfide bonds) as indicated by sodiumdodecyl sulfate (SDS-) PAGE; examples are HMW- and LMW-glutenin subunits (GS), HMW-secalins, D-, C-, and B-hordeins. Because of the different importance of HMW-GS for the bread-making quality of wheat, single subunits have been numbered according to their mobility on SDS-gel electrophoresis (original nos. 1–12), the genome (1A, 1B, or 1D), and the type (x or y); examples of nomenclature are HMW-GS 1Ax1, 1Bx7, and 1Dy10 [62].

The HMW group contains HMW-GS of wheat, HMW-secalins of rye, and D-hordeins of barley (Table 2.4); this type is missing in oats. HMW-GS and HMW-secalins can be subdivided into the x-type and the y-type. The proteins comprise around 600–800 amino acid residues corresponding to MW of 70,000–90,000.

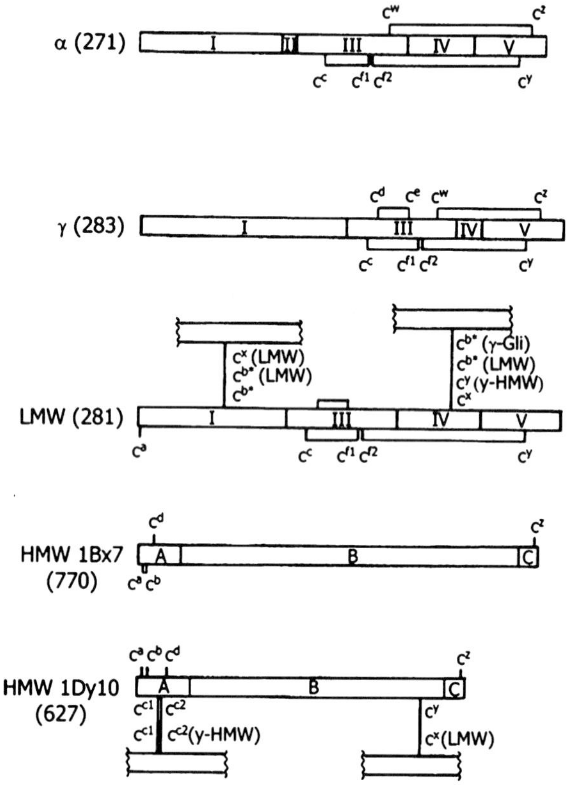

Fig. 2.1 Schematic structure and disulfide bonds of α/β-gliadins, γ-gliadins, LMW-, and HMW-GS (Adapted from [64])

The amino acid compositions are characterized by high contents of glutamine, glycine, and proline. The amino acid sequences [63] can be separated into three structural domains: a nonrepetitive N-terminal domain A of ~100 residues, a repetitive central domain B of 400–700 residues, and a nonrepetitive C-terminal domain C with ~40 residues (Fig. 2.1) [64, 65]. Domain B is dominated by repetitive sequences such as QQPGQG (one-letter-code for amino acids) as a backbone with inserted sequences like YYPTSL, QQG, and QPG with remarkable differences between x- and y-types (Table 2.5). Domains A and C have a more balanced amino acid composition and more amino acid residues with charged side chains. In a native state, the proteins of the HMW group are aggregated through interchain disulfide bonds (Fig. 2.1).

The MMW group consists of the homologous ω1,2-gliadins of wheat, ω-secalins of rye, and C-hordeins of barley including amino acid residues between 300 and 400 and MW around 40,000 (Table 2.4). Additionally, wheat contains unique ω5-gliadins with more than 400 residues and MW around 50,000. This group, likewise, is not present in oats. The proteins of the MMW group have extremely unbalanced amino acid compositions characterized by high contents of glutamine, proline, and phenylalanine, which together account for ~80% of total residues. Most sections of the amino acid sequences are composed of repetitive units such as QPQQPFP or QQQFP. This type of protein occurs as monomers and is readily soluble in aqueous alcohols and, in parts, even in water.

Table 2.5 Partial amino acid sequences of domain B of HMW-GS 1Dx2 (positions 93–338) and of HMW-GS 1Dy10 (positions 106–380) [63]

Position	Sequence[a]	Position	Sequence[a]
93	YYPSVTSPQQVS	106	YYPGVTSPRQGS
105	YYPGQASPQRPGQG	118	YYPGQASPQQPGQG
119	QQPGQG	132	QQPGKW
125	QQSGQGQQG	138	QEPGQGQQW
134	YYP--TSPQQPGQW	147	YYP--TSLQQPGQG
146	QQPEQGQPG	159	QQIGKGQQG
155	YYP--TSPQQPGQL	168	YYP--TSLQQPGQGQQG
167	QQPAQG	183	YYP--TSLQHTGQR
173	QQPGQGQQG	195	QQPVQG
182	QQPGQGQPG	201	QQPEQG
191	YYP-TSSQLQPGQL	207	QQPGQWQQG
204	QQPAQGQQG	216	YYP--TSPQQLGQG
213	QQPGQGQQG	228	QQPRQW
222	QQPGQG	234	QQSGQGQQG
228	QQPGQGQQG	243	HYP--TSLQQPGQGQQG
237	QQPGQG	258	HYL--ASQQQPGQGQQG
243	QQPGQGQQG	273	HYP--ASQQQPGQGQQG
252	QQLGQGQQG	288	HYP--ASQQQPGQGQQG
261	YYP--TSLQQSGQGQPG	303	HYP--ASQQEPGQGQQG
276	YYP--TSLQQLGQGQSG	318	QIPASQ
291	YYP--TSPQQPGQG	324	QQPGQGQQG
303	QQPGQL	333	HYP--ASLQQPGQGQQG
309	QQPAQG	348	HYP--TSLQQLGQGQQT
315	QQPGQGQQG	363	QQPGQK
324	QQPGQGQQG	369	QQPGQG
333	QQPGQG	375	QQTGQG

[a]One-letter-code for amino acids; - deletion

The LMW group consists of monomeric proteins including α/β- and γ-gliadins of wheat, γ-40 k-secalins of rye, γ-hordeins of barley, and avenins of oats, and of aggregative proteins including LMW-GS of wheat, γ-75 k-secalins of rye, and B-hordeins of barley (Table 2.4). They have around 300 amino acid residues and MW ranging from 28,000–35,000, besides γ-75 k-secalins (~430 residues, MW ~50,000) and avenins (~200 residues, MW ~23,000). The amino acid compositions of the LMW group proteins are characterized by relatively high contents of hydrophobic amino acids besides glutamine and proline. The amino acid sequences consist of four (α/β-gliadins five) different sequence sections (Fig. 2.1). The N-terminal section I is rich in glutamine, proline, and phenylalanine forming repetitive units such as QPQPFPPQQPY (α/β-gliadins), QQPQQPFP (γ-gliadins), QQPPFS (LMW-GS), or PFVQQQQ (avenins). Section I of γ-75 k-secalins is prolonged by around 130 residues as compared to γ-40 k-secalins and that of avenins is shortened to around 40 residues. Section II is unique to α/β-gliadins and consists of a polyglutamine sequence (up to 18 Q-residues). Sections III, IV, and V possess more balanced amino acid compositions and most of the cysteine residues that form only intrachain disulfide bonds (monomeric proteins) or both intra- and interchain disulfide bonds (aggregative proteins). The comparison of the amino acid sequences demonstrates that sections III and V contain homologous sequences, whereas section IV is, in part, unique to each type (Table 2.6). γ-Type proteins (γ-gliadins, γ-40 k-secalins [66], γ-75 k-secalins, γ-hordeins) show the highest conformity; α/β-gliadins, LMW-GS, and avenins have the lowest degree of homology within the LMW group. Most oat glutelins are globulin-like proteins and do not show any structural relationship with the HMW-, MMW-, and LMW-type proteins described above [67]. The reasons as to why they are not extractable with a salt solution are not yet clear.

As mentioned earlier, the quantitative composition of storage protein is strongly dependent on both genotype and growing conditions. Nevertheless, some constant data can be observed (Table 2.7) [68–70]: Proteins of the LMW group belong, by far, to the major components. Within this group, monomeric proteins (55–77% of total storage proteins) exceed aggregative proteins (10–25%) in the case of wheat species, whereas rye and barley are characterized by more aggregative (34–48%) than monomeric proteins (~25%). Proteins of the MMW and HMW groups belong to the minor components except ω-secalins (18%) and C-hordeins (36%) or are missing (oats). Within wheat species significant differences can be observed. Common wheat is characterized by the highest values for aggregative proteins (HMW-, LMW-GS) and a low monomeric/aggregated (m/a) ratio, and the "old" wheat species emmer and einkorn by low proportions of HMW-GS and high m/a ratios.

2.3.2.2 Disulfide Bonds

Disulfide bonds play an important role in determining the structure and properties of storage proteins. They are formed between sulfhydryl groups of cysteine residues, either within a single protein (intrachain) or between proteins (interchain). Most information on disulfide bonds is available for wheat gliadins and glutenins. With a few exceptions, ω-type gliadins are free of cysteine and, consequently occur as monomers. Most α/β- and γ-gliadins contain six and eight cysteine residues, respectively, and form three or four homologous intrachain disulfide bonds, present within or between sections III and V (Fig. 2.1) [64]. A small portion of gliadins is different from most gliadins and contains an odd number of cysteine residues. They may be linked to each other or to glutenins by an interchain disulfide bond. Homologous to γ-gliadins, γ-40 k- and γ-75 k-secalins, γ- and B-hordeins as well as avenins contain eight cysteine residues at comparable positions within sections III–V (Table 2.6). Probably they form four intrachain disulfide bonds homologous to those of γ-gliadins (Fig. 2.2). LMW-GS include eight cysteine residues, six of which form three intrachain disulfide bonds homologous to those of α/β- and γ-gliadins [64, 65]. Two cysteine residues located in sections I and IV are unique to LMW-GS; they are obviously not able to form intrachain bonds for steric reasons. They are involved in interchain bonds with residues of different proteins (LMW-GS, modified gliadins, y-type HMW-GS).

Table 2.6 Amino acid sequences of sections III, IV, and V of the LMW group [63]

Type	Positions	Sequences
(a) Section III		
α/β-gliadin	119–186	ILQQILQQQLIPCMDVVLQQHNIVHGRSQVLQQSTY-----QLLQELCCQHLWQIPEQSQCQAIHNVVHAIIL
γ-gliadin	153–224	FIQPSLQQQLNPCKNILLQQCKPASLVSSL-WSIIWPQSDCQVMRQQCCQQLAQIPQQLQCAAIHSVVHSIIM
LMW-GS	101–173	IVQPSVLQQLNPCKVFLQQQCSPVAMPQRLARSQMWQQSRCHVMQQQCCQQLSQIPEQSRYDAIRAITYSIIL
γ-40 k-secalin[a]	–	SIQLSLQQQLNPCKNVLLQQCSPVALVSSL-RSKIFPQSECQVMQQQCCQQLAQIPHHLQCAAIHSVVHAIIM
γ-75 k-secalin	285–356	SIQLSLQQQLNPCKNVLLQQCSPVALVSSL-RSKIFPQSECQVMQQQCCQQLAQIPQQLQCAAIHSVVHAIIM
γ-hordein	135–206	TIQLYLQQQLNPCKEFLLQQCRPVSLLSYI-WSKIVQQSSCRVMQQQCCLQLAQIPEQYKCTAIDSIVHAIFM
B-hordein	112–184	YVHPSILQQLNPCKVFLQQQCSPVPVPQRIARSQMLQQSSCHVLQQQCCQQLPQIPEQFRHEAIRAIVYSIFL
Avenin	43–114	FLQPLLQQQLNPCKQFLVQQCSPVAAVPFL-RSQILRQAICQVTRQQCCRQLAQIPEQLRCPAIHSVVQSIIL
(b) Section IV		
α/β-gliadin	187–222	HQQQKQQQQPSSQVSFQQPLQQYPLGQGSFRPSQQN
γ-gliadin	225–253	QQQQQQQQQGMHIFLPLSQQQQVGQGSL
LMW-GS	174–228	QEQQQGFVQAQQQQPQQSGQGVSQSQQQSQQQLGQCSFQQPQQQLGQQPQQQQQQ
γ-40 k-secalin[a]	–	QQEQREGVQILLPQSHQQLVGQGAL
γ-75 k-secalin	357–381	QQEQREGVQILLPQSHQQHVGQGAL
γ-hordein	207–231	QQGQRQGVQIVQQQPQPQQVGQCVL
B-hordein	185–225	QEQPQQLVEGVSQPQQQLWPQQVGQCSFQQPQPQQVGQQQQ
Avenin	115–155	QQQQQQQFIQPQLQQQVFQPQLQLQQQVFQPQLQQQVFQP
(c) Section V		
α/β-gliadin	223–273	PQAQGSVQPQQLPQF-EEIRNLALQTLPAMCNVYIPPYCTI--APFGIFGTNYR
γ-gliadin	254–308	VQGQGIIQPQQPAQL-EAIRSLVLQTLPSMCNVYVPPECSIMRAPFASIVAGIGGQ
LMW-GS	229–282	VLQGTFLQPHQIAHL-EAVTSIALRTLPTMCSVNVPLYSATTSVPFAVGTGVSAY
γ-40 k-secalin[a]	–	AQVQGIIQPQQLSQFNVGIVLQMLQNLPTMCNVYVPRQCPPSRRHLHAMSLVCGH
γ-75 k-secalin	382–436	AQVQGIIQPQQLSQL-EVVRSLVLQNLPTMCNVYVPRQCSTIQAPFASIVTGIVGH
γ-hordein	232–286	VQGQGVVQPQQLAQM-EAIRTLVLQSVPSMCNFNVPPNCSTIKAPFVGVVTGVGGQ
B-hordein	226–274	VPQSAFLQPHQIAQL-EATTSIALRTLPMMCSVNVPLYRILRGVGPSVGV
Avenin	156–203	QLQQVFNQPQMQGQI-EGMRAFALQALPAMCDVYVPPQCPVATAPLGGF

[a]Gellrich et al. [65]

Table 2.7 Proportions (%) of storage protein types of different wheat species, rye, and barley [68–70]

		Group				
		HMW	MMW	LMW		
Cereal	Variety	a	m	a	m	m/a[a]
Common wheat	Rektor	9.1	10.4	25.1	55.4	1.9
Spelt wheat	Schwabenkorn	6.6	10.4	17.7	65.3	3.1
Durum wheat	Biodur	5.0	6.7	19.3	69.0	3.1
Emmer	Unknown	2.6	10.8	10.0	76.6	6.9
Einkorn	Unknown	3.5	12.8	19.3	64.5	3.4
Rye	Halo	9.0	17.6	48.4	25.0	0.7
Barley	Golden promise	5.0	35.8	34.1	25.1	1.6

[a] *a* aggregative, *m* monomeric

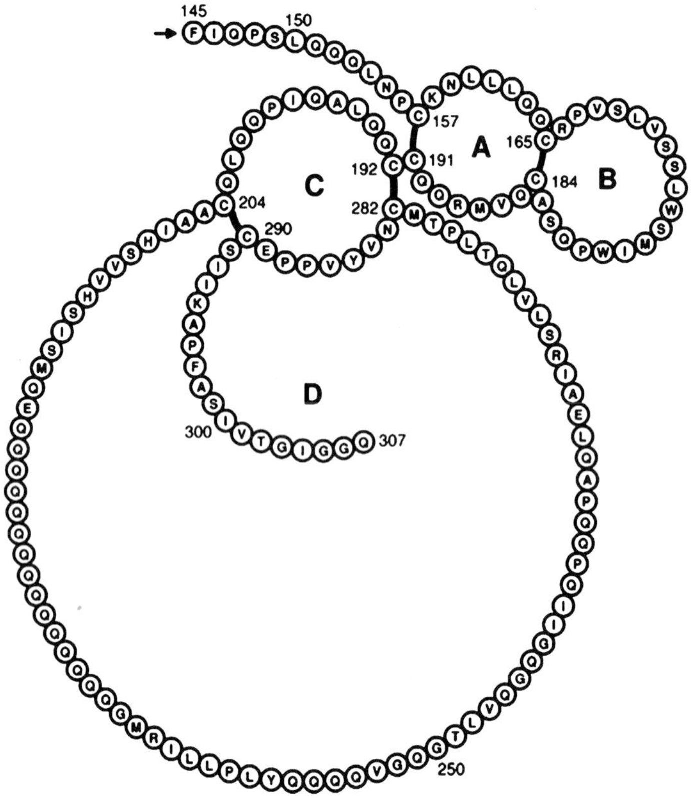

Fig. 2.2 Schematic two-dimensional structures of the C-terminal domain (sections III–V) of γ-gliadins (Taken from [94])

Because HMW-GS do not occur as monomers, it is generally assumed that they form interchain disulfide bonds. The x-type subunits, except subunit 1Dx5, have three cysteine residues in domain A and one in domain C (Fig. 2.1). Cysteines C^a and C^b were found to be linked by an intrachain bond, thus, the others (C^d, C^z) are available for interchain bonds. Subunit 1Dx5 has an additional cysteine residue at the beginning of domain B, and it has been suggested that this might form another interchain bond. Recently, a so-called head-to-tail disulfide bond between HMW-GS has been identified [71]. The y-type subunits have five cysteine residues in domain A and one in each of domains B and C. At present, interchain linkages have only been found for adjacent cysteine residues of domain A (C^{c1}, C^{c2}), which are connected in parallel with the corresponding residues of another y-type subunit, and for cysteine C^y of domain B, which is linked to C^x of section IV of LMW-GS. Thus, HMW- and LMW-GS fulfill the requirement that at least two cysteines forming interchain disulfide bonds are necessary to participate in a growing polymer; they act as "chain extenders." The most recent glutenin model suggests a backbone formed by HMW-GS linked by end-to-end, probably head-to-tail interchain disulfide bonds [65]. LMW-GS form also linear polymers via cysteine residues of sections I and IV; they are linked to domain B of y-type HMW-GS. y-Type HMW-secalins of rye have a second cysteine in domain C, which opens the possibility that an intrachain disulfide bond within domain C is formed inhibiting an interchain bond for polymerization [72]. As far as is known, D-hordeins possess ten (!) cysteine residues [63]; the formation of a regular polymer backbone appears to be impossible.

2.3.2.3 Molecular Weight Distribution

Most information on the quantitative MW distribution (MWD) of native storage (gluten) proteins is available for wheat, because MWD of gluten proteins has been recognized as one of the main determinants of the rheological properties of wheat dough. Native gluten proteins consist of monomeric α/β- and γ-gliadins with MW around 30,000 and monomeric ω5- and ω1,2-gliadins with MW between 40,000 and 55,000. They are alcohol-soluble and amount to ~50% of gluten proteins (Fig. 2.3). Besides monomers the alcohol-soluble fraction contains oligomers with MW roughly ranging between 60,000 and 600,000. They are formed by modified gliadins with an odd number of cysteine residues and LMW-GS via interchain disulfide bonds and account for ~15% of gluten proteins. Composition and quantity of the oligomeric fraction are strongly determined by the conditions of alcohol extraction, for example by temperature and duration. The remaining proteins (~35%) are alcohol-insoluble and mainly composed of LMW-GS and HMW-GS linked by disulfide bonds. Their MW ranges approximately from 600,000 to more than 10 million. The largest polymers termed "glutenin macropolymers" (GMP) are insoluble in SDS solutions and have MW well into the multimillions indicating that they may belong to the largest proteins in nature [73, 74]. Their amounts in flour (20–40 mg/g) are strongly correlated with dough strength and bread volume. GMP is characterized by higher ratios of HMW-GS to LMW-GS and x-type to y-type HMW-GS in

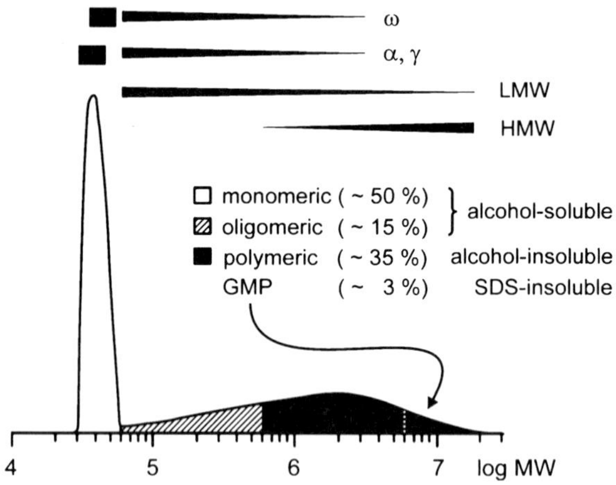

Fig. 2.3 Molecular weight distribution of native wheat storage (gluten) proteins (Modified from [73])

comparison with total glutenins; the HMW-GS combination 1Dx5 + 1Dy10 appears to produce higher GMP concentrations than 1Dx2 + 1Dy12 [73].

Rye storage proteins have a strongly different MWD as compared to wheat. Although rye shows higher proportions of aggregative to monomeric storage proteins than wheat (Table 2.7), the proportions of polymers is much lower (~23%) and the amount of GMP (~5 mg/g flour) strongly reduced [75, 76]. The deficiency of polymeric proteins is balanced by the higher proportion of oligomers (~30%), whereas the proportion of rye monomers (~47%) is similar to that of wheat. Obviously rye storage proteins consist of many more chain terminations (e.g., γ-75 k-secalins, y-type HMW-secalins) and less chain extenders than wheat, which apparently prevents gluten formation during dough mixing. Information about the MWD of native barley and oat proteins is not yet available.

2.3.2.4 Influence of External Parameters

Many studies have substantiated that both structures and quantities of storage proteins are exposed to a continuous change from the growing period of plants to the processing of end products. Because of the importance of wheat as a unique "bread cereal," most investigations have been focused on gluten proteins. In principle, however, the effect of external parameters is similar for all cereal proteins.

Fertilization

The supply with minerals during growing is essential for optimal plant development. Nitrogen (N) fertilization is, in particular, important for common wheat, because a high N supply provides a high flour protein content and thus, increased bread volume. Fertilization with different N amounts demonstrated that the quantities of albumins/globulins are scarcely influenced, whereas those of gluten proteins increase with higher N supply [77]. The effects on gliadins are more pronounced than on glutenins resulting in an elevated gliadin/glutenin ratio. Particularly, the proportions of ω-type gliadins are strongly enhanced by high N supply.

In the past, sulfur (S)-containing fertilizers were not widely used for cereal crops, because air pollution from industry and traffic provided sufficient amounts of S in the soil. The massive decrease in the input of S from atmospheric deposition over the last decades reduced S availability in soils dramatically, and has led to a severe S deficiency in cereals, which exerts a large influence on protein composition and technological properties. In the case of wheat, S deficiency provokes a drastic increase of S-free ω-type gliadins and a decrease of S-rich γ-gliadins and LMW-GS [78]. Moreover, S deficiency has been reported to impair dough and bread properties [79].

Infections

The infection of cereal plants with *Fusarium* strains induces a premature fading of individual spikelets and then, a fading of the whole ears. An outbreak of this infection is often accompanied by mycotoxin contamination of grains and flours, for example by the trichothecene deoxinivalenol. Studies on wheat demonstrated that infection with *Fusarium* caused a distinct reduction in the content of both total glutenins and HMW-GS and impaired dough and bread quality [80]. Glutenins of common wheat have been shown to be more strongly affected than those of emmer and storage proteins of naked barley [81, 82].

Germination

Proteins as well as other constituents are stable in dry grains. Water supply, however, induces germination of grains accompanied by the activation of enzymes, in particular, amylases and peptidases. The latter have been shown, for instance, to cause a fast degradation of prolamins of wheat, rye, barley, and oats during germination [83]. Studies on the Osborne fractions of wheat demonstrated that both monomeric gliadins and polymeric glutenins were strongly degraded during germination for 168 h at 15–30 °C, whereas albumins/globulins were scarcely affected [84]. The degradation of gluten proteins has a drastic negative effect on the bread-making quality of wheat.

Oxidation

Grains contain a considerable amount of LMW thiols such as glutathione, which are known to affect the structures and functional properties of polymeric storage proteins by thiol/disulfide interchange reactions during dough preparation [64]. To prevent this deleterious effect on the bread-making quality of wheat, week-long storage of flour under air (direct oxidation of LMW thiols) and treatment with L-ascorbic acid (indirect oxidation catalyzed by glutathione dehydrogenase) are recommended.

Enzymes

Breads prepared from rye and wheat sourdoughs are of increasing consumer interest due to the improvement of sensorial and nutritional quality, the prolongation of shelf life, and the delay in staling. Wheat storage proteins, however, which are responsible for the viscoelastic and gas-holding properties of dough and for the texture of the bread crumb, are profoundly degraded [85]. Protein degradation during fermentation is primarily due to acidic peptidases present in flour and activated by the lowered pH caused by *Lactobacillus* strains. The strongest decrease was found for the glutenin macropolymer and total glutenins. The extent of the decrease of monomeric gliadins was lower and more pronounced for the γ-type than for the α/β- and ω-types.

Heat

The baking process involves a drastic heat-treatment of proteins, with temperatures of more than 200 °C on the outer layer (crust) and near 100 °C in the interior (crumb) of bread. HPLC analysis of crust proteins from wheat bread indicated serious structural damage of both gliadins and glutenins [86]. With respect to crumb, the extractability of total gliadins with 60% ethanol is strongly reduced compared with those from flours. The single gliadin types are affected differently, ω-type gliadins less and α/β- and γ-types much more. Most gliadins can be recovered in the glutenin fraction after reduction of disulfide bonds suggesting that major heat-induced cross-links of gliadins to glutenins are disulfide bonds.

High Pressure

The effect of hydrostatic pressure is similar to that of heat. Treatment of gluten with pressure in the range of 300–600 MPa at 60 °C for 10 min provokes a strong reduction of gliadin extractability [87]. Within gliadin types, cysteine containing α/β- and γ-type gliadins, but not cysteine-free ω-type gliadins, are sensitive to pressure and are transferred to the ethanol-insoluble glutenin fraction. Cleavage and rearrangement of disulfide bonds have been proposed as being responsible for pressure-induced aggregation.

2.3.2.5 Wheat Gluten

Wheat is unique among cereals in its ability to form a cohesive, viscoelastic dough, when flour is mixed with water. Wheat dough retains the gas produced during fermentation and this results in a leavened loaf of bread after baking. It is commonly accepted that gluten proteins (gliadins and glutenins) decisively account for the physical properties of wheat dough. Both protein fractions are important contributors to these properties, but their functions are divergent. Hydrated monomeric and oligomeric proteins of the gliadin fraction have little elasticity and are less cohesive than glutenins; they contribute mainly to the viscosity and extensibility of dough. In contrast, hydrated polymeric glutenins are both cohesive and elastic, and are responsible for dough strength and elasticity. Thus, gluten is a "two-component glue," in which gliadins can be understood as a "plasticizer" or "solvent" for glutenins [65]. A proper mixture (~2:1) of the two is essential to give desirable dough and bread properties.

Native gluten proteins are amongst the most complex protein networks in nature due to the presence of several hundred different protein components. Even small differences in the qualitative and quantitative protein composition decide on the end-use quality of wheat varieties. Numerous studies demonstrated that the total amounts of gluten proteins (highly correlated with the protein content of flour), the ratio of gliadins to glutenins, the ratio of HMW-GS to LMW-GS, the amount of GMP, and the presence of specific HMG-GS determine dough and bread quality.

Amongst chemical bonds disulfide linkages (Fig. 2.1) play a key role in determining the structure and properties of gluten proteins. Intrachain bonds stabilize the steric structure of both monomeric and aggregative proteins; interchain bonds provoke the formation of large glutenin polymers. The disulfide structure is not in a stable state, but undergoes a continuous change from the maturing grain to the end product (e.g., bread), and is chiefly influenced by redox reactions. These include (1) the oxidation of free SH groups to S-S linkages, which supports the formation of large aggregates, (2) the presence of chain terminators (e.g., glutathione and gliadins with an odd number of cysteine), which stop polymerization, and (3) SH-SS interchange reactions, which affect the degree of polymerization of glutenins. Consequently, oxygen is known to be essential for optimal dough development and oxidizing agents, for example potassium bromide, azodicarbonimide, and dehydroascorbic acid (the oxidation product of ascorbic acid) have been found to be useful as bread improvers [88].

Conversely, reducing agents such as cysteine and sodium metabisulfite are used to soften strong doughs, accompanied by decreased dough development and resistance and increased extensibility. They are specifically in use as dough softeners for biscuits. The overall effect is to reduce the average MW of glutenin aggregates by SH/SS interchange.

Beside disulfide bonds, dityrosine and isopeptide bonds have been described as further covalent cross-links between gluten proteins. Compared with the concentration of disulfide bonds (~10 μmol per g flour) tyrosine-tyrosine cross-links (~0.7 nmol per g flour) appear to be only of marginal importance [89]. Interchain cross-links between lysine and glutamine residues (isopeptide bonds) are catalyzed by the enzyme transglutaminase (TG). Addition of TG to flour results in a decrease in the quantity of extractable gliadins and an increase of the glutenin fraction and the nonextractable fraction [90]. Thereby, dough properties and bread-making quality can be positively influenced, similar to the actions of chemical oxidants.

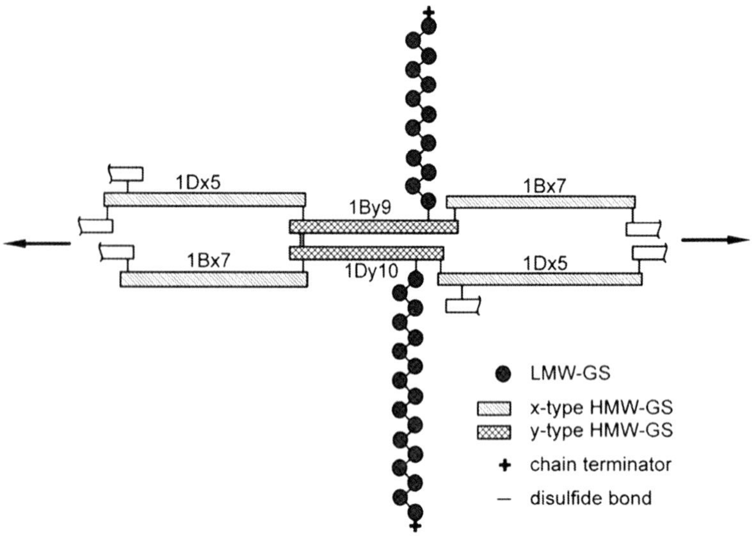

Fig. 2.4 A model double unit for the interchain disulfide structure of LMW-GS and HMW-GS of glutenin polymers (Adapted from [65])

The covalent structure of gluten proteins is complemented by noncovalent bonds (hydrogen bonds, ionic bonds, hydrophobic bonds). Glutamine, predestinated for hydrogen bonds, is the most abundant amino acid in gluten proteins (Table 2.4) and chiefly responsible for the water-binding capacity of gluten. In fact, dry gluten absorbs about twice its own weight of water. Moreover, glutamine residues are involved in frequent protein-protein hydrogen bonds. Though the number of ionizable side chains is relatively low, ionic bonds are of importance for the interactions between gluten proteins. For example, salts such as NaCl are known to strengthen dough, obviously via ionic bonds with glutenins [91]. Hydrophobic bonds can also contribute to the properties of gluten. Because the energy of hydrophobic bonds increase with increased temperature, this type of noncovalent bonds is particularly important for protein interactions during the oven phase.

Both covalent and noncovalent bonds determine the native steric structures (conformation) of gliadins and glutenins. Studies on the secondary structure have indicated that the repetitive sequences of gliadins and LMW-GS are characterized by β-turn conformation, whereas the nonrepetitive sections contain considerable proportions of α-helix and β-sheet structures [92]. The nonrepetitive sections of α/β-, γ-gliadins, and LMW-GS include intrachain disulfide bonds, which are concentrated in a relatively small area and form compact structures including two or three small rings and a big ring (Fig. 2.2) [93]. The nonrepetitive sections A and C of HMW-GS are dominated by α-helix and β-sheet structures, whereas the repetitive section B is characterized by regularly repeated β-turns [94]. They form a loose β-spiral similar to that of mammalian connective tissue elastin; β-spirals have been proposed to transfer elasticity to gluten.

A range of models has been developed to explain the structure and functionality of glutenins. Most recently, the experimental findings on disulfide bonds were transformed into a two-dimensional model [65] (Fig. 2.4). HMW-GS and LMW-GS polymerize separately, both forming linear backbone polymers. Both polymers are cross-linked by a disulfide bond between section IV of LMW-GS and section B of y-type HMW-GS. The backbone of HMW-GS is established by end-to-end, probably head-to-tail linkages. LMW-GS polymers are linked between two sections I and between sections I and IV. The polymerization of HMW-GS and LMW-GS is terminated by chain terminators, either by modified gliadins or LMW thiol compounds.

2.3.3 Storage Proteins of Corn, Millet, Sorghum, and Rice

Overall, the storage proteins of corn, sorghum, millet, and rice are, in part, related and differ significantly from those of wheat, rye, barley, and oats. According to the amino acid composition they contain less glutamine and proline and more hydrophobic amino acids such as leucine [56]. Corn storage proteins, called zeins, can be subgrouped into alcohol-soluble monomeric zeins and cross-linked zeins alcohol-soluble only on heating or after reduction of disulfide bonds. With respect to different structures zeins have been divided into four different subclasses [95]. α-Zeins are the major subclass (71–85% of total zeins), followed by γ- (10–20%), β- (1–5%) and δ-zeins (1–5%), respectively [96]. α-Zeins are monomeric proteins with apparent MW of 19,000 and 22,000 determined by SDS-PAGE. Their amino acid sequences contain up to ten tandem repeats [97]. Proteins of the other subclasses are cross-linked by disulfide bonds and their subunits have apparent MW of 18,000 and 27,000 (γ-zein), 18,000 (β-zein), and 10,000 (δ-zein).

In many ways the storage proteins of sorghum and millet called kafirins are similar to zeins. Sorghum kafirins have also been subdivided into α, β-, γ- and δ-subclasses based on solubility, MW, and structure [98]. α-Kafirins are monomeric proteins and represent the major subclass accounting for around 65–85% of total kafirins. Proteins of the other subclasses are highly cross-linked and alcohol-soluble only after reduction of disulfide bonds. On average, each of them accounts for less than 10% of total kafirins [99, 100]. Within the numerous millet species and varieties the proteins of foxtail millet were studied in detail [101]. SDS-PAGE of unreduced kafirins revealed bonds with apparent MW ranging from 11,000 to 150,000. After the reduction of disulfide bonds two major bands with MW of 11,000 (subunit A) and 16,000 (subunit B) were obtained. Unreduced proteins with higher MW were formed by cross-links of A and/or B subunits. The storage proteins of rice are characterized by the highly unbalanced ratio of prolamins to glutelins (~1:30) [102]. Both fractions show the lowest proline content (~5 mol-%) amongst cereal storage proteins [56]. SDS-PAGE patterns of rice prolamins (oryzins) showed a major band with MW 17,000 and a minor band with MW 23,000 [103]. The apparent MW of glutelin subunits was in a range from 20,000 to 38,000.

2.3.4 *Metabolic Proteins*

Most proteins of the albumin and globulin fractions are metabolic proteins, mainly enzymes and enzyme inhibitors. The corresponding extensive studies have been summarized by Kruger and Reed [104] and recently by Delcour and Hoseney [105]. Many of these proteins are located in the embryo and aleurone layer; others are distributed throughout the endosperm. They have nutritionally better amino acid compositions than storage proteins, particularly because of their higher lysine contents. Those enzymes that hydrolyze carbohydrates and proteins and, thereby, provide the embryo with nutrients and energy during germination, are of most significant importance.

2.3.4.1 Hydrolyzing Enzymes

Carbohydrate-Degrading Enzymes

The many carbohydrate-degrading enzymes include α-amylases, β-amylases, debranching enzymes, cellulases, β-glucanases, and glucosidases. Amylases are enzymes that hydrolyze the polysaccharides in starch granules. They can be classified as endohydrolases, which attack glucosidic bonds within the polysaccharide molecules and exohydrolases, which attack glucosidic bonds at or near the end of chains. The most important enzyme of the endohydrolase type is α-amylase. The enzyme hydrolyzes α-1,4-glucosidic bonds of amylose and amylopectin and produces a mixture of dextrins together with smaller amounts of maltose and oligosaccharides; the pH-optimum is about 5. The other major amylase type is β-amylase, an exohydrolase, which hydrolyzes α-1,4-glucosidic bonds near the nonreducing ends of amylose and amylopectin to produce maltose. Its pH-optimum is similar to that of α-amylase. Both amylase types exist in multiple forms or isoenzymes with different chemical and physical properties. Neither α- nor β-amylase can break α-1,6-glucosidic bonds present in amylopectin. For this kind of hydrolysis debranching enzymes are present in cereal grains. Along with α-glucosidases they assist α- and β-amylases in a more complete conversion of starch to simple sugars and small dextrins. A number of other carbohydrate degrading enzymes exist, their amounts, however, are very low compared to amylases. Examples are α- and β-glucosidases, cellulases, and arabinoxylanases.

Proteolytic Enzymes

Enzymes that hydrolyze proteins are called proteinases, proteases, or peptidases. They attack the peptide bond between amino acid residues and include both endo- and exopeptidases. The latter are divided into carboxypeptidases, when acting from the carboxy terminal and aminopeptidases, when acting from the amino terminal.

The most important proteolytic enzymes are acidic peptidases. They exist in multiple forms having pH-optima between 4.2 and 5.5 and include both endo- and exotypes. On the basis of their catalytic mechanism they can be classified as serine, metallo-, aspartic, and serine peptidases. According to their biological function to provide the embryo with amino acids, their activity is highest during the germination of grains.

Other Hydrolyzing Enzymes

Lipases are the most important enzymes that hydrolyze ester bonds. They attack triacylglycerols yielding mono- and diacylglycerols and free fatty acids. Lipase activity is important, because free fatty acids are more susceptible to oxidative rancidity than fatty acids bound in triacylglycerols. The activity varies widely among cereals with oats and millet having the highest activity. Exogenous lipases are in use to improve the baking performance of wheat flour.

Phytase is an esterase that hydrolyzes phytic acid to inositol and free phosphoric acid. Even partial hydrolysis of phytic acid by phytase is desirable from a nutritional point of view, because the strong complexation of cations such as zinc, calcium, and magnesium ions by phytic acid is significantly reduced.

2.3.4.2 Oxidizing Enzymes

Lipoxygenase is present in high levels in the germ. It catalyzes the peroxidation of certain polyunsaturated fatty acids by molecular oxygen. Its typical substrate is linoleic acid containing a methylene-interrupted, doubly unsaturated carbon chain with double bonds in the *cis*-configuration.

Polyphenoloxidases preferably occur in the outer layers of the grains. They catalyze the oxidation of phenols, such as catechol, pyrogallol, and gallic acid, to quinons by molecular oxygen. Peroxidase and catalase may be classified as hydroperoxidases catalyzing the oxidation of a number of aromatic amines and phenols, for example ferulic acid in arabinoxylans, by hydrogen peroxide. Other oxidizing enzymes are ascorbic acid oxidase and glutathione dehydrogenase.

2.3.4.3 Enzyme Inhibitors

Many investigators have isolated and characterized enzyme inhibitors from germ and endosperm. Most important inhibitors are targeted on hydrolyzing enzymes to prevent the extensive degradation of starch and storage proteins during grain development and to defend plant tissues from animal (insect) or microbial enzymes. Predominant classes are amylase and protease inhibitors concentrated in the albumin/globulin fractions. Amylase inhibitors can be directed towards both cereal and noncereal amylases and protease inhibitors towards proteases from both cereals and animals. Some inhibitors appear to be bifunctional inhibiting amylases as well as proteases.

2.4 Lipids

2.4.1 Lipid Composition

Cereal lipids originate from membranes, organelles, and spherosomes and consist of different chemical structures. Depending on cereal species average lipid contents of 1.7–7% in the grains are present (Table 2.2). Lipids are mainly stored in the germ, to a smaller extent in the aleurone layer and to the lesser extent in the endosperm. In particular oats are rich in lipids (6–8%) in contrast to wheat and rye (1.7%). Cereal lipids have similar fatty acid compositions, in which linoleic acid reaches contents of 39–69%, while oleic acid and palmitic acid make up 11–36% and 18–28%, respectively [106, 107]. Although wheat lipids are only a minor constituent of the flour, they greatly impact the baking performance and have, therefore, been extensively studied.

While triglycerides are the dominating lipid class in the germ and the aleurone layer, phospho- and glycolipids are present in the endosperm (Fig. 2.5). Depending on the extraction rate wheat flour contains 0.5–3% lipids [108]. Extraction with a polar solvent at ambient temperature, i.e., water-saturated butanol, dissolves the nonstarch lipids that make up approximately 75% of the total flour lipids [109]. The residual 25% are the so-called starch lipids. The composition of the nonstarch lipids is given in Table 2.8. They contain about 60% nonpolar lipids, 24% glycolipids, and 15% phospholipids. By extraction with solvents of different polarities they can be further subdivided into a free and a bound fraction. The nonpolar lipids are mainly present in the free lipid fraction, whereas glyco- and phospholipids are part of the bound fraction, in which they can be associated, for example with proteins [106, 107]. The major glycolipid class is the digalactosyldiglycerides. Starch lipids are primarily composed of lysophospholipids, which form inclusion complexes with amylose helices already in native starch [28].

2.4.2 Effects of Lipids on the Baking Performance of Wheat Flour

Only nonstarch lipids affect the rheological properties of wheat doughs. Interactions between starch lipids and starch are sufficiently strong so that this lipid fraction is not available before the starch gelatinizes. Studies with nonstarch lipids have shown that only the polar lipids have a positive effect on baking performance, whereas the

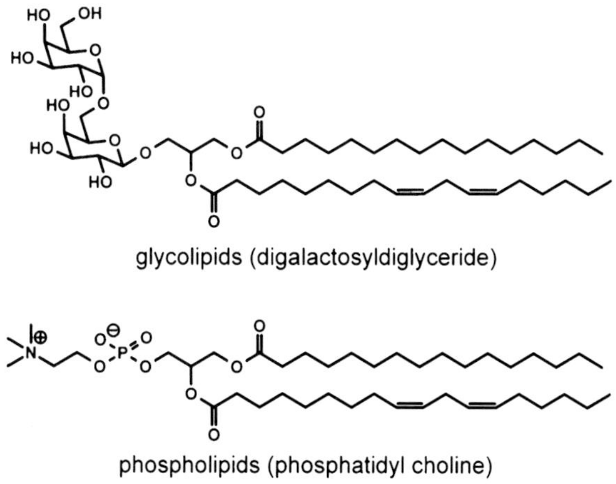

Fig. 2.5 Polar lipids that affect the baking performance of wheat

Table 2.8 Composition of nonstarch lipids of wheat flour [107]. Content (g/100 g) based on total lipid

Nonstarch lipids: 1.70–1.95 g/100 g flour			
Polar	36–42	Nonpolar	58–64
Phospholipids	14–16	Sterol esters	1.9–4.2
Acylphosphatidyl ethanolamine	4.2–4.9	Triglycerides	39.5–49.4
Acyllysophosphatidyl ethanolamine	1.6–2.3	Diglycerides	3.3–5.4
Phosphatidyl ethanolamine/phosphatidyl glycerol	0.7–1.1	Esterified monogalactosyldi-glycerides/monoglycerides	2.7–3.9
Phosphatidyl choline	3.8–4.9	Esterified sterolglycerides	0.8–4.2
Phosphatidyl serine/phosphatidyl inosit	0.4–0.7		
Lysophosphatidyl ethanolamin	0.3–0.5		
Lysophosphatidyl glycerol	0.2–0.3		
Lysophosphatidyl choline	1.4–2.1		
Glycolipids	22–26		
Monogalactosyldiglycerides	5.0–5.9		
Monogalactosylmonoglycerides	0.9–0.4		
Digalactosyldiglycerides	12.6–16.5		
Digalactosylmonoglycerides	0.6–3.4		

nonpolar lipids have the opposite effect [110]. In particular glycolipids have been shown to contribute to the high baking performance of wheat flour [29, 111–113], whereas the functionality of the phospholipids has been found to be less important. If the term "specific baking activity" would be defined, polar lipids would be found to affect the baking performance of wheat flour to a considerably greater extent than proteins. The addition of only 0.13% polar lipids would yield the same increase of loaf volume as a protein content that would be increased by 1%. Polar lipids affect dough properties in many ways, i.e., the dough handling properties are improved and the gas-holding capacity during proofing is increased enabling a prolonged oven spring, increased loaf volume, better crumb resilience, and, in some cases, retardation of bread staling.

2.4.3 Modes of Action of Polar Lipids in Baking

The high baking activities of polar lipids, in particular of the glycolipids, might be explained by modes of action based on the formation of liquid films at the dough liquor/gas cell interface. Possible modes of action are the direct influence of the surfactants on the liquid film lamellae and gas cell interfaces through direct adsorption resulting in an increase of surface activity as suggested by Gan et al. [49, 114] and Sroan et al. [115] as the secondary stabilizing mechanism in the so-called dual film theory. It suggests the presence of liquid lamellae, providing an independent mechanism of gas cell stabilization. As shown recently, the effects of different surface active components may be explained by the type of monolayer that they form [116].

However, in particular the positive effect of some polar lipids such as acylated sterol glucosides and sterol glucosides cannot be explained with this mode of directly stabilizing the liquid film lamellae. Here another mode of action could be the answer, for example the indirect stabilization of the dough liquor/gas cell interface through this type of surfactant [116, 117]. These polar lipid classes have a positive influence on the phase behavior of the endogenous lipids present in the dough liquor in that they lead to an increase in surface activity of the endogenous lipids and hence a better availability and accumulation at the liquid film lamellae/gas cell interface, thus increasing gas cell stabilization, and consequently the bread volume.

Inclusion complexes between amylose helices and polar lipids with one fatty acid residue are responsible for two effects. Complexes present in native starch (starch lipids) increase the temperature of gelatinization and, thus, prolong the oven spring. Inclusion complexes between amylose helices and polar lipids with one fatty acid residue may also form during and after the gelatinization process and are responsible for the anti-staling effect of some polar lipids, for example monoglycerides.

2.5 Minor Constituents

2.5.1 Minerals

The mineral content of cereals ranges from ca. 1.0 to 2.5% (Table 2.2). Compared to other foods this is an intermediate concentration with milk, meat, and vegetables having somewhat lower mineral contents and pulses, which are extraordinarily rich in minerals (mean mineral content ~3.5%). As cereals are among the most important staple foods, and are consumed in high quantities, they are important sources of minerals in the human diet. The major portion of the minerals (>90%) is located in the outer layers of the grains, namely in the bran, the aleurone layer, and the germ. Consequently, products made from whole grains should increasingly be introduced into human nutrition to benefit from the mineral content of cereals.

2.5.2 Vitamins

Cereals contain vitamins in concentrations ranging from below 1 to ca. 50 mg/kg, depending on the compound (Table 2.2). Thus, cereals are a good source of vitamins from the B-group, and, in industrial countries, they cover about 50–60% of the daily requirement of B-vitamins. The most important fat-soluble vitamins are the tocopherols, which are present in concentrations exceeding 20 mg/kg. Like the minerals, vitamins are concentrated in the outer layers of the grains, in particular in the aleurone layer as well as in the germ. Therefore, milling of cereals into white flour will remove most of the vitamins. Consequently, the use of whole-grain products or products enriched in vitamin-containing tissues will be of nutritional benefit for the consumer.

Abbreviations

AX	Arabinoxylans
DSC	Differential scanning calorimetry
GMP	Glutenin macropolymer
GS	Glutenin subunits
HMW	High-molecular-weight
HPLC	High-performance liquid chromatography
LMW	Low-molecular-weight
m/a	Monomeric/aggregated
MMW	Medium-molecular-weight
MW	Molecular weight
MWD	Molecular weight distribution
NSP	Nonstarch polysaccharides
PAGE	Polyacrylamide gel electrophoresis
SDS	Sodium dodecyl sulfate
TG	Transglutaminase
WEAX	Water-extractable arabinoxylans
WUAX	Water-unextractable arabinoxylans

References

1. Food and Agriculture Organization of the United Nations (2012) FAOSTAT Database. http://faostat.fao.org/site/567/default.aspx#ancor. Accessed 18 May 2012
2. Souci SW, Fachmann W, Kraut H (2008) In: Deutsche Forschungsanstalt für Lebensmittelchemie (ed) Food composition and nutrition tables. Deutsche Forschungsanstalt für Lebensmittelchemie. MedPharm Scientific Publishers, Stuttgart
3. Belitz H-D, Grosch W, Schieberle P (2009) Cereals and cereal products. In: Belitz H-D, Grosch W, Schieberle P (eds) Food chemistry, 4th edn. Springer, Berlin, pp 670–675
4. Goesaert H, Brijs C, Veraverbeke WS, Courtin CM, Gebruers K, Delcour JA (2005) Wheat constituents: how they impact bread quality, and how to impact their functionality. Trends Food Sci Tech 16:12–30
5. Zeeman SC, Kossmann J, Smith AM (2010) Starch: its metabolism, evolution, and biotechnological modification in plants. Annu Rev Plant Biol 61:209–234
6. Colonna P, Buléon A (1992) New insights on starch structure and properties. In: Cereal chemistry and technology: a long past and a bright future. In: Proceedings of the 9th international cereal and bread congress, 1992, Paris, France, Institut de Recherche Technologique Agroalimentaire des Céréales (IRTAC), Paris, France, 1992; pp. 25–42
7. Van Hung P, Maeda T, Morita N (2006) Waxy and high-amylose wheat starches and flours-characteristics, functionality and application. Trends Food Sci Tech 17:448–456
8. Jonnala RS, MacRitchie F, Smail VW, Seabourn BW, Tilley M, Lafiandra D, Urbano M (2010) Protein and quality characterization of complete and partial near-isogenic lines of waxy wheat. Cereal Chem 87:538–545
9. Topping DL, Segal I, Regina A, Conlon MA, Bajka BH, Toden S, Clarke JM, Morell MK, Bird AR (2010) Resistant starch and human health. In: Van der Kamp W (ed) Dietary fibre: new frontiers for food and health, proceedings of the 4th international dietary fibre conference, Vienna, Austria, Wageningen Academic Publishers, Wageningen, pp 311–321
10. Hizukuri S, Takeda Y, Yasuda M (1981) Multi-branched nature of amylose and the action of debranching enzymes. Carbohyd Res 94:205–213
11. Shibanuma K, Takeda Y, Hizukuri S, Shibata S (1994) Molecular structures of some wheat starches. Carbohyd Polym 25:111–116
12. Buléon A, Colonna P, Planchot V, Ball S (1998) Starch granules: structure and biosynthesis. Int J Biol Macromol 23:85–112
13. Zobel HF (1988) Starch crystal transformations and their industrial importance. Starch/Stärke 40:1–7
14. Hizukuri S (1986) Polymodal distribution of the chain lengths of amylopectins and its significance. Carbohyd Res 147:342–347
15. Peat S, Whelan WJ, Thomas GJ (1956) The enzymic synthesis and degradation of starch. XXII. Evidence of multiple branching in waxy maize starch. J Chem Soc: 3025–3030
16. French D (1984) Organization of starch granules. In: Whistler RL, BeMiller JN, Paschal EF (eds) Starch chemistry and technology, 2nd edn. Academic, New York, pp 183–212
17. Hejazi M, Fettke J, Paris O, Steup M (2009) The two plastidial starch-related dikinases sequentially phosphorylate glucosyl residues at the surface of both the A-and B-type allomorphs of crystallized maltodextrins but the mode of action differs. Plant Physiol 150:962–976
18. Karlsson R, Olered R, Eliasson A-C (1983) Changes in starch granule size distribution and starch gelatinisation properties during development and maturation of wheat, barley and rye. Starch/Stärke 35:335–340
19. Hizukuri S (1996) Starch: analytical aspects. In: Eliasson A-C (ed) Carbohydrates in food. Marcel Dekker, Inc, New York, pp 347–429
20. Jenkins PJ, Cameron RE, Donald AM (1993) A universal feature in the structure of starch granules from different botanical sources. Starch/Stärke 45:417–420

21. Gallant DJ, Bouchet B, Baldwin PM (1997) Microscopy of starch: evidence of a new level of granule organization. Carbohyd Polym 32:177–191
22. Atwell WA, Hood LF, Lineback DR, Varriano-Marston E, Zobel HF (1988) The terminology and methodology associated with basic starch phenomena. Cereal Foods World 33:306–311
23. Tester RF, Debon SJJ (2000) Annealing of starch—a review. Int J Biol Macromol 27:1–12
24. Waigh TA, Gidley MJ, Komanshek BU, Donald AM (2000) The phase transformations in starch during gelatinisation: a liquid crystalline approach. Carbohyd Res 328:165–176
25. Kalichevsky MT, Ring SG (1987) Incompatibility of amylase and amylopectin in aqueous solution. Carbohyd Res 162:323–328
26. Eliasson A-C, Gudmundsson M (1996) Starch: physicochemical and functional aspects. In: Eliasson A-C (ed) Carbohydrates in food. Marcel Dekker, Inc, New York, pp 431–503
27. Miles MJ, Morris VJ, Orford PD, Ring SG (1985) The roles of amylose and amylopectin in the gelation and retrogradation of starch. Carbohyd Res 135:271–281
28. Morrison WR, Law RV, Snape CE (1993) Evidence for inclusion complexes of lipids with V-amylose in maize, rice and oat starches. J Cereal Sci 18:107–109
29. Selmair PL, Koehler P (2008) Baking performance of synthetic glycolipids in comparison to commercial surfactants. J Agric Food Chem 56:6691–6700
30. Morrison WR, Gadan H (1987) The amylose and lipid contents of starch granules in developing wheat endosperm. J Cereal Sci 5:263–275
31. Blackwood AD, Salter J, Dettmar PW, Chaplin MF (2000) Dietary fibre, physicochemical properties and their relationship to health. J Roy Soc Promot Health 120:242–247
32. Lu ZX, Walker KZ, Muir JG, Mascara T, O'Dea K (2000) Arabinoxylan fiber, a byproduct of wheat flour processing, reduces the postprandial glucose response in normoglycemic subjects. Am J Clin Nutr 71:1123–1128
33. Lu ZX, Walker KZ, Muir JG, O'Dea K (2004) Arabinoxylan fibre improves metabolic control in people with type II diabetes. Eur J Clin Nutr 58:621–628
34. Garcia AL, Steiniger J, Reich SC, Weickert MO, Harsch I, Machowetz A, Mohlig M, Spranger J, Rudovich NN, Meuser F, Doerfer J, Katz N, Speth M, Zunft HJF, Pfeiffer AHF, Koebnick C (2006) Arabinoxylan fibre consumption improved glucose metabolism, but did not affect serum adipokines in subjects with impaired glucose tolerance. Hormone Meta Res 38:761–766
35. Garcia AL, Otto B, Reich SC, Weickert MO, Steiniger J, Machowetz A, Rudovich NN, Mohlig M, Katz N, Speth M, Meuser F, Doerfer J, Zunft HJ, Pfeiffer AH, Koebnick C (2007) Arabinoxylan consumption decreases postprandial serum glucose, serum insulin and plasma total ghrelin response in subjects with impaired glucose tolerance. Eur J Clin Nutr 61:334–341
36. Babio N, Balanza R, Basulto J, Bullo M, Salas-Salvado J (2010) Dietary fibre: influence on body weight, glycemic control and plasma cholesterol profile. Nutr Hospital 25:327–340
37. Izydorczyk MS, Biliaderis CG (1995) Cereal arabinoxylans: advances in structure and physicochemical properties. Carbohyd Polym 28:33–48
38. Meuser F, Suckow P (1986) Non-starch polysaccharides. In: Blanshard JMV, Frazier PJ, Galliard T (eds) Chemistry and physics of baking. The Royal Society of Chemistry, London, pp 42–61
39. Perlin AS (1951) Isolation and composition of the soluble pentosans of wheat flour. Cereal Chem 28:370–381
40. Perlin AS (1951) Structure of the soluble pentosans of wheat flours. Cereal Chem 28:282–393
41. Fausch H, Kündig W, Neukom H (1963) Ferulic acid as a component of a glycoprotein from wheat flour. Nature 199:287
42. Delcour JA, Van Win H, Grobet PJ (1999) Distribution and structural variation of arabinoxylans in common wheat mill streams. J Agric Food Chem 47:271–275
43. Maes C, Delcour JA (2002) Structural characterisation of water-extractable and water-unextractable arabinoxylans in wheat bran. J Cereal Sci 35:315–326

44. Dervilly G, Saulnier L, Roger P, Thibault J-F (2000) Isolation of homogeneous fractions from wheat water-soluble arabinoxylans. Influence of the structure on their macromolecular characteristics. J Agric Food Chem 48:270–278
45. Gruppen H, Hamer RJ, Voragen AGJ (1992) Water-unextractable cell wall material from wheat flour. II. Fractionation of alkali-extractable polymers and comparison with water extractable arabinoxylans. J Cereal Sci 16:53–67
46. Gruppen H, Komelink FJM, Voragen AGJ (1993) Water-unextractable cell wall material from wheat flour. III. A structural model for arabinoxylans. J Cereal Sci 19:111–128
47. Mares DJ, Stone BA (1973) Studies on wheat endosperm. I. Chemical composition and ultrastructure of the cell walls. Aust J Bio Sci 26:793–812
48. Mares DJ, Stone BA (1973) Studies on wheat endosperm. II. Properties of the wall components and studies on their organization in the wall. Aust J Bio Sci 26:813–830
49. Gan Z, Ellis PR, Schofield JD (1995) Mini review: gas cell stabilisation and gas retention in wheat bread dough. J Cereal Sci 21:215–230
50. Atwell WA (1998) Method for reducing syruping in refrigerated doughs. Patent application 1998; WO 97/26794
51. Izydorczyk MS, Biliaderis CG, Bushuk W (1990) Oxidative gelation studies of water-soluble pentosans from wheat. J Cereal Sci 11:153–169
52. Figueroa-Espinoza MC, Rouau X (1998) Oxidative crosslinking of pentosans by a fungal laccase and horseradish peroxidase: mechanism of linkage between feruloylated arabinoxylans. Cereal Chem 75:259–265
53. Vinkx CJA, Van Nieuwenhove CG, Delcour JA (1991) Physicochemical and functional properties of rye nonstarch polysaccharides. III. Oxidative gelation of a fraction containing water-soluble pentosans and proteins. Cereal Chem 68:617–622
54. Courtin CM, Roelants A, Delcour JA (1999) Fractionation-reconstitution experiments provide insight into the role of endoxylanases in bread-making. J Agric Food Chem 47:1870–1877
55. Courtin CM, Gelders GG, Delcour JA (2001) The use of two endoxylanases with different substrate selectivity provides insight into the functionality of arabinoxylans in wheat flour breadmaking. Cereal Chem 78:564–571
56. Wieser H, Seilmeier W, Eggert M, Belitz H-D (1983) Tryptophangehalt von Getreideproteinen. Z Lebensm Unters Forsch 177:457–460
57. Osborne TB (1907) The proteins of the wheat kernel, vol 84. Carnegie Inst, Washington, DC
58. Peterson DM (1978) Subunit structure and composition of oat seed globulin. Plant Physiol 62:506–509
59. Shewry PR, Tatham AS (1990) The prolamin storage proteins of cereal seeds: structure and evolution. Biochem J 267:1–12
60. Wieser H (1994) Cereal protein chemistry. In: Feighery C, O'Farrelly C (eds) Gastrointestinal immunology and gluten-sensitive disease. Oak Tree Press, Dublin, pp 191–202
61. Wieser H, Koehler P (2008) The biochemical basis of celiac disease. Cereal Chem 85:1–13
62. Payne PI, Nightingale MA, Krattinger AF, Holt LM (1987) The relationship between HMW glutenin subunit composition and the bread-making quality of British-grown wheat varieties. J Sci Food Agric 40:51–65
63. Database Uni Prot KB/TREMBL. http://pir.georgetown.edu
64. Grosch W, Wieser H (1999) Redox reactions in wheat dough as affected by ascorbic acid. J Cereal Sci 29:1–16
65. Wieser H, Bushuk W, MacRitchie F (2006) The polymeric glutenins. In: Wrigley C, Bekes F, Bushuk W (eds) Gliadin and glutenin: the unique balance of wheat quality. AACC International, St. Paul, pp 213–240
66. Gellrich C, Schieberle P, Wieser H (2005) Studies of partial amino acid sequences of γ-40 k secalins of rye. Cereal Chem 82:541–545
67. Robert LS, Nozzolillo C, Altosaar I (1985) Characterization of oat (*Avena sativa* L.) residual proteins. Cereal Chem 62:276–279

68. Wieser H (2000) Comparative investigations of gluten proteins from different wheat species. I. Qualitative and quantitative composition of gluten protein types. Eur Food Res Tech 211:262–268
69. Gellrich C, Schieberle P, Wieser H (2003) Biochemical characterization and quantification of the storage protein (secalin) types in rye flour. Cereal Chem 80:102–109
70. Lange M, Vincze E, Wieser H, Schjoerring JK, Holm PB (2007) Suppression of C-hordein synthesis in barley by antisense constructs results in a more balanced amino acid composition. J Agric Food Chem 55:6074–6081
71. Lutz E, Wieser H, Koehler P (2012) Identification of disulfide bonds in wheat gluten proteins by means of mass spectrometry/electron transfer dissociation. J Agric Food Chem 60:3708–3716
72. Köhler P, Wieser H (2000) Comparative studies of high M_r subunits of rye and wheat. III. Localisation of cysteine residues. J Cereal Sci 32:189–197
73. Koehler P (2010) Structure and functionality of gluten proteins: an overview. In: Branlard G (ed) Gluten proteins 2009. INRA, Paris, France, pp 84–88
74. Weegels PL, Hamer RJ, Schofield JD (1996) Functional properties of wheat glutenin. J Cereal Sci 23:1–18
75. Wrigley CW (1996) Giant proteins with flour power. Nature 381:738–739
76. Altpeter F, Popelka JC, Wieser H (2004) Stable expression of 1Dx5 and 1Dy10 high-molecular-weight glutenin subunit genes in transgenic rye drastically increases the polymeric glutelin fraction in rye. Plant Mol Biol 54:783–792
77. Wieser H, Seilmeier W, Kieffer R, Altpeter F (2005) Flour protein composition and functional properties of transgenic rye lines expressing HMW subunits genes of wheat. Cereal Chem 82:594–600
78. Wieser H, Seilmeier W (1998) The influence of nitrogen fertilisation on quantities and proportions of different protein types in wheat flour. J Sci Food Agric 76:49–55
79. Wieser H, Gutser R, von Tucher S (2004) Influence of sulphur fertilisation on quantities and proportions of gluten protein types in wheat flour. J Cereal Sci 40:239–244
80. Zhao FJ, Hawkesford MJ, McGrath SP (1999) Sulphur assimilation and effects on yield and quality of wheat. J Cereal Sci 30:1–17
81. Wang J, Wieser H, Pawelzik E, Weinert J, Keutgen AJ, Wolf GA (2005) Impact of the fungal protease produced by *Fusarium culmorum* on the protein quality and bread making properties of winter wheat. Eur Food Res Tech 220:552–559
82. Eggert K, Wieser H, Pawelzik E (2010) The influence of *Fusarium* infection and growing location on the quantitative protein composition of (part I) emmer (*Triticum dicoccum*). Eur Food Res Tech 230:837–847
83. Eggert K, Wieser H, Pawelzik E (2010) The influence of *Fusarium* infection and growing location on the quantitative protein composition of (part II) naked barley (*Hordeum vulgare nudum*). Eur Food Res Tech 230:893–902
84. Dalby A, Tsai CY (1976) Lysine and tryptophan increases during germination of cereal grains. Cereal Chem 53:222–226
85. Koehler P, Hartmann G, Wieser H, Rychlik M (2007) Changes of folates, dietary fiber, and proteins in wheat as affected by germination. J Agric Food Chem 55:4678–4683
86. Wieser H, Vermeulen N, Gaertner F, Vogel RF (2008) Effects of different *Lactobacillus* and *Enterococcus* strains and chemical acidification regarding degradation of gluten proteins during sourdough fermentation. Eur Food Res Tech 226:1495–1502
87. Wieser H (1998) Investigations on the extractability of gluten proteins from wheat bread in comparison with flour. Z Lebensm Unters Forsch A 207:128–132
88. Kieffer R, Schurer F, Köhler P, Wieser H (2007) Effect of hydrostatic pressure and temperature on the chemical and functional properties of wheat gluten: studies on gluten, gliadin and glutenin. J Cereal Sci 45:285–292
89. Wieser H (2003) The use of redox agents. In: Cauvin SP (ed) Bread making: improving quality. Woodhead Publishing Limited, Cambridge, UK, pp 424–446

90. Hanft F, Koehler P (2005) Quantitation of dityrosine in wheat flour and dough by liquid chromatography-tandem mass spectrometry. J Agric Food Chem 53:2418–2423
91. Bauer N, Köhler P, Wieser H, Schieberle P (2003) Studies on the effects of microbial transglutaminase on gluten proteins of wheat. I. Biochemical analysis. Cereal Chem 80:781–786
92. Kasarda DD (1989) Glutenin structure in relation to wheat quality. In: Pomeranz Y (ed) Wheat is unique. AACC, St. Paul, pp 277–302
93. Tatham AS, Shewry PR (1985) The conformation of wheat gluten proteins. The secondary structures and thermal stabilities of α-, β-, γ- and ω-gliadins. J Cereal Sci 3:104–113
94. Müller S, Wieser H (1997) The location of disulphide bonds in monomeric γ-type gliadins. J Cereal Sci 26:169–176
95. Tatham AS, Miflin BJ, Shewry PR (1985) The β-turn conformation in wheat gluten proteins: relationship to gluten elasticity. Cereal Chem 62:405–412
96. Esen A (1987) A proposed nomenclature for the alcohol-soluble proteins (zeins) of maize (zea-mays-l). J Cereal Sci 5:117–128
97. Wilson CM (1991) Multiple zeins from maize endosperms characterized by reversed-phase HPLC. Plant Physiol 95:777–786
98. Rubenstein I, Geraghty D (1986) The genetic organization of zein. In: Pomeranz Y (ed) Advances in cereal science and technology, vol VIII. AACC, St. Paul, pp 297–315
99. Shull JM, Watterson JJ, Kirleis AW (1991) Proposed nomenclature for the alcohol-soluble proteins (kafirins) of *Sorghum bicolor* (L. Moench) based on molecular weight, solubility, and structure. J Agric Food Chem 39:83–87
100. Watterson JJ, Shull JM, Kirleis AW (1993) Quantitation of a-, b-, and g-kafirins in vitreous and opaque endosperm of *Sorghum bicolor*. Cereal Chem 70:452–457
101. Hamaker BR, Mohamed AA, Habben JE, Huang CP, Larkins BA (1995) Efficient procedure for extracting maize and sorghum kernel proteins reveals higher prolamin contents than the conventional method. Cereal Chem 72:583–588
102. Danno G, Natake M (1980) Isolation of foxtail millet proteins and their subunit structure. Agric Biol Chem 44:913–918
103. Juliano BO (1985) Polysaccharides, proteins, and lipids. In: Juliano BO (ed) Rice chemistry and technology, 2nd edn. AACC, St. Paul, pp 98–142
104. Mandac BE, Juliano BO (1978) Properties of prolamin in mature and developing rice grain. Phytochem 17:611–614
105. Kruger JE, Reed G (1988) Enzymes and color. In: Pomeranz Y (ed) Wheat chemistry and technology, vol I, 3rd edn. AACC, St. Paul, pp 441–476
106. Delcour JA, Hoseney RC (2010) Principles of cereal science and technology, 3rd edn. AACC International, Inc, St. Paul, pp 40–85
107. Eliasson A-C, Larsson KA (1993) Molecular colloidal approach. In: Cereals in breadmaking. Marcel Dekker, Inc, New York
108. Hoseney RC (1994) Principles of cereal science and technology, 2nd edn. AACC, St. Paul, pp 81–101, and 229–273
109. MacMurray TA, Morrison WR (1970) Composition of wheat-flour lipids. J Sci Food Agric 21:520–528
110. Morrison WR, Mann DL, Soon W, Coventry AM (1975) Selective extraction and quantitative analysis of nonstarch and starch lipids from wheat flour. J Sci Food Agric 26:507–521
111. MacRitchie F (1981) Flour lipids: theoretical aspects and functional properties. Cereal Chem 58:156–158
112. Daftary RD, Pomeranz Y, Shogren M, Finney KF (1968) Functional bread-making properties of wheat flour. II. The role of flour lipid fractions in bread making. Food Technol 22:327–330
113. Hoseney RC, Pomeranz Y, Finney KF (1970) Functional (breadmaking) and biochemical properties of wheat flour components. VII. Petroleum ether-soluble lipoproteins of wheat flour. Cereal Chem 47:153–160
114. Gan Z, Angold RE, Williams MR, Ellis PR, Vaughan JG, Galliard T (1990) The microstructure and gas retention of bread dough. J Cereal Sci 12:15–24

115. Sroan BS, Bean SR, MacRitchie F (2009) Mechanism of gas cell stabilization in bread making. I. The primary gluten-starch matrix. J Cereal Sci 49:32–40
116. Sroan BS, MacRitchie F (2009) Mechanism of gas cell stabilization in breadmaking. II. The secondary liquid lamellae. J Cereal Sci 49:41–46
117. Selmair PL, Koehler P (2009) Molecular structure and baking performance of individual glycolipid classes from lecithins. J Agric Food Chem 57:5597–5609

Chapter 3
Technology of Baked Goods

Maria Ambrogina Pagani, Gabriella Bottega, and Manuela Mariotti

3.1 Introduction

Baked goods are a heterogeneous market category of products. The high diversification that distinguishes these products makes it difficult to find one single, general and satisfactory definition. It is, however, possible to identify baked goods based on the common commodities matrix since they are all foods derived from cereal flour [1]. Other similarities within this group are the basic ingredients used, mainly wheat flour or, less commonly, rye flour, water and leavening agents. Although the technological processes may differ, each one comprises mixing, leavening and baking. These successive stages allow the transformation of flour, or a mixture of different types of flour, into an appetizing and digestible food, which, at the macroscopic level, has a complex structure differentiated by a friable crust and an internal alveolar structure.

The classification of baked goods may be based on different criteria [2]. One criterion, which is widely used in the baking trade but is not governed by legislation, is the presence of sugar in the formula which can be perceived by taste and corresponds to at least 10% of the weight of the flour. Baked products with more simple formulas, therefore, constitute the various types of bread. Despite its importance, not only sensorial but also nutritional and textural, the amount of sugar in the dough is not sufficient for classifying this complex category in a satisfactory way. As proposed by Cauvain and Young [1], the distinction of the basic products comprising the formula may be made by considering the fat:flour ratio.

M.A. Pagani (✉) • G. Bottega • M. Mariotti (✉)
Department of Food, Environmental and Nutritional Sciences (DeFENS),
University of Milan, via Celoria 2, 20133 MILANO, Italy
e-mail: ambrogina.pagani@unimi.it; gabriella.bottega@unimi.it;
manuela.mariotti@unimi.it

M. Gobbetti and M. Gänzle (eds.), *Handbook on Sourdough Biotechnology*,
DOI 10.1007/978-1-4614-5425-0_3,

A further criterion for differentiating baked goods is the evaluation of their lightness and softness, and so of the specific volume (correlated with lightness) and humidity (which determines the softness). These two structural characteristics are the result of complex phenomena that occur during each stage of the technological process. Indeed, the final characteristics of the product not only depend on leavening but also on mixing and baking (see Sect. 3.4). A light and soft baked good usually has a specific volume higher than 2.5–3.0 mL/g and humidity higher than 15–18%. A dry and friable baked good has a specific volume between 1.3 and 2.5 mL/g and humidity values lower than 5–10% [2]. On the basis of these two criteria, four categories of baked products can be identified. Moreover, each category can be further subdivided according to the leavening method used (see Sect. 3.3.3).

3.2 Wheat, a Preferred Raw Material for Bread and Baked Goods

The choice of ingredients is of fundamental importance for the production of leavened baked goods that satisfy consumer tastes. Although barley and rye flours provide a workable dough, bakers realized a long time ago that the best results in terms of volume development were always obtained with wheat flour. The superior quality of this cereal, especially its most evolved species *Triticum durum* and *T. aestivum*, is of a purely technological nature. Only wheat flour provides a cohesive and homogenous dough, where the single and original particles of flour are no longer recognizable.

The property that makes wheat dough unique is its viscoelasticity. This rheological property enables the mass to be stretched and deformed without rupturing. At the same time, the dough is elastic and tenacious, capable of maintaining its shape even when subjected to physical stress. After baking, gluten proteins denature and loose viscoelasticity, ensuring the maintenance of the final shape of baked goods. The reason for this versatile behaviour does not depend on differences in the quantity of components. Indeed, the protein content of the numerous varieties of wheat extends over quite a wide interval, from 9 to 16% of the weight of the grain [3]. This variability coincides with that of other cereals. The technological superiority of wheat is related to complex qualitative differences at the level of protein fractions. Wheat storage proteins are generally classified based on molecular weight (MW), solubility and conformation. Storage proteins, gliadins and glutenins, account for ca. 80% of the entire protein fraction and have a particular amino acid composition. Gluten proteins have a high percentage of glutamine, about one third of all amino acids, and proline, and low levels of lysine (Chap. 2, [4]). This composition is responsible for the low nutritional value of wheat protein. However, it also accounts for the protein–protein interactions that lead to the formation of gluten, the three-dimensional network which is continuous and homogenous throughout the mass. Both non-covalent bonds, such as hydrophobic interactions and hydrogen bonds (guaranteed by the large amounts of glutamine), and covalent bonds are involved, the most significant of which are the disulfide bonds between cysteine residues.

Gliadins and glutenins have different roles for the viscoelastic characteristics of gluten [5–7]. Gliadins can be easily deformed and stretched, thanks to their viscous properties that are typical of fluids, because of their globular shape and interconnections via disulfide bonds. In contrast, glutenins form long chains by means of inter-protein disulfide bonds, which resist deformation and form an elastic and tenacious mass. Although durum wheat semolina is used for bread making, especially in the Mediterranean regions, the common wheat (*T. aestivum*) flour may be the optimal raw material because it gives the best leavening results [8]. Soluble proteins in wheat flour are mostly proteins with enzymatic activity. Most of these enzymes are hydrolases (amylases, proteases, lipases), which specifically act on the reserve macromolecules. Also oxidative enzymes (lipoxygenases, peroxidases) are present. For many wheat processes, the enzymatic activities have a central and strategic role which has to be carefully monitored (see also Chap. 2).

3.2.1 The Milling Process

About 70% of the worldwide wheat production is used for food [9], mainly by bakery industries. The essential preliminary step is the milling process. The objective of milling is twice. On the one hand, it separates the starchy endosperm, the area containing gluten proteins, from germ, pericarp and seed coats, that form bran and other by-products. On the other hand, it reduces the particle size of the endosperm to values lower than 150–200 μm for common wheat flour and 500 μm for durum wheat semolina. The fine granulometry of the flour gives an optimal workability to the resulting dough and contributes to its processing into a palatable and appetizing food. The removal of bran improves both the hygienic properties of the flour, since the peripheral parts of the grain are often contaminated by chemical residues and biotic pollutants, and the technological characteristics of the flour. In fact, non-starch polysaccharides and enzymes, which are abundant in the bran, worsen the rheology properties of the dough [10]. The separation of the oil-rich germ prevents rancidity, which compromises the storage of flour [11]. However, the elimination of these two parts, which are rich in several functional components, decreases the nutritional value of the flour.

Overall, milling is much more complex than mere grinding. Because of the particular morphological structure of wheat grain, which is characterized by the crease or ventral furrow, an introflexion along the length of the kernel, which hides a double layer of teguments, the milling has to promote the extraction of the flour from grains. Therefore, the first step consists of opening the grain. Then, proceeding from the inside towards the outside, the endosperm is recovered via repeated sequences of size reduction and separation stages, excluding the more external areas (bran, aleurone layer, etc.), which are known as tailing products. This procedure fully justifies the definition of "flour extraction yield", defined as the quantity of flour produced from 100 parts of cleaned and conditioned wheat grains [12]. This is the only solution for preventing the passage of the bran layers from the furrow, in which

Table 3.1 Chemical composition of wheat regions (Adapted from [15]) (data expressed as g/100 g dm)

Region	Kernel (%)	Starch and soluble sugars (%)	Proteins (%)	Lipids (%)	NSP[a] (cellulose, etc) (%)	Ash (%)
Fruit coat (*pericarp*)	5	14÷16	10÷14	1÷3	60÷74	3÷5
Seed coat	2	9÷11	13÷19	3÷5	53÷63	9÷15
Aleurone	8	10÷14	29÷35	7÷9	35÷41	5÷15
Endosperm	82	80÷85	8÷14	2÷3	1÷3	0.5÷1.5
Germ	3	19÷21	36÷40	13÷17	20÷24	4÷6

[a]Non-starch polysaccharides

25–30% of total bran is hidden, into the flour of the bran layers [13]. Milling is generally simpler for those cereals without furrow.

The physical separation of the various parts of the wheat grain is made possible by the different composition of the three morphological areas of the kernel, that determine different behaviour during processing (Table 3.1). The separation of the endosperm from the bran is not quantitative, since parts of the endosperm are lost into the milling by-products and small percentages of bran fragments are inevitably present in the flour. Therefore, the milling process has to reach a compromise between "extraction yield" and "grade of refinement" (accuracy of elimination of bran) of the flour. The main criterion for flour classification is based on the accuracy of teguments separation. The approach, used by legislation in many countries, is based on the threshold value of a number of parameters, especially ash and proteins.

The current technology for milling considers the following four stages: (1) receiving, pre-cleaning and storage of the incoming wheat; (2) cleaning and conditioning; (3) milling; (4) storage of the flours. Wheat coming to the mill is usually transported in bulk in trucks, trains or ships and it is unloaded into large hoppers with a grilled opening to facilitate the elimination of large foreign matter. This operation is preceded by rapid analytical inspections of samples representative of the entire batch, to assess quality parameters. Preliminary cleaning operations made before loading the wheat into the silos are intended to remove mainly coarse foreign materials, in order to provide for better storage. The cleaning of the wheat is carried out immediately before the milling process. This involves a sequence of operations, each performed by a special machine, with the aim of removing impurities, foreign matters and powders. Differences in size, shape and density compared to the whole and sound wheat grains are used to achieve this aim [14]. The conditioning or tempering of kernels is decisive to achieve an optimal milling. This operation includes grain humidification and the successive resting time, to increase the water content of 2–4%. The conditioning step toughens the bran, favouring its break off in form of large particles, and mellows the endosperm, thereby facilitating the separation between these two parts. The conditioning time (from 6 to 24–36 h), the quantity of water used, and the ways to add water (one or two tempering steps) depend on the initial humidity and on the hardness and vitreousness of the wheat kernels.

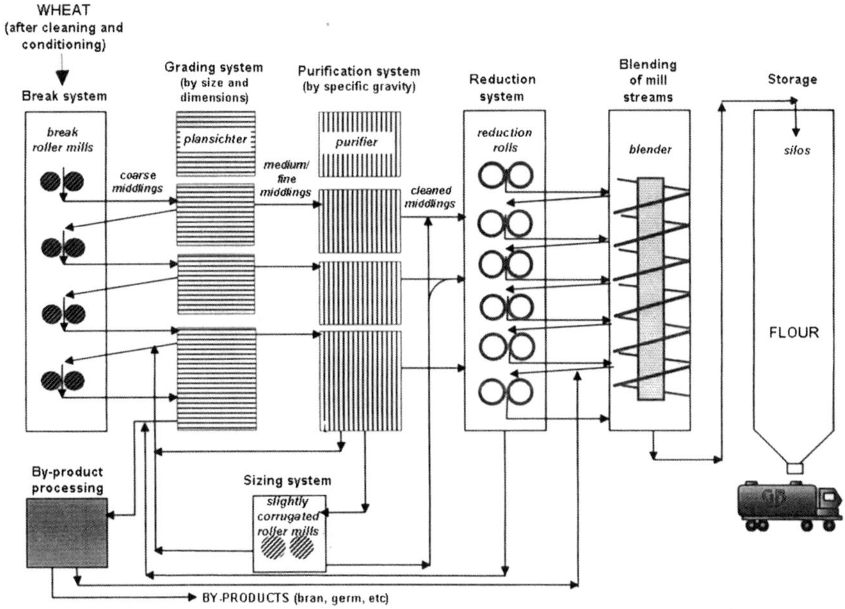

Fig. 3.1 Simplified schema of wheat milling operations (Adapted from [15])

After this stage, caryopses have 15.5–17% humidity and are in the best state for milling [14]. The milling process is complex in terms of both the type and number of operations involved and the methods used (Fig. 3.1). The process is classified into different systems: (a) break system, which separates the endosperm from bran and germ; (b) sizing or purification system, which separates particles according to the presence of bran pieces; and (c) reduction system, where the large particles of endosperm are reduced to flour [15]. First, the kernel is subjected to breaking by means of roller mills that are made up of pairs of cast iron cylinders, each at a set distance from each other, with corrugated surfaces and turning at differential speeds. The breaking system has the function to open, cut and flake the grains: it separates the endosperm from teguments and leaves bran in the form of large and flat flakes to facilitate their removal. Grains have to be ground gradually (four to five subsequent breaking steps), thus limiting the formation of flour and the disintegration of bran [12]. Each breaking step is followed by sieving, carried out by plansichters or sifters. They are large machines, within which numerous sieves are stacked with a mesh granulometry suitable for the material to be sifted (Fig. 3.2). The coarse particles of endosperm are called "semolina". Some of them, referred to as middlings, have various degrees of attached bran layers, whereas others are clean middlings, composed of pure endosperm. These two fractions are separated according to their specific gravity and size. The sizing rolls, formed by smooth or slightly corrugated rolls, detach the bran pieces attached to the middlings. These operations are immediately followed by classification through plansichters. The clean middlings are sent

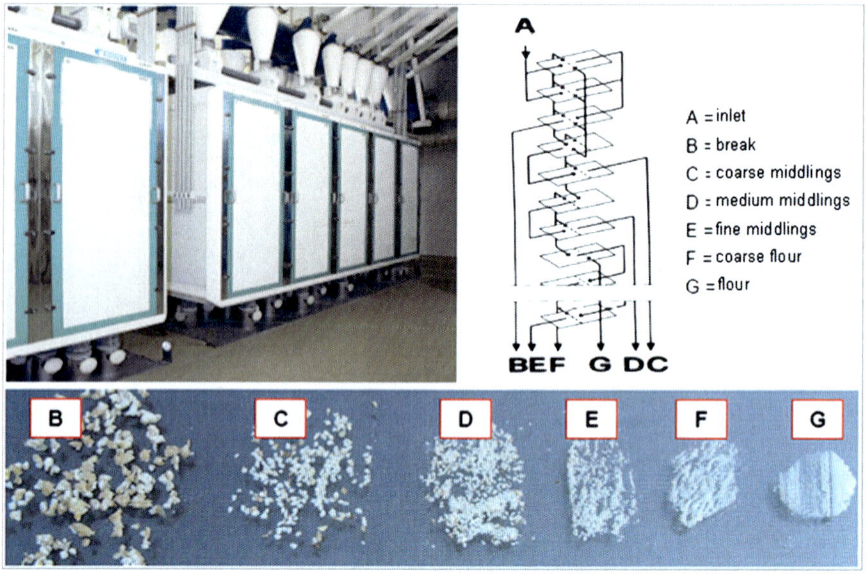

Fig. 3.2 Sieving or grading section: plansichters (*above*) and middling fractions separated by sieving action (*below*) (Courtesy of Bühler AG, Switzerland)

to the reduction system, the final stage of the milling process, with the objective to reduce the size of clean middlings to flour. It consists of a sequence of several smooth roll mills (up to eight to ten, according to the size and the expected starch damage) and sifters. The milling diagram comprises a number of the above steps, to ensure that the majority of the endosperm is converted into flour and that most of the teguments are removed as by-products.

The flour extraction yield varies between 74 and 76%. Since the bran and the germ together represent around 20% of the weight of the wheat grain, the flour extraction yield is lower than the theoretical value. More refined flours have lower extraction rates, since most of the external layers of the endosperm are eliminated with teguments. Milling of durum wheat requires different diagrams, which are characterized by a higher number of purifiers to improve the separation of bran particles. Nevertheless, the yield is lower (68–72%), since semolina is mostly formed by particles larger than 250–300 μm and contains only a minimum amount of fine particles (lower than 200 μm).

The separation of the more external layers and germ of the caryopsis inevitably causes a marked change of the chemical composition and, consequently, of the nutritional value of the flour. It has a lower concentration of ash, proteins, vitamins and soluble sugars than the caryopsis and a higher starch content. This difference depends on the efficiency of the separation of the more external layers of the caryopsis from the endosperm. Consequently, refined flours, corresponding to an extraction yield of approximately 75%, contain only 5% of fibre, 45% of fats, 30% of

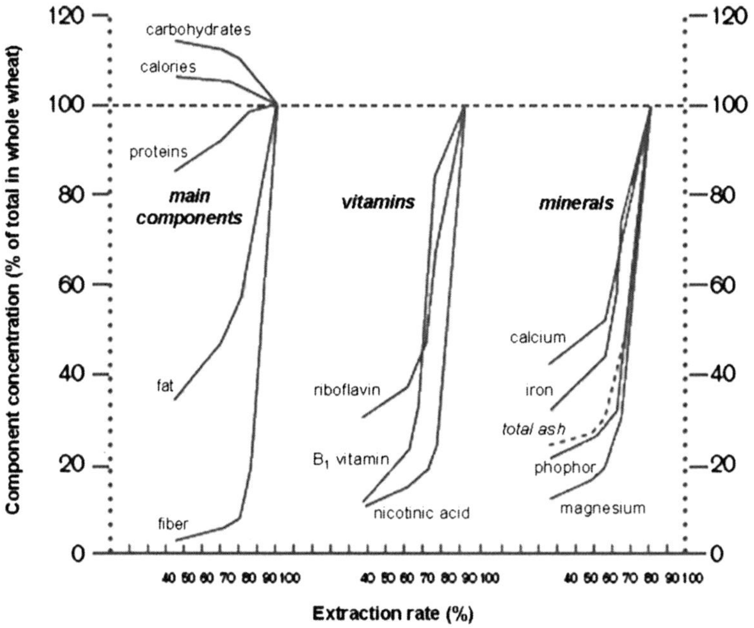

Fig. 3.3 Composition of flour according to the extraction rate: comparison with whole kernel (*dotted line* refers to wheat kernel) (Adapted from [15])

minerals and between 15 and 40% of different vitamins in comparison to the native caryopsis (Fig. 3.3).

3.2.1.1 Milling Optimization and Innovation

During the last decade, new food safety-related regulations, for example HACCP, ISO Standards 9001; 22000 and 22005, traceability, labelling health claim and use of GMO, (FAO/WHO, 1997) have supported the improvement of the process control in milling and bakery industries [16, 17]. In particular, technological solutions that not only consider the production and economic aspects, but also the hygiene and nutritional characteristics of flour were proposed.

Cleaning

The most recent innovation proposed for this stage is the use of an optical sorter (www.buhlergroup.com), a device that efficiently removes all types of contaminants and foreign materials in one step. The product stream is fed inside the sorting machine and different high-resolution cameras detect and recognize defects based on colour, shape and other optical properties. Specific sensors and high-speed ejectors carry out a precise ejection, enabling the accurate separation of contaminants in a very short time.

Debranning and Pearling

Debranning technology, also referred to as pre-processing or pearling, is currently used in rice milling and barley processing for removing hulls that have no technological and nutritional value. Although this practice is necessary in the case of covered cereals, the sequential removal of the outer bran layers prior to milling is quite uncommon in naked grain processing.

The original debranning systems for wheat are the Tkac process and the PeriTec process marketed by Satake Corporation. In both processes, two distinct machines assure the bran removal by friction (kernel to kernel) and abrasion (kernel to stone) actions [18, 19]. New debranning equipments for wheat were developed to carry out abrasion and friction on the same machine. The Vertical Debranner VCW (Satake Corporation; www.satake-group.com) includes two separate working chambers in the same equipment: the upper one has an abrasive zone where rotating abrasive rings work the grain against a peripheral slotted screen to eliminate the outer bran layers; partially debranned grains then enter the lower chamber, where friction completes the debranning process. Applying the Tkac system to several durum wheat varieties, Dexter et al. [20] found an increase in semolina yield and a higher semolina refinement. Consequently, the colour of pasta was improved. Other interesting results were also found at laboratory and industrial scales [21]. Because of the positive results obtained with durum wheat, attention was shifted to common wheat. Except for the decrease of the microbial contaminations [22] and bunt infections [23], the results are still contradictory. Currently, the pre-milling treatment of common wheat consists of a peeling process carried out with a machine (DC-Peeler Bühler AG – www.buhlergroup.com or the DHA Vertical Debranner from Ocrim S.p.A.—www.ocrim.it) which promotes a mild removal of the outermost layer of the kernel (maximum 1.5–2% of the grain weight) and, at the same time, enables one to reduce contaminations (bacteria, mycotoxins and heavy metals) by 50%.

3.2.2 Improvement of Flour Performance

The natural aptitude of wheat to be processed into varieties of baked goods is further improved by selective breeding research. These studies mainly concern changes of the quality/quantity of seed storage proteins. Nonetheless, the performance of wheat during processing is sometimes different from expectations due to characteristics that are markedly influenced by environmental and agronomic parameters [24]. The practice of improving the technological functionality of the flours by the addition of improvers is, therefore, significantly widespread. Among the most commonly used improvers there are enzymes (e.g. amylase, hemicellulase, lipase and protease) and emulsifiers (e.g. mono- and di-glycerides), which increase the volume of the dough and the gas retention and crumb softness of baked goods. It is interesting to note that many of these improvers are naturally present in the wheat grain but they are removed with the by-products (bran and germ) during milling.

3.3 Other Raw Materials

3.3.1 Water

Water plays a key role both during processing and for the shelf-life and sensory properties of baked goods. On the basis of the definition used for studying polymers, the adjective "plasticizing" is often used for water. This identifies a material incorporated into a polymer to increase workability, flexibility and extensibility [25]. The water is subjected to significant changes during processing, both in terms of absolute quantity (total humidity) and availability (relative humidity).

The amount of water added to convert flour to dough has to ensure the hydration of all hydrophilic components, especially proteins. The addition of this solvent determines the radical change of the three-dimensional conformation of proteins. In 1973, following their observations with an optic microscope, Bernardin and Kasarda described an "explosion" of flour particles when in contact with excess water, and the rapid formation of protein strings [26]. This spontaneous rearrangement is caused by the immediate exposure to water of the hydrophilic areas of proteins which are rich in polar amino acids, while the more hydrophobic areas are hidden inside. The interaction among different protein chains is ensured through the formation of disulfide bonds and through stabilization via hydrogen bonds and hydrophobic interactions. Gluten proteins undergo a kind of glass transition when they absorb water, passing from the hard and glassy state to the soft and rubbery state [27]. Cuq et al. [28] described the changes that occur during the bread making using state diagrams and phase changes. The glass transition is the period of marked increase in molecular mobility that involves amorphous polymers (e.g. proteins) or the amorphous areas of semi-crystalline polymers (e.g. starch). Amorphous polymers are in the glass state at low temperatures and/or with low water content. An increase of temperature or of the water content induces amorphous polymers to become soft and viscoelastic. When the water content is higher than 15–20%, the glass transition of proteins occurs below environmental temperature [28]. During mechanical mixing, the relative mobility of the protein molecules and their high reactivity cause the formation of intermolecular covalent links that are responsible for the formation of the continuous and homogenous gluten network.

The level of water needed to obtain an optimal consistency of the dough is not always easy to quantify. Overall, hydration less than 35% does not give an optimal and homogenous hydration of gluten [29]. The absorption of water varies according to the degree of refinement, granulometry, level of damaged starch, quantity and quality of proteins and humidity of the flour. The hydration capacity of the flour is usually calculated based on the farinographic absorption index (see Sect. 3.5.1.1). Although it is very practical for routine applications, this approach does not describe the distribution of water within the components and the competition among the hydrophilic compounds. The dough is a highly complex system where numerous aqueous stages coexist and each one is variously rich in chemical components [30]. The dough is a sophisticated metastable dispersed system, where water moves from one phase to another and where thermodynamic incompatibility among different polymers occurs [31]. The time needed to obtain homogeneous and uniform hydration of the flour particles is strictly related to the dimension of the particles, their hardness and vitreosity, the presence of non-starch polysaccharides and intensity of mixing.

Water is not only necessary for the formation of gluten, but it also performs a solvent action for the other ingredients present in the formula (e.g. salt and sugars) and it allows the enzymatic activities to take place. Another function played indirectly by water is the control of the dough temperature. After mixing, ca. 45% of the total water in dough is associated with starch, 30% with proteins and 25% with non starch polysaccharides (pentosans) [32]. Water allows the swelling of the starch granules during baking and their gelatinization, a key phenomenon for the physical and nutritional properties of baked goods [33].

3.3.2 Other Ingredients: Salt, Sugar and Fats

Flour, water and leavening agents are the indispensable ingredients for making baked goods. Often the formula requires the addition of salt, which influences the sensory and rheological properties of the dough, and additives or improvers. The use of salt in leavened baked products generally refers to sodium chloride. Salt is an ingredient that is almost always present in the formulation of bread and other bakery products. The role of salt is related to its ability to enhance the aroma of the product and to mask off-flavours such as a bitter and metallic taste, but salt addition also strengthens the structure of the dough. This effect on dough structure is caused by the positive effects on both hydrogen bonds and hydrophobic interactions between protein macromolecules. Salt optimizes the mixing time and the kneading, and controls the speed of yeast fermentation [34]. In dough without salt, the high speed of CO_2 production may be responsible for the deterioration in the product structure. On the contrary, NaCl allows the leavening step to be controlled and optimized by slowing the rate of gas production [35]. However, this positive effect is strongly influenced by the amount of salt added [36]. In particular, a great increase in volume was observed in bread prepared by adding 1.5–2% of NaCl, while quantities exceeding this threshold were associated with a strong decrease in the volume of the product.

Other ingredients such as sugar or fats are optional. The addition of sugar or fat over a certain value markedly changes the rheological properties of the dough. The presence of sugars influences each stage of processing: it gives more consistency to the dough, in some cases promotes the fermentation, facilitates the browning of the crust during baking and ensures a soft crumb during storage. Fats such as olive oil, lard or butter are also used. If the concentration of fat is lower than 5%, the main technological role is as a lubricant: the gluten increases the extensibility before rupture, favouring a higher dough volume [37]. If the formulation is very rich in fat, the dough becomes short and loses its extensibility. This effect justifies the definition of shortening for these components. Polar lipids such as mono- and di-glycerides stabilize the air bubbles formed during mixing and provide a crumb with a finer and more regular alveolar structure [38]. During storage, lipids prevent the interaction among starch macromolecules, slowing down their reorganization into ordered and crystalline structures (retrogradation), as well as the migration of water between starch and proteins, slowing down the phenomena of staling and aging [39].

3.3.3 Leavening Agents

Essentially three types of leavening agents are used for making baked goods: baker's yeast, chemical agents and sourdough. Since sourdough will be covered in detail in all the other chapters, this paragraph only aims at shortly describing the main features of the other leavening agents.

3.3.3.1 Baker's Yeast

Baker's yeast refers to *Saccharomyces cerevisiae* strains. After a brief initial respiratory activity due to the oxygen dispersed in the dough, baker's yeast ferments glucose, fructose, maltose and sucrose from the flour to CO_2 and ethanol. Nowadays, the types of baker's yeast available on the market have different shelf life (yeast cream, compressed yeast, dried yeast), osmotolerance features (suitable for baked goods with elevated levels of sugars) and activity at low temperatures (frozen dough) (see Chap. 6). The use of baker's yeast as a leavening agent is very often an alternative to the use of sourdough, especially for industrial bakeries. During processing, baker's yeast mainly determines the leavening of the dough due to the production of CO_2; it also synthesizes some volatile compounds that positively affect the flavour and taste of baked goods.

3.3.3.2 Chemical Agents

The production of CO_2 within the dough can also be obtained by a reaction between sodium bicarbonate and an acid (e.g. tartaric acid). Chemical agents are not used for bread making but for sweet baked goods, both dry and light (such as biscuits, sponge-cakes, etc.). The use of chemical agents for leavening is recommended when CO_2 has to be produced rapidly in doughs rich in sugars and fats, ingredients that slow down and/or inhibit the metabolism of the biological agents. Sodium bicarbonate lacks toxicity, is cheap and easy to use, and it has a high solubility at room temperature. On the basis of the rapidity of the reaction, the chemical agents are classified as: fast or immediate, slow or delayed, and double acting powders. The former develop gas as soon as they are introduced into the dough. The delayed acting powders determine the formation of negligible quantities of gas during mixing, which only develops at high temperatures during baking in the oven. The double effect powders react in part at room temperature and in part during baking, and in this case two acids are involved, one soluble and one insoluble. The reactivity of a chemical agent is expressed as the "neutralization value" (g of NaOH that are neutralized by 100 g of acid salt).

3.4 Baked Goods Making Process

A very large number of baked goods are manufactured worldwide. Breads are the most diverse and several differences are found, for instance, between those manufactured in the Mediterranean areas and those from the Anglo-Saxon market [40]. Overall, in the Mediterranean areas no significant quantities of sugar or other high hydrophilic substances are added and pans are generally not used in the leavening and baking stages.

Apart from the large variety of baked goods, the technological process may be summarized in a sequence of operations that require long periods of time and which have the primary objective of aerating the dough and making it porous (Table 3.2).

3.4.1 Discontinuous Processes (Straight-Dough and Sponge-and-Dough)

Bread making, both at artisan and industrial levels, is traditionally a discontinuous process since the various stages of mixing, leavening and baking are carried out on limited quantities of materials and in separated facilities. Discontinuous bread-making processes are performed using the straight-dough or the sponge-and-dough methods. Bread making with sourdough could be considered as a particular sponge-and-dough method. Bread characteristics are influenced not only by processing but also by flour and formula. In the straight-dough method, the fastest and easiest to manage, all the ingredients are mixed together simultaneously to form a dough which is then left to rise (Table 3.3). Fermentation has to be carried out at least in two stages. The first leavening is generally made with large quantities of dough, and for variable times (from 30 min to 3 h), depending on the process. The primary objective of this operation is not to get a volume increase, but to induce changes in the rheological properties of the dough [41]. The first leavening provides a greater workability to the dough, which achieves the capacity to maintain shape during the second leavening (or proofing). In this stage, individual pieces of dough, corresponding to the final size,

Table 3.2 Bread-making process: aim and modifications associated with the main operations

Step/phase/operation	Aim	Modifications
Mixing	Homogeneous distribution of ingredients (including minor components)	Hydration and solubilization of the water compounds
	Formation of uniform and "coherent" structure	Formation of soluble gluten
	Inclusion of air bubbles	Inclusion of air microbubbles
Leavening/proofing	*Increase in volume of dough*	*Formation of gas (CO_2)*
	Development of typical flavour	Production of fermentation metabolites important for developing flavour and able to change the macromolecules solubility
Shaping	Giving shape to dough Division of dough into final pieces	*Subdivision of gas bubbles and inclusion of new air*
Cooking	Giving the product its typical aspect	*Increase in volume due to evaporation of gases: 20–30% of the volume is obtained during baking (oven-spring)*
		Formation of crust and crumb
	Decrease of water content	Protein denaturation
	Stabilization of leavened and shaped dough	Starch gelatinization
	Making product appetizing and digestible	Development of flavour
	Completing leavening of the dough	Evaporation of water and ethanol
Cooling	Product packaging	Change of solubility of sugars
		Hardening of fats

Table 3.3 Characteristics of bread obtained according to the nature of the flour (straight-dough process using 100 g flour)

	Bread			
Cereal flour bread	Volume (mL)	Height (mm)	Weight (g)	Specific volume (mL/g)
100% wheat flour	900	121	140	6,4
100% einkorn flour	440	74	142	3,0
70% wheat flour + 30% rye flour	485	85	148	3,3
70% wheat flour + 30% maize flour	440	75	146	3,0

are maintained at controlled temperature and humidity conditions for approximately 1 h until the maximum volume is reached.

With the sponge-and-dough method, the ingredients are added at different times, during the refreshments of the dough. Using baker's yeast, the sponge starts by mixing compressed yeast with part of the flour and water (from 50 to 70% of the total dough) required for the formula. After a leavening phase (from 3 to 4 h up to 10 h or more), the remaining parts of flour and water are added, together with any other ingredient required. During the maturing stage, the yeast adapts to the environment and reaches the optimal fermentative capacities when all the ingredients are added. The final dough is cut into pieces, shaped and left to rise again for about 1–2 h, before being baked. The long leavening time required to get the sponge ensures an alveolar structure with a large number of bubbles in the product, some of them of considerable size [42]. This structure guarantees extreme lightness, as indicated by the high specific volume and porosity of the crumb (Table 3.4).

According to the consistency of the sponge, different definitions of the process are given. Polish, Poolish or Viennese methods use highly fluid dough. This process was initially introduced in Poland and subsequently widespread in Vienna at the beginning of the last century [41]. The Poolish premix is generally prepared using 50–75% of the water required by the recipe, to which an equivalent amount of flour is added. This ratio gives a mass of low consistency that is left to ferment from 3 to 8 h, according to the quantity of baker's yeast added. This method gives sensory advantages and allows the formation of a delicate aroma.

Leavening with sourdough requires the use of a starter, represented by a piece of dough from a previous batch which is fermented and stored under controlled conditions of temperature and humidity. Sourdough is generally used at 5–20% of the formula. Refreshing is of fundamental importance for dough development and maintains the microbial species typical of the sourdough [43, 44]. The intense acidification markedly influences the sensory and shelf-life features of the baked goods [43]. In particular, making bread with sponge and dough methods, and especially using sour dough, ensures a higher initial lightness compared to the straight dough method (Table 3.4). The bread is characterized by a more highly developed and irregular alveolar structure, since the longer fermentation times determine a slow and progressive production of CO_2, accompanied by coalescence phenomena. The bread also has a characteristic taste and smell, caused by the formation of volatile compounds that are mainly formed during baking, following the Maillard reaction between glucose and free amino acids, which are released during fermentation.

In Anglo-Saxon countries, and in particular in Great Britain, the most widespread bread-making process is the straight dough method known as the Chorleywood Bread Process (CBP). It was developed in the 1960s by the Flour Milling and Baking Research Association at Chorleywood in England [38, 45]. The leavening period is substantially reduced thanks to the high-speed mixers used for mixing the dough (see Sect. 3.4.3.1). CBP uses an intense mechanical mixing of the dough that lasts just a few minutes and ensures the incorporation of large quantities of air inside the mass. The decrease of time limits the cost of the process but the special conditions used during mixing require enriched flours and formulas with elevated concentration of baker's yeast (higher than 2.5–3.0%). Emulsifiers and fats with a high fusion point are also indispensable for the oxidizing properties and to stabilize the numerous tiny alveoli that develop from the microscopic air bubbles included during the mixing stage. Another characteristic of this process is the reduction of the pressure during mixing to adjust the size of the alveoli.

3.4.2 Continuous Processes

These processes were introduced in the USA during 1950. The two best-known types are the Do-Maker process, developed in 1954, and the Amflow process, developed in 1960 by AMF Incorporated [46]. These technologies took hold in the 1970s, and were widespread especially in the United States and Great Britain.

Contrarily to the discontinuous process working in batches, the continuous process is characterized by a substantial reduction of time, more compact machinery and better durability of the characteristics of the baked goods [47]. These processes are based on the possibility of eliminating the long leavening times by using yeast cultures or pre-ferments propagated separately without or with small quantities of flour. The subsequent high-speed mixing, with the simultaneous addition of all the ingredients, favours volume development even without long leavening times. As for the Chorleywood process, the intense mechanical stress during the high-speed mixing can be "supported" by the dough only if strong oxidizing improvers are added; emulsifying lipids are also indispensable.

3.4.3 The Main Stages of the Process

3.4.3.1 Mixing

As shown in Table 3.2, during mixing ingredients are distributed and blended within the mass, and gluten is formed. These phenomena are described as dough development. Many variables are involved. One of these is the quantity of water added to the flour, which may be indicated as the "level of absorption" or "hydration". In some processes the level of absorption does not correspond to the optimal quantity as determined by the farinograph but it is mainly related to the handling characteristics of the dough. Stiff dough with hydration levels between 40 and 45% (dough humidity: 38–41%) has reduced extensibility; consequently, the baked goods have a limited porosity with a very fine alveolar structure. Soft or slack doughs have hydration levels higher than 60% (dough humidity: about 50%). They are difficult to handle due to their low consistency, which is responsible for the long and irregular shape such as the Italian Ciabatta [2]; the crumb presents large alveoli, often long in shape, which result from the coalescence of smaller bubbles.

Table 3.4 Physical characteristics of flour, dough and bread according to the bread-making process

	Specific volume (mLg^{-1})	Porosity (%)	
Flour	1.5	–	
Dough	0.85	–	
Proofed dough	0.9–1.0	–	
Wheat bread			
– Straight and dough	2.5 ÷ 3.0	20 ÷ 25	
– Sponge and dough	3.0 ÷ 10.0[a]	25 ÷ 75[a]	
– Sourdough	3.2 ÷ 3.7	35 ÷ 40	

[a]According to the parameters of the technological process

Mixing during traditional processes is carried out at least in two different times with different speeds. In the first stage (the French *frasage*), the low speed lasts about 5 min. Water is distributed among ingredients to allow the hydration of the protein macromolecules and the development of gluten. Air bubbles are entrapped into the mass and their sizes vary between 30 and 100 μm [48]. This phenomenon is completed during the high-speed stage (the French *malaxage* and *soufflage*). At the microscopic level, the sequence of events is associated with significant rheological changes and the mass is rather wet and sticky until hydration is completed during the cleanup stage (Fig. 3.4). Development of the dough is completed when the mass clears away from the walls and blades of the mixer bowl and starts to crackle in the bowl. Furthermore, stretching deforms the dough without breaking until it becomes a semi-transparent film [47], thanks to the viscoelastic properties of the gluten.

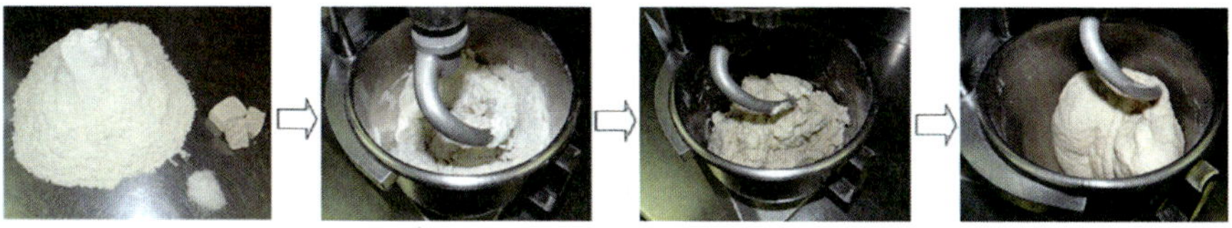

Fig. 3.4 Raw materials and "evolution" of dough aspect during dough-making

The air included at the end of the mixing stage represents ca. 8–10% of the volume of the dough [48]. From this point onwards, any further mixing causes irreversible changes of the rheological properties. The over-mixed dough becomes sticky and loses elasticity.

The Mixers

The mixers are usually classified into two categories: vertical and horizontal mixers. The former consist of machines with rotating bowls of various capacities ranging from 100 to 3,000 kg and more. The first machines, known as Artofex, have two reciprocating arms whose movement, both circular and vertical rotation at moderate speed, simulates the movement of the baker's arms. The action performed by the dual-arms is very delicate. Consequently, the time necessary to completely develop the dough is long and the productivity of the machine is low. Currently, they are almost completely substituted by spiral mixers. The mechanical action against the dough is usually completed by a rotating central post, a device whose function is to hold and expose the dough to the action of the spiral tool, decreasing cutting effects with a smoothly mixing action. The different types of mixer vary depending on the capacity and rotation speed of the bowl, the arrangement and rotation speed of the mixing tools, the possibility of cooling down the mass via insufflation of CO_2 and the possibility of extracting the bowl to guarantee the easy movement within the working premises.

As mentioned previously, high-speed mixers, capable of completing the process in a few minutes and ensure the retention of large volumes of air, were developed to increase the productivity of the bread-making process. The horizontal bowl, normally made of stainless steel, presents a single mixing shaft with several transversal bars, whose profile varies in function of the process involved. Auxiliary equipment includes the microprocessor controls for monitoring all mixer functions. These devices ensure constant and uniform characteristics in all dough batches. In the Chorleywood method, the Tweedy mixers are preferred (www.bakerperkinsgroup.com). These high-intensity mixers have an impact mixer plate at the bottom of the bowl. Some baffles are present at the sidewalls of the bowl to direct the dough towards the mixer plate. The rotation speed of the mixing device may exceed 300 rpm, while the bowl remains fixed. These conditions are suitable to fully develop dough in only 4–5 min.

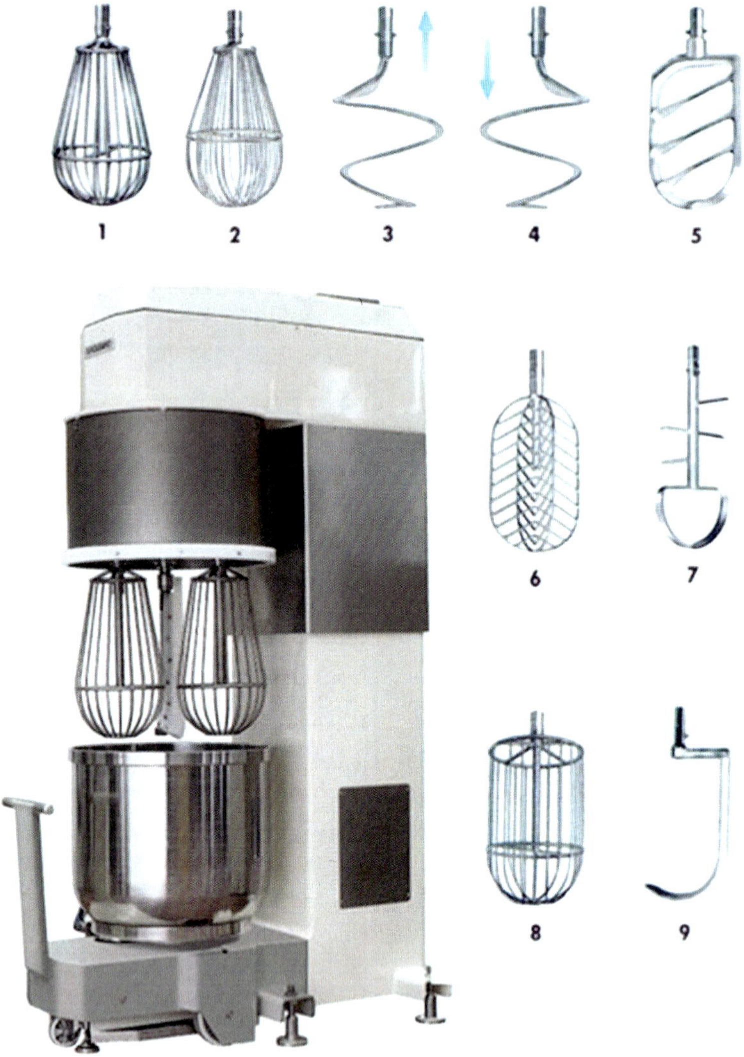

Fig. 3.5 Planetary mixer and examples of mixing tools (Courtesy of Sancassiano S.p.A., Italy)

The quantity of air retained in the dough varies based on the mixer characteristics, flour strength and the addition of specific ingredients. To obtain batters (highly aerated dough for sugar- and fat-based cakes), the planetary mixers are used. Their mixing tool is similar to a whisk (Fig. 3.5) and performs a whipping action on the mass. The presence of emulsifiers is indispensable to stabilize this very low density foamy mass and to produce an extremely fine grain and uniform texture.

According to the mixing tool shape and the speed applied, the time needed to obtain a well-developed dough or a light batter varies from 30 to just a few minutes. Consequently, the temperature of the mass remains the same or just a little higher

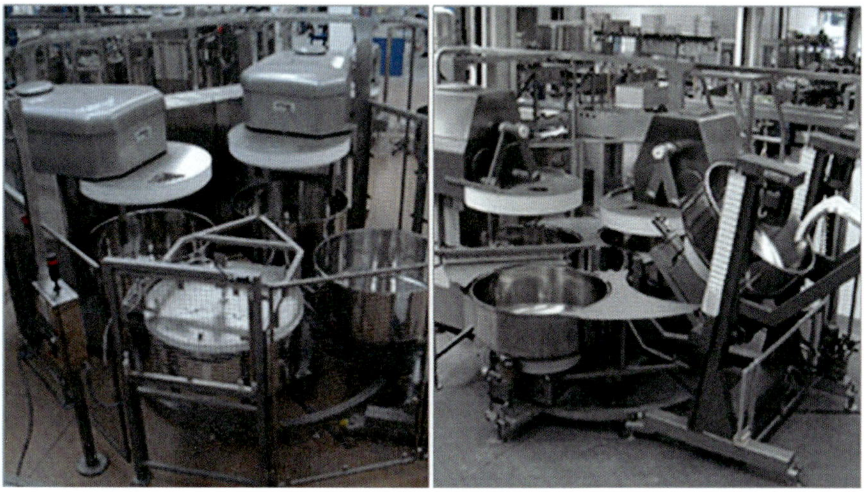

Fig. 3.6 Carousel system (Courtesy of Sancassiano S.p.A., Italy)

than that of the ingredients (mixers with dual arms or spirals, 20–60 rpm) or may increase even by 10–12°C in the case of ultra-high-speed mixers, working at 500–1,000 rpm.

The mixers previously described work discontinuously. The developed dough must be removed and the bowl emptied ready for the next batch. This system has a limited effect on the organization of the work at artisanal level but at industrial plants it creates serious problems. One proposed solution consisted of the carousel, a modular system which, although it did not shorten the mixing times, guaranteed the availability of a well-developed dough at set times, with standardization and an almost continuous feeding of the machines. As shown in Fig. 3.6, the carousel (www.sancassiano.com) is formed by a number of stations that move automatically (controlled by a Programmable Logic Control, PLC), rotating and occupying different positions at set times. It passes from the first position (raw materials are dosed into the first bowl) to intermediate mixing and kneading positions up to the last position where the dough is discharged and poured out into the feeding hopper of the next machine. The time between the first and the last position corresponds to that necessary to develop a mass with the desired rheology and texture characteristics.

In recent years, the companies working in this sector have put forward solutions for further improvement of the automation and versatility of this stage of the industrial process. The different stages of dough development and leavening are controlled by PLC in robotized plants. A variable number of bowls are handled and moved by a robot shuttle which takes each bowl from a parking area and subjects it to the various work stations, on the basis of the sequence of the production cycle. The advantages of this plant are many, including high flexibility, capacity to satisfy all types of technological cycles (direct, indirect, etc.), possibility of feeding several lines in parallel (even with different types of dough) and high levels of hygiene and cleaning.

3.4.3.2 Leavening

The production of any leavened baked goods concerns the transformation of the semi-solid mass of the dough, a kind of emulsion with a continuous phase represented by hydrated gluten that surrounds the starch granules, and a dispersed phase, consisting of microbubbles of air, into a foam, where the continuous phase retains considerable volumes of gas (Table 3.2).

Leavening is the stage associated with the significant expansion of the original volume of the mass. This is possible thanks to the viscoelastic properties of the dough, and in some cases also to the presence of emulsifiers. The number of alveoli retained in the mass upon completion of the mixing is estimated to be between 10^2 and 10^4 per mm^3 [50]. Volume expansion can be obtained using biological and chemical leavening agents and also through the physical approach. In this latter case, the inclusion of air follows intense mechanical actions during mixing. Mixed leavening is also considered. Microalveoli incorporated during mixing (physical leavening) are further expanded following the chemical leavening in the oven. Danish pastry, used for making particular sweet products, is obtained from mixed leavening: the dough is first biologically leavened, then formed with a lamination process to distribute the fat in thin and alternate layers within the dough. Although CO_2 is considered the major gas responsible for the development of dough volume, other gases and low-boiling substances may interfere with the overall volume of the dough. For instance, ethanol is solubilized in the aqueous stage of the mixture and forms an azeotrope with a boiling point of 78°C and water vapour.

The most obvious phenomenon associated with leavening is the volume expansion. The CO_2 produced is solubilized firstly in the aqueous stage of the dough. Once saturation is reached, the gas settles in the bubbles entrapped in the dough gradually dilating and expanding them, without any breakages. The pressure inside the alveoli increases but the dough reacts by stretching thanks to gluten viscoelasticity. The high diameter of the bubble makes it possible to balance the overpressure that is created. The film (ca. 1 μm) created by surfactants, soluble proteins, polar lipids, or pentosans, on the surface of the alveolus plays the principal role in this phenomenon [45, 50, 51]. The leavened dough is therefore a foam consisting of a semi-solid aqueous phase where gas bubbles are distributed. The coalescence of these gas bubbles is delayed as long as the lipoprotein film is able to expand, reducing its thickness. Its breakage is associated with the merger with adjacent bubbles.

Acidification during sourdough fermentation also influences the rheological properties of the dough. As shown by extensographic analyses, the acidification determines the full maturation of the dough. The extensibility of the dough is modified so that it can better support the dividing and final moulding stages. A fully maturated dough will break clean and sharp with minimum resistance to pull [47]. The reasons for this behaviour are numerous, complex, and only partly understood.

They are variously attributed to the continuous and progressive hydration of the proteins, to the presence of metabolites (CO_2, organic acids) that determine a new organization of gluten, and to changes of the aqueous phases where polymers are immersed. In some cases, the rheological changes are hidden by amylase and protease activities that occur simultaneously during leavening but with opposite effects.

Fermentation and Proofing Rooms

At the artisanal level, fermentation is carried out in chambers or cabinets (for limited daily production) where the temperature and humidity are kept constant. The recommended temperature ranges between 27°C and 37°C and the environmental humidity has to be above 75–80% [47]. Lower relative humidity causes the formation of a surface skin that impedes dough development during baking. Optimal temperature and humidity conditions are maintained thanks to air conditioning systems. The baked goods are usually arranged on trays located on mobile trolleys that facilitate movement. Long fermentation times, especially in the sponge-and-dough methods, require one to work overnight. Today, equipment called the "retarder" or "sponge conditioner" allows the optimization of the operations by controlling the kinetic of fermentation at low temperatures [41]. These machines are thermostatic chambers that allow one to control and monitor the temperature (from −10°C to +35°C), through the presence of refrigerator groups and resistors, and the relative humidity. After mixing, the dough is placed inside the chamber where the temperature is lowered and then slowly raised to achieve the fermentation. This optimization of the process reduces or eliminates the night shifts, and distributes the work during the day, guarantying fresh baked goods also in the evening.

At the industrial level, the proofing chambers usually consist of a tunnel with dynamic transport and automated controls of the environmental variables. The dough is placed on trays or belts and proceeds towards the exit, thanks to catenaries. The tunnels are sized so that the leavening time coincides with the time needed for the dough to travel inside.

3.4.3.3 Dough Makeup Operations

As briefly described before, the two leavening stages are alternated by dividing, and rounding and moulding operations which were previously carried out manually. Currently, these operations are performed by automated machines with different working capacities.

Generally, dividing is carried out within the shortest possible time on a volumetric basis. The fermented dough is inserted into a chamber with a piston whose course is directly proportional to the volume to be divided. The individual piece of dough is cut off by a knife, then ejected from the chamber and shaped. This operation is usually alternated with a period of rest necessary to allow the dough to recover from physical stresses associated with compression, followed by cutting and shearing. The working of the divider is completed by rounding and moulding the individual pieces with different machines. Generally, three steps are carried out: the sheeting of dough pieces into a uniform layer, the rolling into a cylinder by means of a rounding or curling machine, and compression to obtain the desired shape.

3.4.3.4 Baking

Baking is considered the most important stage of the entire cycle. During baking, the dough heats up and loses humidity. Heating occurs from the outside towards the inside, water loss occurs in the opposite direction. These two phenomena cause multiple changes, differing in their physical, chemical and biochemical nature and intensity, according to the temperature and the area of the dough. The change from the foam state to the sponge state [52], and the diversification between crust and crumb are observed.

The sequence of changes (Table 3.5) is different according to bread area. The temperature inside the dough is always below 100°C, while on the surface it reaches 180–200°C. As soon as the leavened dough is inserted into the oven, the fermentative activity increases [38] until microbial death occurs at temperatures higher than 50°C. Heating causes a further significant volume expansion (oven spring) of about 40% of the volume compared to the leavened dough, corresponding to an increase of the surface area of 10% [49]. The volume occupied by gases, CO_2, ethanol vapours, water vapour, increases as the temperature increases. Starting from 70°C, the chemical and biochemical transformations of the macromolecules stabilize the complex. A porous network of interconnected alveoli separated from each other by a solid matrix with very fine walls is formed [52]. During this passage, proteins and starch achieve new properties. The gluten is denatured, completely loses its extensibility and achieves elasticity. The starch swells up and gelatinizes. The intensity of these two phenomena depends on the distance from the geometric centre of the dough. Baking is completed when, even in the most internal part of the dough, the temperature has reached values that promoted the structural consolidation. The temperature at the centre point has to be in the range of 90–95°C to prevent collapse due to a non-rigid structure [52].

Because of the temperature gradient that is created in the dough during baking, the surface, which is exposed to the oven temperature from the beginning, reaches the sponge state much more quickly than the internal part. The surface areas, therefore, become more and more dehydrated and permeable, facilitating evaporation and the release of the water vapour that is generated and accumulated within. Once baking is completed, the crust has a humidity of less than 5% [2]. This value ensures friability and crispiness. The internal crumb retains higher humidity and remains soft and light.

The complex chemical reactions that occur during baking are of marked importance for the aroma and taste of baked goods. The starch in the crust is degraded into dextrins between 110 and 140°C. Caramelization starts at 140–150°C and continues, producing pyrodextrins, at higher temperatures. Proteins irreversibly react with

Table 3.5 Changes during baking and storage of bread[a]

	During baking		During storage	
	Macroscopic level	Molecular level	Macroscopic level	Molecular level
Crust	Evaporation of gases	Protein/sugar interaction	Loss of crispness for increase in crust moisture	Water migration from crumb to crust macromolecules
	Progressive drying	Dextrinization/caramelization		
	Non enzymatic browning	Maillard reaction		
Crumb	Water migration towards crust	Protein coagulation	Increase in hardness	Starch retrogradation
	Structure strengthening	Starch gelatinization	Crumbling tendency	Water migration exchange at inter and intra macromolecular degree
		Water retention by non-starch polysaccharides	Loss of typical flavour	Interaction among aromatic compounds/macromolecules
		Change of fat structure	Appearance of "stale bread" flavour	Oxidative phenomena

[a]According to bread size

sugars in the Maillard reaction, forming a number of compounds responsible for the colour and typical aroma of the crust [53]. If the intensity of the Maillard reaction does not exceed certain limits, these effects contribute to the sensory properties. On the contrary, the protein–sugar interactions lead to unavailable lysine and, in advanced stages, cause the synthesis of toxic compounds. The *International Maillard Reaction Society* site (http://imars.case.edu) provides more information on the nutritional and sensory properties of baked goods.

The intensity of the colour of the crust is strictly related to baking temperature, while the thickness of the crust is influenced by baking time. Baked goods undergo a loss of weight during baking, a key step of the process. Usually, bakers aim to obtain the highest yield according to the values of humidity which are allowed by national regulations. In that sense, the practice of including vapour in the oven during the initial baking is justified. The vapour condenses on the surface of the dough, accelerates heat transfer to the dough, slows down evaporation, and decreases weight loss [45]. As a further positive effect, the viscoelastic properties are retained for a longer time on the surface of the dough, allowing a higher development of the baked goods.

In the oven, the heat transfer towards the dough occurs by radiation from the walls and by air convection. Inside the dough, the transfer occurs by conduction. Because the temperature is usually between 200 and 230°C, the kinetic of baking is quite slow and provides the consolidation of the internal structure without unacceptable scorching of the crust.

Ovens

Regardless of the characteristics of the oven, the baking floor is called the sole or deck and the upper part of the chamber the crown.

Ancient ovens, often made of stone or bricks, had a single chamber where combustion and baking occurred. This type of baking is referred to as a "direct firing system" because the combustion gases are in contact with the dough. The initial temperature reaches 350–400°C and decreases during baking to 160–170°C. These ovens, today used only for special traditional or typical breads (e.g. Italian *Altamura* bread or Arabic bread), were replaced by ovens with an "indirect firing system", where the combustion area is separated from the baking chamber and the combustion gases do not come in contact with the dough but circulate in tubes above and below the baking surface [47]. This configuration provides a higher uniformity of heating and guarantees hygiene. The heat transfer may be improved using forced air systems inside the chamber (ventilated ovens) or forcing the circulation of the combustion fumes into tubes via a ventilator (Cyclotherm ovens). This method also improves the heat exchange thanks to the repeated passage of the combustion fumes within the tubes positioned above and below the conveyor. Heat is transferred to the

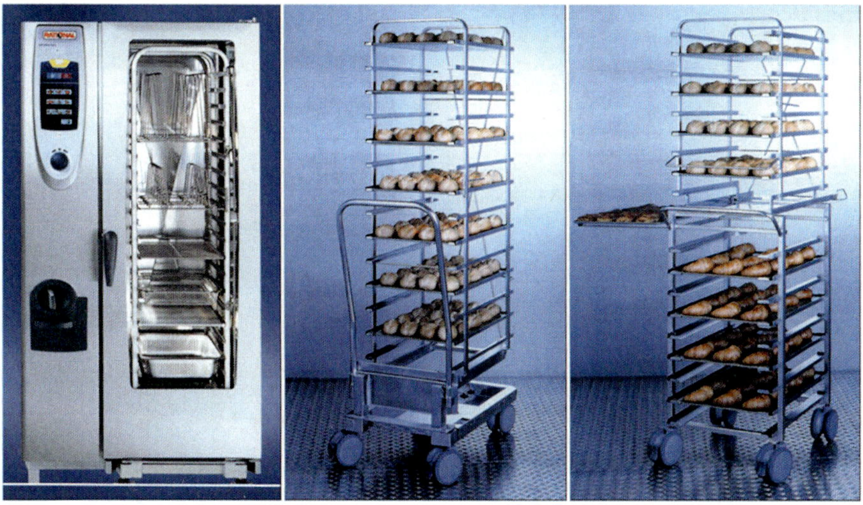

Fig. 3.7 Modern rack oven (Courtesy of Rational AG, Germany)

dough mainly by radiation. Several oven doors are installed, which change the flow of the fumes towards the upper or bottom area of the chamber to optimize baking based on the dough characteristics. Systems that provide a better heat yield, with an energy saving that may reach 30% compared to those from the conventional thermal cycle ovens, are also available.

In artisanal bakeries, the most common ovens have fixed decks with separate chambers which are arranged vertically or consist of a cabinet equipped with a rotating rack carrying trays or frames (Fig. 3.7). In the former ovens, baked goods have to be loaded and unloaded by hand using long peels or special loading devices (Fig. 3.8). This operation requires time and skill. More recent solutions make it possible to vary baking conditions for each chamber. In the rack ovens, baked goods are placed on the pans or trays located on the rack (often during the leavening stage) that is inserted into the oven and is rotated to give a better uniformity during baking.

In industrial bakeries, baking is a continuous operation performed in a long horizontal tunnel with different sections or zones, each one having its own burner and where the temperature is variable. Shutters control the evacuation of the water vapour which accumulates in the chamber towards extraction flues. Baking time is determined by the speed of the belt that transports baked goods and by the length of the oven. According to the heating system, ovens are heated by gas, fuel oil or electricity. Microwave ovens combined with traditional ovens are also proposed, but their application is suitable only for specific industrial purposes.

Fig. 3.8 Artisanal bakery (Courtesy of Wizard Bakery Company, Germany)

3.4.4 *The Production Units*

3.4.4.1 Artisanal Production

The equipment of a modern artisanal baking laboratory is shown in Fig. 3.8. Although it covers a limited surface area, the solution is highly rational as it combines machineries based on the different stages of the technology process. The mixing areas (Fig. 3.8a) are adjacent to the shaping areas (Fig. 3.8b), while the leavening section (Fig. 3.8c, d) is next to the oven area (Fig. 3.8e, f) to facilitate the movement of the dough and its introduction into the oven.

3.4.4.2 Industrial Production

At the industrial level, processing is carried out continuously. All the different stages, which are identical to those of the artisanal process, are connected via conveyor belts that carry the semi-finished product to the next stage of the cycle. Rational solutions (see www.wpib.de/; www.itecaspa.com/; www.esmach.it/) provide long horizontal systems with linear transportation to avoid bends or turns. Modern plants consist of completely automated systems of mixing area, controlled by a robotized centre that exclude manual work for dosage of the ingredients, mixing and movement of the bowls to the successive stages. After cutting and shaping, the dough comes to the continuous proofing chamber, whose horizontal or vertical development is proportional to the time needed to complete this stage. Baking under a continuous belt oven is followed by a long cooling stage, which is controlled by circular net transporters.

3.5 Quality Assessment of Dough Baking Properties

Apart from information on the composition of the raw material, a complete knowledge of the changes that occur during the whole technological process is necessary to assess the quality of baked goods. Many instruments and techniques have been developed for this purpose.

3.5.1 Rheology and Descriptive Empirical Measurements

Rheology is the study of the flow and deformation of materials in response to the application of mechanical force. The force is usually defined in terms of stress, the amount of force applied per unit area, with strain being the resulting deformation. The rheological features of the dough are important throughout the bread-making process and determine the quality of baked goods [54]. Rheological measurements are carried out to obtain a quantitative description of the mechanical properties of the materials as well as information related to their molecular structure and composition [54]. Usually, rheological techniques are classified based on the type of strain imposed (e.g. compression, extension, shear, torsion) and on the relative magnitude of the imposed deformation (e.g. small or large deformation). The main techniques used for measuring the properties of cereals are descriptive empirical techniques and fundamental measurements.

3.5.1.1 Quality Assessment of Flour and Descriptive Rheology of Dough

After mixing, subjective manual assessments of the dough were used for a long time to indicate whether it was suitable for processing and baking. Over time, a significant use of descriptive empirical measurements of rheological properties was observed. Devices such as the Penetrometer, Texturometer, Consistometer, Amylograph, Farinograph, Mixograph, Extensigraph, Alveograph and Rheofermentometer were developed [55] (Fig. 3.9). Different users may have different requirements. For instance, plant breeders require rapid and automated tests which use small amounts of sample, while millers and bakers require rapid and reliable tests to assay the baking quality [56]. All the tests mimic the complex responses (deformations) in the dough when subjected to various stresses (mixing, overpressure during leavening and baking, etc.). Nevertheless, the behaviour of the dough matrix is non-linear since the deformation is not proportional to the strength that determines it. For instance, the dough appears harder and stiffer when deformed slowly [57]. Therefore, the results from these tests provide useful information only if the nature and intensity of the deformations are similar to those that occur during in situ processing. According to these considerations, the definition is of imitative or empirical or descriptive rheology [45].

Empirical rheology is divided into two main classes [57]. The first class includes instruments that measure the viscosity variation of the dough during mixing/torsion at room temperature (e.g. Brabender Farinograph, Micro-DoughLAB, Mixograph) or according to a gradient of temperature (e.g. Mixolab). The Farinograph (www.brabender.com) measures the water absorption, dough development time, stability and softness. The Mixograph (www.national-mfg.com) determines the mixing time, maximum resistance and tolerance [57]. However, to determine the mixing properties of the dough, mechanical parameters such as mixer speed, mixing bowl capacity, and mixer geometry have to be precisely controlled. The Mixolab (www.chopin.fr) is a relatively new instrument which resembles the Farinograph and Viscoamylograph in terms of performance. It determines in real time the torque (Nm) produced by the dough between the two blades and, once the dough is formed, the device measures its behaviour as a function of time, mixing development and temperature. The second class includes equipment with devices that record and quantify the elastic characteristics of the dough, which are correlated to the resistance that the dough opposes to stretching that is protracted until rupture occurs. The most used are the Alveograph (www.chopin.fr) and the Extensograph (www.brabender.com). The Extensograph performs an extensional test where a cylindrical dough sample is clamped horizontally in a cradle and stretched by a hook which is placed in the middle of the sample and moves downwards until the rupture occurs. The Alveograph uses air pressure to inflate a thin sheet of dough and measures the resistance to expansion and the extensibility of the dough by providing the measurement for maximum over pressure, average abscissa at rupture, index of swelling and deformation energy of dough. Another interesting instrument is the Rheofermentometer (www.chopin.fr). It enables the measurement of gas production and retention, dough permeability, and volume and tolerance during leavening. It is used to determine the quality of the flour, the fermentative activity, the quality of the protein network, and the activities of additives to be selected. Through height and pressure sensors, it measures the fermentation and development of the dough under a weight imposed on it. During recent years, the Texture Analyzer (www.stablemicrosystems.com) has found

Fig. 3.9 Representation of the main tests used for the quality assessment of dough baking properties

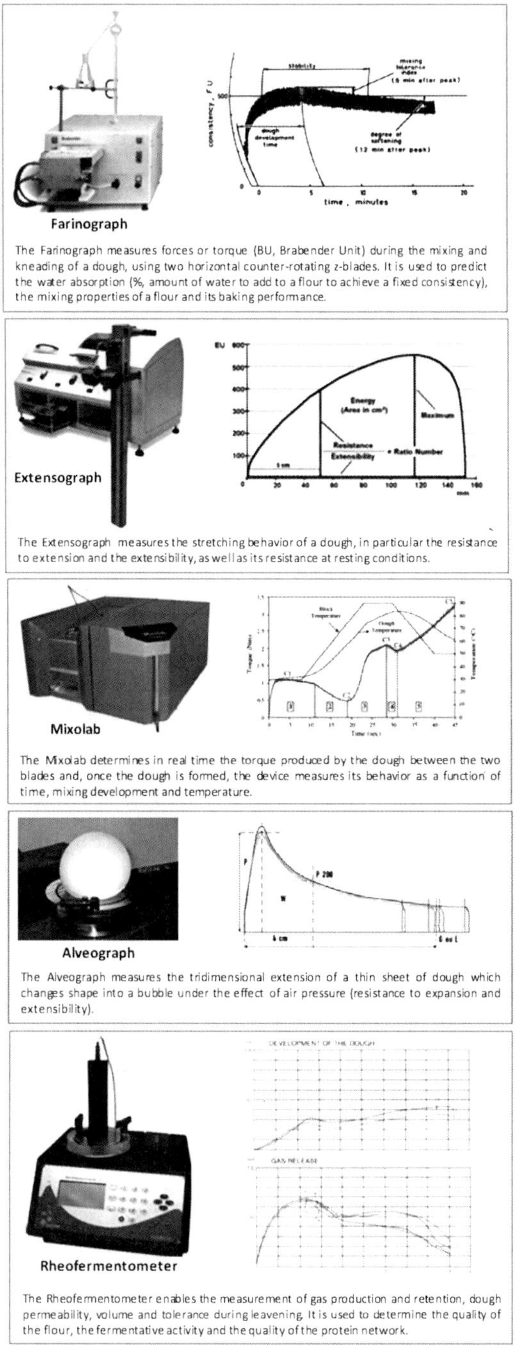
time , minutes
Farinograph
The Farinograph measures forces or torque (BU, Brabender Unit) during the mixing and kneading of a dough, using two horizontal counter-rotating z-blades. It is used to predict the water absorption (%, amount of water to add to a flour to achieve a fixed consistency), the mixing properties of a flour and its baking performance.
Extensograph
The Extensograph measures the stretching behavior of a dough, in particular the resistance to extension and the extensibility, as well as its resistance at resting conditions.
Mixolab
The Mixolab determines in real time the torque produced by the dough between the two blades and, once the dough is formed, the device measures its behavior as a function of time, mixing development and temperature.
Alveograph
The Alveograph measures the tridimensional extension of a thin sheet of dough which changes shape into a bubble under the effect of air pressure (resistance to expansion and extensibility).
Rheofermentometer
The Rheofermentometer enables the measurement of gas production and retention, dough permeability, volume and tolerance during leavening. It is used to determine the quality of the flour, the fermentative activity and the quality of the protein network.

increasing application. The instrument provides information about a long list of textural properties: hardness, brittleness, elasticity, cohesiveness, stickiness, gumminess, springiness, consistency and fracturability. Kieffer et al. [59, 60] developed a micro-method for extension tests. The Kieffer Extensibility Rig is a type of miniature of the Brabender Extensograph and is used in combination with various mechanical testing machines. The cylindrically shaped dough is clamped at both ends and is extended uniaxially by a hook that moves upwards. The Dough Inflation System was introduced in the early 1990s and it was developed based on the concept of the Alveograph. It measures the stress and strain relationships based on the inflation of a sheet of dough through a biaxial extension test and operates at strain rates lower than the Alveograph. The deformations involved in biaxial extension tests are preferred as they are more relevant to the type of deformation of the dough around an expanding gas bubble during proofing and baking.

All the above instruments provide a great deal of information on the quality and performance of baked goods. Nevertheless, they are purely descriptive, frequently disruptive, and they depend on the type of instrument used, the size and geometry of the test sample and the specific conditions under which the test is carried out [45, 61].

3.5.1.2 Fundamental Measurements

The study of the effective mechanical properties of dough concerns fundamental rheology. It requires sophisticated equipment and considerable experience. It measures absolute and not relative parameters, providing results that are independent from the particular instrument used. A material is considered solid when it does not change shape continuously when subjected to stress. A material is considered liquid when it changes shape continuously when subjected even to the slightest stress. Dough cannot be classified univocally since it behaves as solid or liquid according to the experimental conditions, and because it shows both solid and liquid features simultaneously. These materials are defined as viscoelastic. The majority of foods show a linear behaviour below a certain deformation value and a non-linear behaviour above. When this limit is exceeded the results depend on the experimental conditions and do not describe the fundamental characteristics of the material [62]. Therefore, the measurements should be carried out during the linear viscoelastic interval of the material. Nevertheless, studies on doughs have also recently focused on non-linear and time-dependent behaviour [63].

Usually, the properties of viscoelastic materials are measured by creep and recovery, stress relaxation or dynamic oscillatory tests. These measurements are usually made on samples placed between two plates of a rheometer. In the creep and recovery test, an instantaneous stress is applied to the sample at rest and the change in strain (creep) is observed over time. When the stress is released, some recovery may be observed as the material attempts to return to its original shape. In the stress-relaxation test, the sample is subjected to an instantaneous strain and the stress required to maintain the deformation is observed as a function of time [62].

In the oscillatory tests, samples are subjected to deformation or stress which varies harmonically with time. Sinuosoidal and simple shear is typical [62]. This testing procedure is the most common dynamic method for studying the viscoelastic behaviour of food. Results are very sensitive to the chemical composition and physical structure [62]. Using a sinusoidally oscillating deformation of known magnitude and frequency, the phase lag angle between stress and strain is measured and used to calculate the elastic (storage modulus or G') and viscous (loss modulus or G'') components of a complex viscosity η^*.

The rheological behaviour of the dough is determined by protein–protein interactions at large deformations, while starch–starch interactions dominate at small deformations. Therefore, empirical tests correlate well with the results of the baking test [64], as the deformation that occurs is reasonably large compared with the deformation applied during the creep and dynamic rheological tests. In contrast, fundamental tests provide well-defined basic rheological information (viscosity and elasticity) and provide better defined experimental conditions of stress and strain, which allow results to be interpreted in fundamental units. Although various types of tests and instruments have been developed to describe dough performance during processing, it is fair to say that no single technique could completely describe its rheological behaviour.

3.5.2 *Innovative Approaches*

3.5.2.1 Image Analysis

Recently, image analysis has been introduced to evaluate the quality of foods, including baked goods. This technique uses protocols based on image digitalization at the macro- and micro-structural level through different systems (e.g. scanners, video cameras and microscopes). Image analysis provides a rapid and objective definition of the morphological and densitometrical characteristics of single objects or complex structures (Fig. 3.10). It makes it possible to study and model the phenomena that occur during processing continuously or even on-line [65, 66].

The analysis of an image requires a number of passages: (1) image acquisition in a digital format (a pixel image); (2) image pre-processing, to improve the image while maintaining its original dimensions; (3) image segmentation, to divide the digital image into separate, non-overlapping areas (e.g. to better distinguish the objects from the rest of the image, such as the alveoli in a slice of bread); (4) measurement of objects, to determine their different characteristics (size, shape, colour, texture); and (5) classification, to identify the objects by classifying them into different classes [67]. When used for baked good processing, image analysis allows the determination of several parameters such as the increase of volume, changes of shape, time needed to complete the dough development, extent and distribution of the alveolar structure during leavening, initial increase and successive contraction of volume, and gelatinization of the starch and surface browning during baking [68].

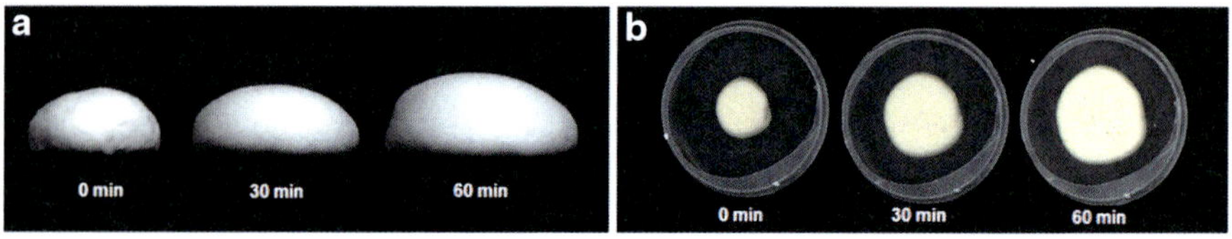

Fig. 3.10 The use of the image analysis approach in the evaluation of the leavening phase of dough

3.5.2.2 Microscopy

The macroscopic behaviour of the dough depends on its microstructure. The latter is affected by composition, spatial arrangements of the components and types of bond existing [69].

Microscopy is often used to determine the optimal mixing time of the dough, the extent of the development of the gluten and the nature of the gluten matrix [70–72]. Many details are explored via electronic microscopy, both by transmission (TEM, transmission electron microscopy) and scanning (SEM, scanning electron microscopy). To minimize the influence of sample preparation, atomic force microscopy (AFM) may be used. This technique provides high-resolution images of the surface of the starch granules [73]. Confocal laser scanning microscopy (CLSM) has recently found application in the analysis of foods. It offers the possibility of optically dissecting the material and reconstructing the 3D image [74], and to observe dynamic processes such as the formation and growth of air bubbles in the dough during leavening and baking [75]. The different components of the dough may also be simultaneously identified and located, using specific fluorescent markers. Electronic microscopy was also used to study the rupture of the gluten network following freezing and thawing [76]. A very useful instrument for the observation of frozen matrices is the cryo-SEM. This instrument shows the ultrastructure of the starch–protein associations and the state of the gluten fibrils forming the protein network [77].

3.5.2.3 Spectroscopy

The request for quick, reliable and easy to use methods that provide automation or on-line applications is increasingly frequent. Near-infrared spectroscopy (NIR) satisfies these requirements. This technique is based on the acquisition of information from a sample via the interactions that occur between its molecules and the electromagnetic waves in the near infrared. NIR offers the possibility of analyzing the matrix in a non-destructive way, does not require the use of reagents, and is highly informative. NIR spectra allow the simultaneous quantification of the various components or information regarding the mutual relations between them [78]. Recently, the NIR technique was used as a potential on-line sensor to monitor processing [79, 80]. To completely exploit the potential of NIR, advanced chemometric techniques are needed for the interpretation of spectral data which are arranged in wide bands with overlapping peaks that originate from the different components present in the matrix [78]. NIR spectroscopy is largely used to quickly determine the chemical composition of caryopses and flours. Other studies reported its application to determine the technological quality of the flours [81–84], to evaluate the molecular interactions between the dough components (water-protein-starch) [85, 86], and to monitor mixing [87], leavening, and staling [78].

Recent developments concerned the acquisition of information on dough via interactions that occur between the molecules and the electromagnetic waves in the infrared medium (MIR, mid-infrared spectroscopy). Nuclear magnetic resonance (NMR) was also used for baked good processing. It was applied to monitor the distribution and mobility of the water, to investigate the structure of the product and track the staling phenomenon [88–90].

References

1. Cauvain SP, Young LS (2006) Current approaches to the classification of bakery products. In: Cauvain SP, Young LS (eds) Baked products: science, technology and practice, 1st edn. Blackwell, Oxford, UK, pp 1–13
2. Pagani MA, Lucisano M, Mariotti M (2006) Italian bakery. In: Hui YH, Corke H, De Leyn I, Nip WK, Cross N (eds) Bakery products, 1st edn, Science and technology. Blackwell Publishing Professional, Ames, pp 527–560
3. Chung OK, Park SH, Tilley M, Lookhart GL (2003) Improving wheat quality. In: Cauvain SP (ed) Bread making. Improving quality, 1st edn. CRC-Woodhead, Cambridge, UK, pp 536–561
4. Shewry PR (2007) Improving the protein content and composition of cereal grain. J Cereal Sci 46:239–250
5. Shewry PR, Popineau Y, Lafiandra D, Belton P (2001) Wheat gluten subunits and dough elasticity: findings of the Eurowheat project. Trends Food Sci Technol 11:433–441
6. Cornell H (2003) The chemistry and biochemistry of wheat. In: Cauvain SP (ed) Bread making. Improving quality, 1st edn. CRC-Woodhead, Cambridge, UK, pp 31–70
7. Hamer RJ, MacRitchie F, Weegels P (2009) Structure and functional properties of gluten. In: Khan K, Shewry PR (eds) Wheat: chemistry and technology. AACC, St. Paul, pp 153–178
8. Hoseney RC (1989) The interaction that produce unique products from wheat flour. In: Pomeranz Y (ed) Wheat is unique, 1989th edn. American Association of Cereal Chemists, St. Paul, pp 595–606
9. FAO – Food Outlook; www.fao.or/docrep
10. Katina K (2003) High-fibre baking. In: Cauvain SP (ed) Bread making. Improving quality, 1st edn. CRC-Woodhead, Cambridge, UK, pp 487–499
11. Paradiso VM, Summo C, Trani A, Caponio F (2008) An effort to improve the shelf-life of breakfast cereals using natural mixed tocopherols. J Cereal Sci 47:322–330
12. Webb C, Owens GW (2003) Milling and flour quality. In: Cauvain SP (ed) Bread making. Improving quality, 1st edn. CRC-Woodhead, Cambridge, UK, pp 200–219
13. Posner ES (2000) Wheat. In: Kulp K, Ponte JG Jr (eds) Handbook of cereal science and technology, second edition, revised and expanded. Headquarters, Marcel Dekker, New York, pp 1–29
14. Posner ES, Hibbs AN (1997) In: Posner ES, Hibbs AN (eds) Wheat flour milling. American Association of Cereal Chemists, St. Paul, pp 1–341

15. Lucisano M, Pagani MA (1997) Cereali e derivati. In: Daghetta A (ed) Gli alimenti: aspetti tecnologici e nutrizionali. Istituto Danone, Milano, pp 7–67
16. Werner A (2002) Traceability of raw material in bakeries. Getreide Mehl und Brot 56:358–360
17. Sperber WH (2007) Role of microbiological guidelines in the production and commercial use of milled cereal grains: a practical approach for the 21st century. J Food Prot 70:1041–1053
18. Tkac JJ (1992) US Patent 5 082 680
19. Willis M, Giles J (2001) The application of a debranning process to durum wheat milling. In: Kill RC, Tumbull K (eds) Pasta and semolina technology. Blackwell Science, Oxford, UK, pp 64–85
20. Dexter JE, Martin DG, Sadaranganey GT, Michaelides J, Mathieson N, Tkac JJ, Marchylo BA (1994) Preprocessing: effects on durum wheat milling and spaghetti-making quality. Cereal Chem 71:10–16
21. Bottega G, Cecchini C, D'Egidio MG, Marti A, Pagani MA (2009) Debranning process to improve quality and safety of wheat and wheat products. Tec Molitoria Int 60:67–78
22. Laca A, Mousia Z, Dıaz M, Webb C, Pandiella SS (2006) Distribution of microbial contamination within cereal grains. J Food Eng 72:332–338
23. Sekhon KS, Singh N, Singh RP (1992) Studies of the improvement of quality of wheat infected with Karnal bunt. I. Milling, rheological and baking properties. Cereal Chem 69:50–54
24. Wrigley C, Batey I (2003) Assessing grain quality. In: Cauvain SP (ed) Bread making. Improving quality, 1st edn. CRC-Woodhead Publishing Ltd, Cambridge, UK, pp 71–96
25. Sears JK, Darby JR (1982) The technology of plasticizers. In: Sears JK, Darby JR (eds), Wiley Interscience, New York, pp 1–1166
26. Bernardin JE, Kasarda DD (1973) Hydrated protein fibrils from wheat endosperm. Cereal Chem 50:529–536
27. Hoseney RC, Zeleznak K, Lai CS (1986) Wheat gluten: a glassy polymer. Cereal Chem 63:285–286
28. Cuq B, Abecassis J, Guilbert S (2003) State diagrams to help describe wheat bread processing. Int J Food Sci Tech 38:759–766
29. Feillet P, Guinet R, Morel MH, Rouau X (1994) La Pâte. Formation et développement. In: Guinet R, Godon B (eds) La panification française. Lavoisier, Paris, pp 226–279
30. Schiraldi A, Fessas D (2003) Classical and Knudsen thermogravimetry to check states and displacements of water in food systems. J Therm Anal Cal 71:225–235
31. Tolstoguzov VB (2000) Food as dispersion system. Thermodynamic aspects of composition-property relationships in formulated food. J Therm Anal Cal 61:397–409
32. Bushuk W (1966) Distribution of water in dough and bread. Baker's Dig 40:38–40
33. Eliasson AC (2003) Starch structure and bread quality. In: Cauvain SP (ed) Bread making. Improving quality, 1st edn. CRC-Woodhead, Cambridge, UK, pp 145–167
34. Eliasson AC (2003) Starch structure and bread quality. In: Cauvain SP (ed) Bread making. Improving quality, 1st edn. CRC-Woodhead Publishing Ltd, Cambridge, UK, pp 145–167
35. He H, Roach RR, Hoseney RC (1992) Role of salt in baking. Cereal Food World 53:4–6
36. Holmes JT, Hoseney RC (1987) Effect of pH and certain ions on bread-making properties. Cereal Chem 64:343–348
37. Desgrets R (1994) Les ingredients spécifiques: panification fine, viennoiserie. In: Guinet R, Godon B (eds) La panification française. Lavoisier, Paris, pp 132–151
38. Campbell GM (2003) Bread aeration. In: Cauvain SP (ed) Bread making. Improving quality, 1st edn. CRC-Woodhead, Cambridge, UK, pp 352–374
39. Gray JA, Bemiller JN (2003) Bread staling: molecular basis and control. Comprehensive Reviews. Compr Rev Food Sci Food Safety 2:1–21
40. Kulp K, Ponte JG Jr (1981) Staling of white pan bread: fundamental causes. CRC Crit Rev Food Tec 11:1–48
41. Chargelegue A, Guinet R, Neyreneuf O, Onno B, Pointrenaud B (1994) La fermentation. In: Guinet R, Godon B (eds) La panification française. Lavoisier, Paris, pp 280–325

42. Brown J (1993) Advances in breadmaking technology. In: Kamel BS, Stauffer CE (eds) Advances in baking technology. Blackie Academic and Professional, New York, pp 38–87
43. Corsetti A, Farris GA, Gobbetti M (2010) Uso del lievito naturale. In: Gobbetti M, Corsetti A (eds) Biotecnologia dei prodotti lievitati da forno. CEA, Milano, pp 171–187
44. Vogel RF, Müller M, Stolz P, Ehrmann M (1996) Ecology in sourdough produced by traditional and modern technologies. Adv Food Sci 18:152–159
45. Dobraszczyk BJ, Campbell GM, Gan Z (2001) Bread: a unique food. In: Dendy DAV, Dobraszczyk BJ (eds) Cereals and cereal products: chemistry and technology. Aspen, Gaithersburg, pp 182–232
46. Millar S (2003) Controlling dough development. In: Cauvain SP, Cauvain SP (eds) Bread making. Improving quality, 1st edn. CRC-Woodhead, Cambridge, UK, pp 401–423
47. Pyler EJ (1988) Fundamental of baking technology. In: Pyler EJ (ed) Baking science and technology. Part three, Vol. II. Sosland, Merriam
48. Campbell GM, Rielly CD, Fryer PJ, Sadd PA (1993) Measurement and interpretation of dough densities. Cereal Chem 70:517–521
49. Spicher G (1983) Baked goods. In: Reed G (ed) Biotechnology. Food and feed production with microorganisms, vol 5. Verlag Chemie, Weinheim, pp 1–80
50. Gan Z, Ellis PR, Schofield JD (1995) Gas cell stabilization and gas retention in wheat bread dough. J Cereal Sci 21:215–230
51. MacRitchie F (1976) The liquid phase of dough and its role in baking. Cereal Chem 53:318–326
52. Cauvain SP, Young LS (2003) Water control in baking. In: Cauvain SP (ed) Bread making: improving quality, 1st edn. CRC-Woodhead, Cambridge, UK, pp 447–466
53. Richard-Molard D (1994) Le goût du pain. In: Guinet R, Godon B (eds) La panification française. Lavoisier, Paris, pp 452–476
54. Lefebvre J (2006) An outline of the non-linear viscoelastic behaviour of wheat flour dough in shear. Rheol Acta 45:525–538
55. Dobraszczyk BJ, Morgenstern MP (2003) Rheology and the breadmaking process. J Cereal Sci 38:229–245
56. Dobraszczyk BJ, Salmanowicz BP (2008) Comparison of predictions of baking volume using large deformation rheological properties. J Cereal Sci 47:292–301
57. Bloksma AH (1990) Rheology of the breadmaking process. Cereal Foods World 35:228–236
58. Walker CE, Hazelton JL (1996) Dough rheological tests Cer Foods World 41:23–28
59. Kieffer R, Kim JJ, Belitz HD (1981) Zugversuche mit Weizenkleber im Mikromaβstab. Z Lebensm Unters Forsch 172:190–192
60. Kieffer R, Garnreiter F, Belitz HD (1981) Beurteilung von Teigeigenschaften durch Zugversuche im Mikromaβstab. Z Lebensm Unters Forsch 172:193–194
61. Dobraszczyk BJ (2003) Measuring the rheological properties of dough. In: Cauvain SP, Cauvain SP (eds) Bread making. Improving quality, 1st edn. CRC-Woodhead, Cambridge, UK, pp 375–400
62. Steffe JF (1996) In: Steffe JF (ed) Rheological methods in food engineering. Freeman Press, East Lansing
63. Lefebvre J (2009) Nonlinear, time-dependent shear flow behaviour, and shear-induced effects in wheat flour dough rheology. J Cereal Sci 49:262–271
64. Bloksma AH, Bushuk W (1988) Rheology and chemistry of dough. In: Pomeranz Y (ed) Wheat chemistry and technology, vol 11, 3rd edn. AACC, St. Paul, pp 131–152
65. Russ JC, Stewart WD, Russ JC (1988) The measurement of macroscopic images. Food Technol 42:94–102
66. Riva M (2003) Analisi dell'immagine dei prodotti da forno. Tecnologie Alimentari 14:30–33
67. Du C-J, Sun D-W (2004) Recent developments in the applications of image processing techniques for food quality evaluation. Trends Food Sci Tech 15:230–249
68. Riva M, Pagani MA, La Prova M (2004) Un approccio innovativo per lo studio della lievitazione di impasti da pane: l'Analisi d'Immagine. Tec Molitoria 55:629–650

69. Létang C, Piau M, Verdier C (1999) Characterization of wheat flour–water doughs. Part I: rheometry and microstructure. J Food Eng 41:121–132
70. Moss R (1974) Dough microstructure as affected by the addition of cysteine, potassium bromate, and ascorbic acid. Cereal Sci Today 19:557–562
71. Parker ML, Mills ENC, Morgan MRA (1990) The potential of immunoprobes for locating storage proteins in wheat endosperm and bread. J Sci Food Agr 52:35–45
72. Sidi H, Moss R (1991) Light microscopy observations on the mechanism of dough development in Chinese steamed bread productions. Food Struct 10:289–293
73. van de Velde F, van Riel J, Tromp RH (2002) Visualisation of starch granule morphologies using confocal scanning laser microscopy (CSLM). J Sci Food Agr 82:1528–1536
74. Vodovotz Y, Vittadini E, Coupland J, McClements DJ, Chinachoti P (1996) Bridging the gap: use of confocal microscopy in food research. Food Technol 50:74–82
75. Blonk JCG, van Aalst H (1993) Confocal scanning light microscopy in food research. Food Res Int 26:297–311
76. Esselink EFJ, van Aalst H, Maliepaard M, Duynhoven JPM (2003) Long-term storage effect in frozen dough by spectroscopy and microscopy. Cereal Chem 80:396–403
77. Yi J, Kerr WL (2009) Combined effects of freezing rate, storage temperature and time on bread dough and baking properties. Food Sci Tech (LWT) 42:1474–1483
78. Sinelli N, De Dionigi S, Pagani MA, Riva M, Belloni P (2004) Spettroscopia FT-NIR nel monitoraggio on-line di prodotti da forno: lievitazione e raffermamento. Tec Molitoria 55:1075–1093
79. Hoyer H (1997) NIR on-line analysis in the food industry. Process Contr Qual 9:143–152
80. Cimander C, Carlsson M, Mandenius CF (2002) Sensor fusion for on-line monitoring of yoghurt fermentation. J Biotechnol 99:237–248
81. Sirieix A, Downey G (1993) Commercial wheat flour authentication by discriminant analysis of near infrared reflectance spectra. J Near Infrared Spec 1:187–197
82. Allosio N, Boivin P, Bertrand D, Courcoux P (1997) Characterisation of barley transformation into malt by three-way factor analysis of near infrared spectra. J Near Infrared Spec 5:157–166
83. Pagani MA, Lucisano M, Mariotti M (2002) Valutazione del grado di gelatinizzazione dell'amido mediante tecnica NIR. Tec Molitoria 53:1218–1223
84. Manley M, Van Zyl L, Osborne BG (2002) Using Fourier transform near infrared spectroscopy in determining kernel hardness, protein and moisture content of whole wheat flour. J Near Infrared Spec 10:71–76
85. Osborne BG (1996) Near infrared spectroscopic studies of starch and water in some processed cereal foods. J Near Infrared Spec 4:195–200
86. Kays SE, Barton FE, Windham WR (2000) Predicting protein content by near infrared reflectance spectroscopy in diverse cereal food products. J Near Infrared Spec 8:35–43
87. Wesley IJ, Larsen N, Osborne BG, Skerritt JH (1998) Non-invasive monitoring of dough mixing by near infrared spectroscopy. J Cereal Sci 27:61–69
88. Assifaoui A, Champion D, Chiotelli E, Verel A (2006) Characterisation of water mobility in biscuit dough using a low-field ^{1}H NMR technique. Carbohydr Polym 64:197–204
89. Ishida N, Takano H, Naito S, Isobe S, Uemura K, Haishi T, Kose K, Koizumi M, Kano H (2001) Architecture of baked breads depicted by a magnetic resonance imaging. Magn Reson Imaging 19:867–874
90. Vodovotz Y, Vittadini E, Sachleben JR (2002) Use of H-1 crossrelaxation nuclear magnetic resonance spectroscopy to probe the changes in bread and its components during aging. Carbohydr Res 337:147–153

Websites

91. www.bakerperkinsgroup.com
92. www.brabender.com
93. www.buhlergroup.com
94. www.chopin.fr
95. www.esmach.it
96. www.fao.org/docrep (FAO - Food Outlook)
97. www.imars.case.edu
98. www.itecaspa.com
99. www.national-mfg.com
100. www.ocrim.it
101. www.sancassiano.com
102. www.satake-group.com
103. www.stablemicrosystems.com
104. www.wpib.de

Chapter 4
Technology of Sourdough Fermentation and Sourdough Applications

Aldo Corsetti

4.1 Definition of Sourdough

Sourdough technology is widely used in bread making and cake production as it confers distinctive characteristics, high sensory properties and shelf life to the resulting products (Table 1). Sourdough is "a mixture of wheat and/or rye flour and water, possibly with added salt, fermented by spontaneous (from flour and environment) lactic acid bacteria and yeasts which determine its acidifying and leavening capability. These activities are obtained and optimized through consecutive refreshments (or re-buildings, replenishments, backslopping)" (3–5). The term refreshment deals with the technique by which a dough made of flour, water and possibly other ingredients ferments spontaneously for a certain time (possibly at a defined temperature) and it is subsequently added as an inoculum to start the fermentation of a new mixture of flour and water (and possibly other ingredients).

When applied for a defined interval of time such a process provides a sourdough with constant and repeatable leavening and acidifying performances reliant on the growth of lactic acid bacteria and yeasts that are well adapted to the environment. After the preparation of the sourdough, the refreshment technique is aimed at maintaining the metabolic activity of the microbial communities at all times (6).

When the sourdough is added to a mixture of water and flour to start consecutive propagations (refreshments) to obtain the final mass or full sour to be used as the leavening agent, it can be designated the "mother sponge" (2, 4). Generally, a sourdough contains a variable number of lactic acid bacteria and yeasts, ranging from 10^7 to 10^9 cfu/g and 10^5 to 10^7 cfu/g, respectively, with a ratio of about 100:1 (7).

A. Corsetti (✉)
Department of Food Science, University of Teramo, Teramo, Italy
e-mail: acorsetti@unite.it

M. Gobbetti and M. Gänzle (eds.), *Handbook on Sourdough Biotechnology*,
DOI 10.1007/978-1-4614-5425-0_4,

Table 4.1 Characteristics of sourdough bread versus baker's yeast bread (adapted from (1, 2))

Characteristics	Sourdough bread	Baker's yeast bread
pH	3.8–4.6	5.3–5.8
Lactic acid	0.4–0.8%	0.005–0.04%
Acetic acid	0.10–0.40%	0.005–0.04%
Bread volume	0.22–0.30	≤0.20
Flavour	Complex aroma and flavour	
Staling	Slow	Rapid
Shelf life	Good protection against microbial contaminations	High sensibility to bacteria and mould spoilage
Some nutritional aspects	Optimal phytase activity and hydrolysis of phytic acid responsible for ion (Ca^{2+}, Fe^{2+}, Mg^{2+} etc.) binding	Low phytase activity, decalcifying effect
	Free amino acid concentration increase	Free amino acid concentration similar to that of flour
	Decrease of glycemic index	

The main role of lactic acid bacteria (mainly obligately and facultatively heterofermentative lactobacilli) is in the acidification process while yeasts mainly account for the leavening of the dough by releasing CO_2 (4).

4.2 Sourdough Preparation and Storage

Sourdough (mother sponge) preparation can be fulfilled through many different protocols. The main objective is to obtain a leavening agent that contains well-adapted resident microorganisms. Such microorganisms have to produce sufficient CO_2 to leaven the dough, and organic acids and other metabolites to provide rye or wheat bread with good texture and sensory properties, and extended shelf life (3). Two classical procedures, for example the French and American systems, are discussed below.

4.2.1 The French System

Mother sponge preparation for obtaining the French "*pain au levain*" begins with a quite firm wheat flour dough (dough yield, DY of 150–152, see Sect. 4.6.2) with addition of salt and malt. This dough undergoes a first fermentation step lasting ca. 24 h. This corresponds to the early fermentative activity of flour-resident yeast and lactic acid bacteria, which results in a low CO_2 and organic acid release (2). The decrease of pH induces the activity of flour endogenous proteases which together with bacterial hydrolytic enzymes act on gluten and lead to a lower dough firmness. The second step begins with the first refreshment which is aimed at introducing oxygen and new fermentable carbohydrates into the mixture to stimulate microbial growth and activity. The refreshment is obtained by adding a quantity of flour corresponding to the weight of the previous fermented dough and a quantity of water to bring the DY to a value (e.g. 148) lower than the previous one. This dough ferments quickly and represents the starting dough for the next refreshment. By applying such a procedure a sourdough with a steady fermentative and leavening capability is obtained. During the last step, each refreshment is carried out at a regular interval of time (e.g. 7–8 h), with the aim of maintaining an equilibrium in the ratio between microbial communities (2). According to Calvel (2), when the dough volume increases by three to fourfold with respect to the initial dough, a new refreshment should be performed. The mother sponge (*levain chef*), which is obtained following the above procedure, represents the dough used to prepare the full sour needed to leaven the bread-dough.

4.2.2 The American System

The American system relies for the mother sponge preparation on a mix of water and wheat and/or rye flours. In contrast to the French system, the value of DY ranges from 225 (liquid dough with a ratio water:flour of 1.25:1) to 250 (liquid dough with a ratio water:flour of 1.5:1). These values of DY remain unaltered throughout the consecutive refreshments (8). The time and temperature of fermentation are strictly controlled during each phase. In the first step, the water-flour mixture ferments at 32–35°C for 24 h in order to acidify the dough. At the end of this step, the first refreshment is obtained by adding flour and water to the previous fermented dough, without changing the value of DY. After 8 h of fermentation at 32–35°C, the second refreshment is carried out and the dough ferments for another 16 h. After the above procedure is applied, refreshments are carried out every 8–16 h between which the dough is allowed to ferment at 24–27°C. With the aim of calculating the amount of water and flour to be added at each replenishment stage a multiplicative factor of 4 is considered. Under these conditions, the weight of the fermented dough is multiplied by 4 every two refreshments. This allows one to determine the quantity of water and flour to be used for the next refreshment and the ratio between the two ingredients is maintained to 1.25:1. By using this system the value of DY remains constant until a sourdough with a pH value of 3.6–3.8 and TTA (total titratable acidity) of 16–20 ml NaOH/20 g of dough (as calculated following the American system) is obtained. A last fermentation of 8 h at 24–27°C is needed to confer to the sourdough the sensory and leavening performances of the mother sponge. Generally, a sourdough with the above characteristics is obtained in approximately 5 days, after which it is maintained in an active state by storage at low temperature (e.g. 4°C) and subjected to refreshment at least once a day (8).

4.2.3 Sourdough Storage

The liquid sourdough (DY 225–250) that is frequently used by bread manufacturers in the United States is stored at 1–2°C, after rapid refrigeration, or at 4–5°C. It is used to start a new fermentation within 2–3 days, without the refreshment step. In the case of prolonged storage (10 days) one or two refreshments are needed to activate the metabolism of the lactic acid bacteria and yeasts. A prolonged storage of the sourdough for a few months at 4–5°C is possible when the ratio between water and flour is reduced by adding flour at ca. 0.43:1 (30% water: 70% flour). In this case, a firm sourdough, with DY of 143, is produced. Such a storage type necessarily requires sourdough reactivation (at least two refreshments) before use (8). In many artisan bread preparations, different and often empirical storage techniques are applied. Generally, as a consequence of the daily schedule of bread manufacture, a portion of the sourdough is refreshed at least one time before its use. Nevertheless, by applying a separate storage protocol, part of the mother sponge can be stored at low temperature (4–6°C) for some weeks after putting it in a cloth bag tied with string (9). Refreshments are needed before reusing such a sourdough in a bread-making protocol. In some cases sourdough can be frozen and reused after refreshment.

4.3 Classification of Sourdoughs

Sourdough bread making is an ancient biotechnological process and various protocols for its use are applied in many countries. On the basis of the technology applied, sourdoughs have been grouped into three types (10), to which a fourth type, named sponge-dough, can be added.

4.3.1 Type I Sourdough

Traditional sourdoughs whose microorganisms are kept metabolically active through daily refreshments are included in this group. Type I sourdoughs are generally suitable for achieving dough leavening without addition of baker's yeast; the dough propagation described above for French and U.S. sourdoughs are examples of Type I sourdoughs. Generally, a three-stage protocol is applied relying on three refreshments over 24 h in order to obtain the leavened dough to bake. Each step is characterized by a given DY as well as fermentation temperature and time. At the end of the last step of fermentation the sourdough is used as the leavening agent; thus it can be considered as a natural starter culture containing many microbial strains (11). In wheat and/or rye flour sourdoughs, dominating strains belong to the species *Lactobacillus sanfranciscensis* which can co-exist with other obligately heterofermentative lactic acid bacteria such as *L. pontis*, *L. brevis*, *L. fermentum*, *L. fructivorans* and with the yeasts *Candida milleri*, *C. holmii*, *Saccharomyces cerevisiae* and *S. exiguus* (recently renamed *Kazachstania exigua*).

4.3.2 Type II Sourdough

Sourdoughs obtained through a unique fermentation step of 15–20 h followed by storage for many days belong to this group. Type II sourdoughs are generally not suitable for achieving dough leavening but are used for dough acidification, and as dough improvers. These sourdoughs are generally liquid (DY of ca. 200) and they are produced at the industrial level using bioreactors or tanks at a controlled temperature that exceeds 30°C. Such a protocol aims at shortening the fermentation process (12). During storage, a portion of the mature sourdough can be used as the inoculum with the aim of acidifying the dough and enriching it with aroma and flavour compounds which are characteristics of sourdough baked goods. On the basis of the long fermentation time, high DY, and temperature of fermentation, lactobacilli such as *L. panis*, *L. reuteri*, *L. johnsonii*, and *L. pontis*, which are resistant to low pH, dominate these sourdoughs (10, 13). Spontaneous flour yeasts are inhibited and, consequently, the leavening of the final dough is obtained by adding commercial baker's yeast.

Through a combined approach consisting of culture-dependent and culture-independent systems, Meroth et al. (14, 15), by using an experimental model based on a laboratory scale-fermentation, showed a different prevalence of lactic acid bacteria and yeasts in type I and type II rye-based sourdough, started with a mixture of commercially available sourdough starters and baker's yeast. In particular, *L. sanfranciscensis*, *L. mindensis* and *C. humilis* dominated the type I sourdough. *L. crispatus*, *L. pontis*, *L. fermentum* and *S. cerevisiae* prevailed in the type II sourdough propagated at 30°C; the same sourdough propagated at 40°C showed a dominance of *L. crispatus*, *L. panis* and *L. frumenti* as well as the disappearance of all the yeasts that had been added with the starter mixture. Finally, *L. johnsonii*, *L. reuteri* and *C. krusei* characterized a type II sourdough fermented at high temperature (40°C), but produced with rye bran instead of rye flour.

4.3.3 Type III Sourdough

Type II liquid sourdoughs, which are dried/stabilized after preparation, are named type III sourdoughs (12). They are mainly used at the industrial level as their quality is more constant compared to type I sourdough, and they are simpler to manage and less time consuming. On the basis of the preparation method, type III sourdoughs are dominated by drying-resistant lactic acid bacteria such as *Pediococcus pentosaceus*, *L. plantarum* and *L. brevis* (10). A detailed description of this type of sourdough and its industrial application are reported in Sect. 4.8.

4.3.4 Sponge Dough

The sponge dough is aimed at acclimatizing baker's yeast (*S. cerevisiae*) and improving swelling of the flour components, loaf volume, taste and flavour of the bread and its shelf life. It is an indirect process which, from a technological and microbiological point of view, could be considered as an intermediate procedure between straight-dough (or a direct process in which just baker's yeast is used to start the fermentation) and sourdough (an indirect process in which fermentation starts without the addition of baker's yeast). Sponge dough is obtained in two steps: in the first dough (pre-dough) the baker's yeast is mixed with a part of the flour and water of the recipe, while the second dough is obtained by adding the rest of the flour, water and possibly other ingredients to the fully fermented pre-dough. Depending on the type of bread, the length of pre-dough fermentation can vary from 3 to 20 h and as a consequence, various percentages of flour and yeast, besides different combinations of DY and temperature can be applied in this step (16, 17). As in the case of longer fermentations lactic acid bacteria present as contaminants from either baker's yeast or flour grow in the dough reaching typically more than 10^8 cfu/g and contribute to the overall quality of baked goods (4), the sponge dough can be included in the category of sourdoughs.

4.4 Examples of Sourdough Applications

Once produced, the sourdough is generally submitted to various refreshments either to activate the microbial metabolism or to increase the dough mass (Fig. 4.1).

Different types of sourdough (e.g. type I, II or III) are used for the manufacture of various leavened baked products, which range from traditional Italian or French wheat bread, white pan bread, rye bread, San Francisco bread to traditional cakes famous throughout the world such as Panettone. Examples of those sourdough applications are given below.

4.4.1 A Traditional Italian Sourdough Bread: The Altamura Bread

The Altamura bread is the first European bread that received the PDO (Protected Denomination of Origin) status. It is manufactured in the Apulia region (Italy) using specific cultivars of durum wheat (*Triticum durum*) flour and the technology is based on type I sourdough. The full sour is obtained by a three-stage procedure in order to gradually increase the amount of leavened dough. At each step, water and durum wheat flour are mixed with a previously fermented dough, which is added at the proportion of ca. 20% based on flour weight. On the basis of the ratio between

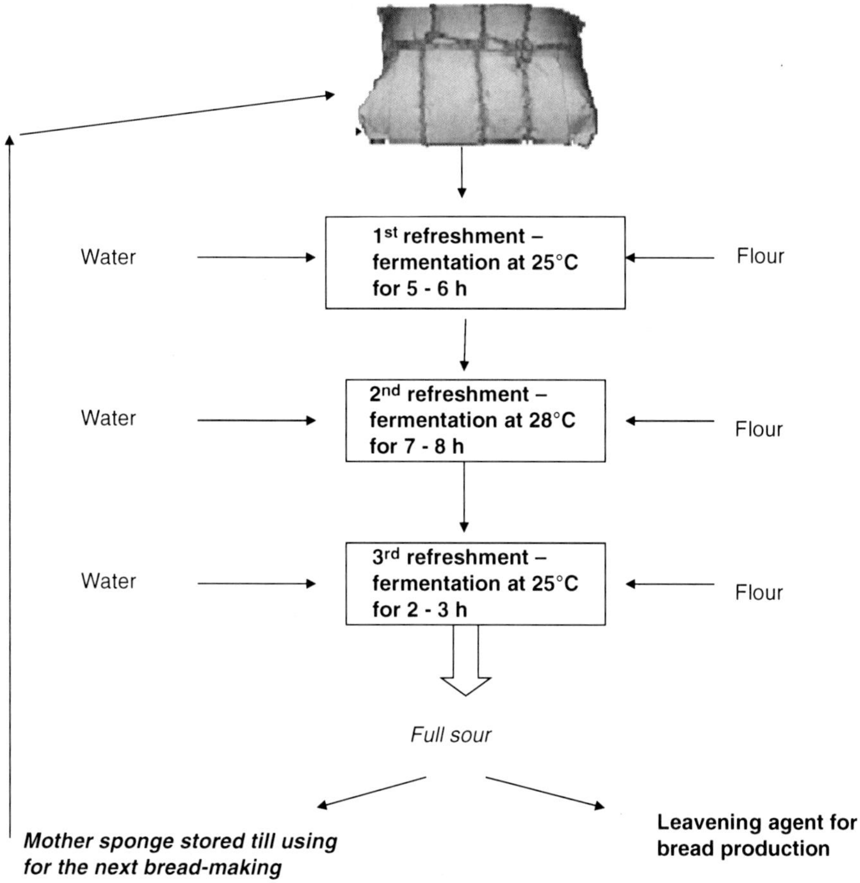

Fig. 4.1 Representative flow sheet to obtain a *full sour* from a *mother sponge* using a three-step procedure. Fermentation time and temperature given in the example can change depending on bread-making protocol (Adapted from (1))

the ingredients that are indicated in the original recipe (100 kg of durum wheat flour + 20 kg of sourdough + 2 kg NaCl + 60 l of water) and assuming a full sour with a DY of 160, which contains 12.5 kg of flour, the final dough should have a DY of 162. As in many traditional productions, the time and temperature used at each stage are not defined but variously determined based on the bread-makers experience.

4.4.2 French Bread (Pain au levain)

Although various protocols based on one or two steps are used to obtain an adequate quantity of full sour (*levain tout point*), a three-stage system is usually applied to prepare the traditional wheat flour sourdough French bread, according to the type I sourdough procedure. The first step consists of the mixing of a part of the mother sponge (*levain chef*) with flour and water to achieve around four times the initial mass. This dough (DY 160) ferments for 1.5–2 h at ca. 25°C and is named *levain de première* (fresh sour). It is used as a starter to obtain a dough with a DY of ca. 185, which ferments for 7–8 h at a temperature slightly higher than the previous one. Those parameters, including the long mixing time to oxygenate the dough, stimulate the growth of yeasts and the leavening capability of the *levain de seconde* (basic sour). By using this dough as the starter for the third refreshment, the last dough or *levain tout point* (full sour) is obtained after a short fermentation time (ca. 2 h) with the aim of controlling the hydrolytic activities of the dough and of preserving the gas-retaining and bread-making capabilities. The *levain tout point* is used at the proportion of 25% with respect to the mass of the final dough and the dough is ready for baking after 30 min of leavening (2).

4.4.3 Rye Sourdough Bread

As reported for French bread, the rye or wheat/rye flour sourdough bread is obtained using single or multi-step protocols, depending on the type of bread making and product. The "three-stage sourdough process, basic-sour overnight" requires ca. 15–20 h and, as described by Spicher and Pomeranz (17), represents a good reference model for type I sourdough application in rye-flour-based bread making. The first stage leads to the *anfrischsauer* (fresh sour), which, in turn, is started by the *anstellgut* (mother sponge), a commercial starter or a part (ca. 10–15% of the total dough weight) of a full sour (*vollsauer*) from a previous bread making. The *anfrischsauer* has a DY of 200–250 and ferments for ca. 6 h at 25–26°C. Afterwards, it contains a high number of yeasts and it is used to start a new dough based on rye flour and water (DY of about 160–180). After 5–8 h of overnight fermentation at 26–30°C, lactic acid bacteria grow and the *grundsauer* (basic sour) is obtained. A further refreshment (3 h at 30–33°C), using the basic sour as the starter for a mix of rye flour and water (DY 180–200), leads to the *vollsauer* (full sour), which has values of pH and TTA of 4.0 and 10.5 ml NaOH 0.1 N/10 g of dough, respectively. A part (40%) of this sourdough represents the natural starter for the final dough (DY 170), which contains rye flour and an equivalent amount of wheat flour in the case of a rye/wheat flour bread preparation. The dough is baked after dough resting of 10 min at 28°C, followed by proofing.

4.4.4 White Pan Bread

The white pan bread is one of the most diffused breads in the United States and represents a good example of bread that is manufactured using type II sourdough. As reported by Kulp (8), the manufacture of this bread relies on a "one-stage system" where just one refreshment is needed to mix the liquid full sour with the other ingredients of the recipe. The quite soft dough (DY 175) ferments in two separate steps: the first at room temperature (ca. 25–26°C) for 25–30 min and the second at 35°C for ca. 75–100 min. Before baking, the dough shows values of pH of 4.3–4.4 and values of TTA of 9–11 ml NaOH 0.1 N/20 g of dough. Overall, the liquid starter (full sour) represents 37% of the final dough but different percentages can be used, which depend on the recipe and industrial or artisan levels. In the last case, a lower quantity of starter is used, which requires a longer fermentation period and different working schedule. Because of the high acidity of the sourdough starter, a short mixing time is required due to the increased solubility of the gluten proteins, which decreases the time for development of the gluten network.

4.4.5 Panettone Cake

Panettone is a traditional Italian cake consumed over Christmas and famous throughout the world. Panettone has a soft structure, with regular holes, and characteristic flavour, which is derived from dough ingredients (water, flour, butter, sugar, eggs, salt, and others) and processing. Both at artisan and industrial levels, the sourdough biotechnology is traditionally based on many refreshments (e.g. type I sourdough) and an increased concentration of sugar is added during the last steps (9, 12). Comparable to most other type I sourdoughs, sourdough microbiota are predominantly composed of *L. sanfranciscensis*. During traditional manufacture, a sourdough with a low fermentation quotient (FQ, see Sect. 4.6.3) is used and time and temperature of fermentation are strictly controlled. In particular, the dough temperature does not exceed 30°C and the temperature of water and other ingredients has to be not lower than 20–22°C. Various studies have attributed the prolonged softness and shelf life of Panettone cake (it is manufactured several months before consumption) to the presence of dextran producing *Leuconostoc mesenteroides* in the sourdough, which uses sucrose as an exo-polysaccharide (EPS) precursor (12). On the basis of such a finding a new protocol relying on type III sourdough has recently been proposed as an alternative to the traditional manufacture (see Sect. 4.8). A sourdough containing 25% (on dry matter) of dextran, which is stabilized by refrigeration, pasteurization or drying, has been used to shorten the time to obtaining a product with similar characteristics to the traditional Panettone cake (18).

4.4.6 San Francisco Bread

San Francisco bread is traditionally manufactured using the very famous San Francisco sourdough, which possesses a typical microbial community, mainly consisting of *L. sanfranciscensis* and *S. exiguus* (renamed to *K. exigua*). Traditional bread making relies on the type I sourdough. Various refreshments and long fermentation times at low temperature are used to increase the concentration of acetic acid synthesized by the obligately heterofermentative *L. sanfranciscensis*. Recently, a liquid San Francisco sourdough, which is stabilized through pasteurization, has been introduced to the market. Thus, also the type III sourdough is used as an acidifying and flavouring agent for the manufacture of the San Francisco bread in ca. 3 h (no-time-dough process) (see Sect. 4.7) (12).

4.5 Stability of the Sourdough Microbiota During Propagation and Use

In order to keep microorganisms active, type I sourdoughs are, generally, propagated daily. Thus, problems of microbiota stability during propagation and in the use of such a type of "biological starter" arise.

In general, it is recognized that in a sourdough ready to be used (e.g. the mother sponge—see Sect. 4.2) for bread and other baked goods production the following basic characteristics occur:

- The prevalence of a microbial community different from that of the flour used as raw material (3, 19, 20);
- A stable ratio among lactic acid bacteria and yeasts (to the order of 100:1 or higher) (7);
- A predominance of facultatively and, especially, obligately heterofermentative lactobacilli over other lactic acid bacteria (3).

As reported by Vrancken et al. (21), laboratory sourdoughs based on wheat, rye, or spelt that are backslopped daily, reach equilibrium through a three-step process: (1) prevalence of sourdough-atypical lactic acid bacteria, (2) prevalence of sourdough-typical lactic acid bacteria, and (3) prevalence of highly adapted sourdough-typical lactic acid bacteria (21). In a recent paper dealing with the comparative analysis of seven mature sourdoughs of type I backslopped for 80 days at artisan and laboratory levels under constant technological parameters, Minervini et al. (22) showed that, while the cell density of presumptive lactic acid bacteria and related biochemical features were not affected by the environment of propagation, all sourdoughs harboured a certain number of species and strains, which were dominant throughout time and, in several cases, varied depending on the environment of propagation (21). Moreover, besides stable species and strains, other lactic acid bacteria temporarily contaminated the sourdoughs with differences between artisan and laboratory levels. Interestingly, this occasional succession of strains and species only slightly affected the kinetic of sourdough acidification; nevertheless, it could influence the sensory properties of the resulting leavened baked product. Overall, as a stable microbiota during sourdough propagation and use is essential in order to obtain standard and repeatable final products, the problem of microbial stability in terms of species and

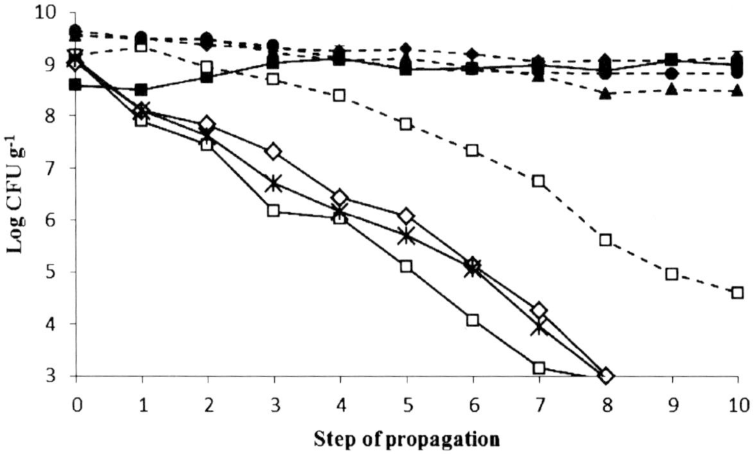

Fig. 4.2 Persistence of strains of *Lactobacillus plantarum* (DB200, --♦--; 12H1, --●--; 2MF8, --□--; G10C3, --Δ--) and *Lactobacillus sanfranciscensis* (LS6, –□–; LS41, –■–; LS48, —◊—; LS3, —X—) after sourdough propagation at 30°C for 6 h during ten subsequent days (Adapted from (24) and from (23))

strain composition in type I sourdough propagated by applying different endogenous (e.g. type of flour, quantity of water) and exogenous (e.g. temperature/time of fermentation) parameters should be carefully addressed, both in terms of identification and typing of dominant and sub-dominant microorganisms (see Sect. 4.6.1) as well as to select robust, well-adapting and competitive starter strains (23). The robustness of sourdough lactobacilli varies depending on the species and on the strains. While the majority of *L. sanfranciscensis* strains showed quite a low robustness during daily backslopping performed at the laboratory level (24), selected strains of *L. plantarum* seemed to share several phenotypic traits that determined the capacity to outcompete the contaminating lactic acid bacterium biota (23) (Fig. 4.2).

4.6 Methods to Evaluate the Performance of the Sourdough

Both microbiological and physico-chemical parameters are used to evaluate the performance of a sourdough. The level of complexity is different and depends on the purpose of the analyses. The microbiological aspect essentially deals with the assessment of the community of lactic acid bacteria and yeasts, as those microorganisms are dominant in a good quality sourdough and are generally present at the ratio of approx. 100:1 (7). Nevertheless, an array of both phenotypic and genotypic methods is necessary to identify the species/strain composition of the dominant and sub-dominant microbiota of the sourdough. An overview of those systems is given below.

4.6.1 Determination of the Number of Lactic Acid Bacteria and Yeasts

The standard plate count is used to estimate the cell density of both lactic acid bacteria and yeasts. Suitable culture media to assess the number of sourdough lactic acid bacteria include SDB (Sour Dough Bacteria) medium, as originally described by Kline and Sugihara (25); or MRS (de Man, Rogosa, Sharpe) medium (51), with modifications concerning the pH and the nutrient composition, particularly to include maltose and fructose as carbon sources and electron acceptors, respectively, and cysteine as reducing agent (26–28); Homohiochii medium, especially recommended to enumerate obligately heterofermentative lactic acid bacteria (29); SFM (San Francisco medium), containing wheat or rye bran especially useful for lactic acid bacteria as well as lactobacilli (30).

To all the above media cycloheximide (100 ppm) can be added to inhibit the growth of yeast when requested.

Yeast are generally enumerated on WL medium (31); YPDA medium (32, 33); Sabouraud dextrose agar (34, 35); YGC agar (14); and Malt extract agar (36). To inhibit bacterial growth, all the above media generally contain 50–100 ppm of chloramphenicol.

4.6.2 Phenotypic and Genetic Analyses to Identify and Type Lactic Acid Bacteria and Yeasts

Sourdough lactic acid bacteria and yeasts are generally identified using a polyphasic approach, combining both phenotypic and genotypic assays.

Besides morphological and physiological tests, specific biochemical assays are available to identify at species level *Weissella* spp., *Pediococcus* spp. and *Enterococcus* spp. as well as the predominant sourdough lactic acid bacteria, which are often members of the genus *Lactobacillus* (19). In that case, the miniaturized commercial test (e.g. API 50 CHL, Biomerieux, France) or automated assay (e.g. Biolog system) are frequently applied after a preliminary evaluation based on cell morphology, Gram stain, catalase test, growth at 15°C and 45°C, CO_2 production from glucose, and NH_3 release from arginine (37). Nevertheless, the high biochemical diversity in the above-mentioned genus, means that the identification scheme must include some chemotaxonomic analysis (e.g. SDS-PAGE, *Sodium Dodecyl Sulphate-PolyAcrylamide Gel Electrophoresis* of total or cell-wall associated proteins) (38, 39) or, mainly, genotypic tests such as 16S rRNA gene sequencing (30). Moreover, on the basis of specifically designed primers targeting species-specific sequences, many rapid identification systems relying on PCR (e.g. multiplex-PCR, Intergenic Spacer Region-based PCR) have been recently applied to sourdough lactic acid bacteria identification (27, 40, 41). Among genotypic systems, PCR-RAPD *(Random Amplified Polymorphic DNA)* as well as RFLP *(Restriction Fragment Length Polymorphism)* or PFGE (*Pulsed Field Gel Electrophoresis*) have been successfully applied for the evaluation of intra-specific differences (e.g. strain typing) (38, 42, 43).

On the basis of a similar identification scheme as described above, sourdough yeasts can be preliminarily identified by a phenotypic approach consisting of cell and colony morphology, presence and number of spores, carbohydrates, organic acids, alcohols and nitrite fermentation/assimilation tests using commercial kits (e.g. ID32C, API 20C AUX, RapID *Yeast Plus System*, *Rapid Yeast Plus*), growth at different temperatures and salt concentrations (44). As for lactic acid bacteria, chemotaxonomy based on fatty acids, proteins and polysaccharides can be successfully applied to yeast identification. The genotypic approach is often based on the sequencing of various ribosomal genes (e.g. 18S, 5,8S, 26S, 5S, and spacer fragments) (45), as well as on the PCR-RFLP of the rDNA 5,8S-ITS (*Internal Transcribed Spacers*) (34). As for lactic acid bacteria, yeast strain typing mainly relies on the PCR-RAPD technique.

Both lactic acid bacteria and yeasts can be identified and monitored during sourdough fermentation by culture-independent systems, for example by PCR-DGGE (*Denaturing Gradient Gel Electrophoresis*), generally directed toward the amplification of specific variable regions of 16S or 26S rRNA genes of lactic acid bacteria and yeasts, respectively. A similar approach was applied by Meroth et al. (14, 15) to evaluate the effect of type I and type II sourdough fermentation on microbial population dynamics.

4.6.3 Determination of the Physico-Chemical Parameters

The balance between the communities of lactic acid bacteria and yeasts and the composition of the microbial communities in terms of species and strains markedly influences the sourdough performances and the overall quality of related leavened baked goods. The microbial community is influenced by some sourdough characteristics and, in turn, it modifies the chemical composition and physical parameters of the sourdough, which is reflected in the overall quality of the products (Fig. 4.3). An overview of the physico-chemical characteristics of the sourdough, which are related to its performance, is determined through the evaluation of the following parameters:

- Dough Yield (DY)
- Dough Acidity (pH and Total Titratable Acidity, TTA)
- Fermentation Quotient (FQ).

Dough Yield (DY). The ratio between water and flour in the dough is indicated as dough yield and it deals with the dough consistency. Considering that different flours have different capabilities to absorb water, doughs of various consistency are obtained having the same DY. DY is calculated as follows: DY = (flour weight + water weight) × 100/flour weight. To consider other ingredients in the formula, the expression is modified as follows: DY = dough weight × 100/flour weight.

Overall, a firm wheat flour sourdough has a DY of 150–160. A liquid sourdough shows values close to 200. Intermediate values of DY indicate soft dough (12). It has been observed that both DY and temperature of fermentation markedly influence

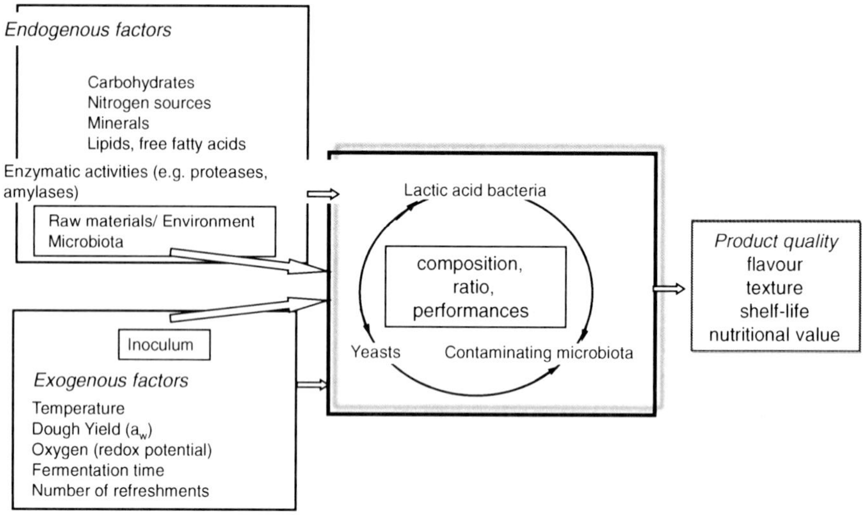

Fig. 4.3 Endogenous and exogenous factors impacting on sourdough characteristics and performances as well as on the overall quality of the final product (Adapted from (4))

the aroma of the sourdough and, especially, the molar ratio between lactic and acetic acids (FQ). Overall, more acetic acid is present in firm dough fermented at 25–30°C, while more lactic acid is found in soft dough fermented at 35–37°C (12, 46). Obviously, on the basis of microbial adaptation to the various environmental factors, the combination of DY and temperature during refreshments markedly influences sourdough microbiota and its performance.

Acetic acid plays an important role by influencing many bread properties (flavour, rope inhibition, shelf life), and its content in sourdough can be increased by fructose addition or by aeration in the presence of heterofermentative lactobacilli (47). In many cases, the effect of temperature on the acetic acid content of type I sourdoughs (generally propagated at a temperature of about 25°C), can strictly depend on the yeast activity: at low temperature, yeasts grow and hydrolyze kestose and other fructo-oligosaccharides, with the fructose being available as electron acceptors for a higher production of acetic acid by *L. sanfranciscensis* via the acetate-kinase pathway (47). At a high temperature (more than 32°C), growth of yeasts is inhibited, fructo-oligosaccharides remain un-hydrolyzed, and *L. sanfranciscensis* has less fructose available for acetate production (48).

pH. The final pH, which ranges from 3.5 to 4.3, is usually considered as an index of a well-developed sourdough fermentation (50). The pH of the sourdough influences the values of pH of the final dough and bread, depending on the amount of full sour that is used as the inoculum. In the case of a standard inoculum of 20% (referred to the dough weight), values of pH that range from 4.7 to 5.4 are usually found in the final dough (50).

TTA (Total Titratable Acidity). The value of TTA is a measure of the total organic acids synthesized during sourdough fermentation. TTA is expressed as ml of NaOH

0.1 N/10 g of dough. The values of TTA range from 30 to 150 ml NaOH 0.1 N/10 g for liquid to 40–220 NaOH 0.1 N/10 g for dried sourdoughs. Nevertheless, the optimal value of TTA for the sourdough depends on the type of bread. Overall, sourdough with high TTA are preferred for bread making with rye flour (48).

Fermentation Quotient (FQ). FQ indicates the molar ratio between lactic and acetic acids during sourdough fermentation. After the determination of the concentration of lactic and acetic acids by enzymatic or chromatographic methods, the FQ is calculated as follows: FQ=(g of lactic acid in 100 g of dough/molecular weight of lactic acid): (g of acetic acid in 100 g of dough/molecular weight of acetic acid). This parameter is strictly related to the type of lactic acid bacteria dominating the fermentation and markedly varies depending on the balance between homo- and hetero-fermentative lactobacilli. In turn, this balance depends on exogenous and endogenous factors which prevail during fermentation (e.g. fermentable sugar and oxygen concentration, DY, time and temperature, etc.).

4.7 The Use of Stabilized Sourdoughs Versus Active Sourdoughs

On the basis of the research progress in the field of bread-making biotechnology, novel types of sourdough have recently been developed with the main aim of improving the flavour of leavened baked products without using traditional protocols of manufacture based on active sourdough. Currently, various commercial preparations of stabilized sourdough are available on the market. They are used for making traditional breads and are based on a mix of tailor-made aroma compounds. Different types of stabilized sourdough are obtained starting from an accurate selection and mix of raw materials (e.g. flour from *T. aestivum* or *T. durum* wheat, rye or other cereals) and microbial strains. Dried sourdough is a type of stabilized product useful for this purpose. Different drying protocols can be applied, for example freeze-drying, spray granulation, fluidized bed drying, spray drying and drum drying, among which the latter two can be cited here as the most common for type III sourdough production (12, 48). In both cases, the higher the DY of the starting type II sourdough the higher the resulting TTA value of the derived type III sourdough, which also increases due to water evaporation. In the spray-drying process, the liquid sourdough is dried by using a warm air flux that removes water until the humidity becomes less than 10%. In the drum-drying system, the vapour of the warmed drums removes water from the thin-layered liquid sourdough during contact. On the basis of various combinations of time and temperature and on the extent of the Maillard reaction, different type III sourdoughs are obtained, which show different degrees of caramelization or toasting and, as a consequence different flavouring activities. Since many volatile compounds (especially acetic acid) are missing due to the evaporation, even if to a different extent depending on the drying technique, pasteurization, cooling or salting can be applied to obtain a stabilized liquid or pasty sourdough (12, 48). With the exception of cooling, all the other stabilization systems lead to microbiota inactivation and stop gas and/or acid production (48), giving a sourdough which can be classified as type III. Especially liquid sourdough represents an advantage for industrial applications because it can be pumped and easily dosed, while showing constant quality (12). Generally, the use of a stabilized sourdough is quite simple. It can be stored at room temperature for a long time (30–60 days) and directly added to the final dough at a proportion of 5–10%. Because of the low cell density of lactic acid bacteria and yeasts, baker's yeast is generally added to leaven the dough for bread production. Examples of applications of type III sourdough in the modern bakery industry are the use of a stabilized liquid rye sourdough for rye bread production; dextran-containing sourdough (produced by the use of selected *L. mesenteroides* strains) stabilized by cooling, pasteurization, or drying for the production of Panettone, rye bread and toast bread; pasteurized San Francisco sourdough in liquid form for the production of San Francisco bread (12).

The selection of microbial strains represents the current challenge for sourdough bread-making industries and relies on several metabolic traits which are related to the most important flavour compounds. Overall, strains isolated from raw materials or sourdoughs are in vitro and in situ selected and evaluated mainly based on the kinetics of acidification, proteolytic activity and nitrogen metabolism and synthesis of aromatic compounds. Recently, fermentation processes based on selected lactic acid bacteria and liquid sourdough have been evaluated with the aim of improving the biosynthesis of specific chemical compounds (49).

4.8 Use of Pure Cultures of Sourdough Starters Versus Stabilized Sourdoughs

As for other fermented food biotechnologies (dairy, oenology, meat) the interest in using a commercial starter culture is rapidly increasing for sourdough breadmaking. In this context, some researches have been devoted to the use of freeze-dried or frozen preparations of lactic acid bacteria (4, 20). Nevertheless, only a few applications of selected lactic acid bacteria starters are currently documented both at artisan and industrial levels. Difficulties are mainly related to the possibility of obtaining highly concentrated lactic acid bacteria preparations as well as to the performance of the selected lactic acid bacteria in situ. From an applicative point of view starter cultures should be pre-cultivated in a mix of flour and water to obtain metabolically active strains before use as an inoculum. Nevertheless, it has been found that many strains, which were in vitro selected for interesting traits, show some difficulties in dominating the sourdough ecosystem and overcoming the endogenous lactic acid bacteria throughout refreshments. Recently, Siragusa et al. (24) demonstrated that, after ten consecutive refreshments, only three out of nine *L. sanfranciscensis* strains, which were in vitro selected for the optimal rate of acidification, proteolytic activity and release of flavour compounds, were able to persist during sourdough fermentation due to the dominance of autochthonous *L. plantarum* strains. Thus, the capability to adapt to the sourdough ecosystem seems to be an essential trait for selecting strains in order to obtain a sourdough with a constant microbial composition and performance. Currently, this seems the main limitation in the use of pure cultures as sourdough starters and the most innovative applications mainly deal with the use of stabilized sourdoughs.

References

1. Corsetti A, Farris GA, Gobbetti M (2010) Uso del lievito naturale. In: Gobbetti M, Corsetti A (eds) Biotecnologia dei prodotti lievitati da forno. Casa Editrice Ambrosiana, Milano, cap. 9
2. Onno B, Roussel P (1994) Technologie et microbiologie de la panification au levain. In: De Roissart H, Luquet FM (eds) Bactéries lactiques, vol 11. Lorica, Uriage, France, cap. V-5
3. Corsetti A, Settanni L (2007) Lactobacilli in sourdough fermentation: a review. Food Res Int 40:539–558
4. Hammes WP, Gänzle MG (1998) Sourdough breads and related products. In: Woods BJB (ed) Microbiology of fermented foods, vol 1. Blackie Academic and Professional, London, cap. 8
5. Wood BJB (2000) Sourdough bread. In: Robinson RK, Batt CA, Patel PD (eds) Encyclopedia of food microbiology, vol 1. Academic, London
6. De Vuyst L, Vancanneyt M (2007) Biodiversity and identification of sourdough lactic acid bacteria. Food Microbiol 24:120–127
7. Gobbetti M (1998) The sourdough microflora, interactions of lactic acid bacteria and yeasts. Trends Food Sci Tech 9:267–274
8. Kulp K (2003) Baker's yeast and sourdough technologies in the production of U.S. bread products. In: Kulp K, Klaus L (eds) Handbook of dough fermentations. Marcel Dekker, New York, cap. 5
9. Ottogalli G, Galli A, Foschino R (1996) Italian bakery products obtained with sour dough: characterization of the typical microflora. Adv Food Sci 18:131–144
10. Böcker G, Stolz P, Hammes WP (1995) Neue Erkenntnisse zum Ökosysyem Sauerteig und zur Physiologie des sauerteig-typischen Stämme *Lactobacillus sanfrancisco* und *Lactobacillus pontis*. Getreide Mehl und Brot 49:370–374
11. Hammes WP (1991) Fermentation of non-dairy food. Food Biotech 5:293–303
12. Decock P, Cappelle S (2005) Bread technology and sourdough technology. Trends Food Sci Tech 16:113–120
13. Wiese BG, Strohmar W, Rainey FA, Diekmann H (1996) *Lactobacillus panis* sp. Nov., from sourdough with a long fermentation period. Int J Syst Bacteriol 46:449–453
14. Meroth CB, Walter J, Hertel C, Brandt MJ, Hammes WP (2003) Monitoring the bacterial population dynamics in sourdough fermentation processes by using PCR-denaturing gradient gel electrophoresis. Appl Environ Microbiol 69:475–482
15. Meroth CB, Hammes WP, Hertel C (2003) Identification and population dynamics of yeasts in sourdough fermentation processes by PCR-denaturing gradient Gel electrophoresis. Appl Environ Microbiol 69:7453–7461
16. Pagani MA, Lucisano M, Mariotti M (2007) Traditional Italian products from wheat and other starchy flours. In: Hui YH (ed) Handbook of food products manufacturing. Wiley Interscience New York, USA, pp 327–388
17. Spicher G, Pomeranz Y (1985) Bread and other baked products. In: Ullmann's Encyclopedia of industrial chemistry, vol A4, 5th edn. Verlag VCH, Weinheim
18. Vandamme EJ, Renard CEFG, Arnaut FRJ, Vekemans NMF, Tossut PPA (1997) Process for obtaining improved structure build-up of baked products. EP 0 790 003 B1
19. Corsetti A, Settanni L, Valmorri S, Mastrangelo M, Suzzi G (2007) Identification of subdominant sourdough lactic acid bacteria and their evolution during laboratory-scale fermentations. Food Microbiol 24:592–600
20. De Vuyst L, Neysens P (2005) The sourdough microflora: biodiversity and metabolic interactions. Trends Food Sci Tech 16:43–56

21. Vrancken G, Rimaux T, Weckx S, Leroy F, De Vuyst L (2011) Influence of temperature and backslopping time on the microbiota of a type I propagated laboratory wheat sourdough fermentation. Appl Environ Microbiol 77:2716–2726
22. Minervini F, Lattanzi A, De Angelis M, Di Cagno R, Gobbetti M (2012) Influence of artisan bakery- or laboratory-propagated sourdoughs on the diversity of lactic Acid bacterium and yeast microbiotas. Appl Environ Microbiol 78:5328–5340
23. Minervini F, De Angelis M, Di Cagno R, Pinto D, Siragusa S, Rizzello CG, Gobbetti M (2010) Robustness of *Lactobacillus plantarum* starters during daily propagation of wheat flour sourdough type I. Food Microbiol 27:897–908
24. Siragusa S, Di Cagno R, Ercolini D, Minervini F, Gobbetti M, De Angelis M (2009) Taxonomic structure and monitoring of the dominant population of lactic acid bacteria during wheat flour sourdough type I propagation using *Lactobacillus sanfranciscensis* starters. Appl Environ Microbiol 75:1099–1109
25. Kline L, Sugihara TF (1971) Microorganisms of the San Francisco sour dough bread process. II. Isolation and characterization of undescribed bacterial species responsible for the souring activity. Appl Microbiol 21:459–465
26. Corsetti A, Lavermicocca P, Morea M, Baruzzi F, Tosti N, Gobbetti M (2001) Phenotypic and molecular identification and clustering of lactic acid bacteria and yeasts from wheat species *Triticum durum* and *Triticum aestivum*) sourdoughs of Southern Italy. Int J Food Microbiol 64:95–104
27. Corsetti A, Settanni L, Van Sinderen D, Felis GE, Dellaglio F, Gobbetti M (2005) *Lactobacillus rossii* sp. nov. isolated from wheat sourdough. Int J Syst Evol Microbiol 55:35–40
28. Meroth CB, Hammes WP, Hertel C (2004) Characterisation of the microbiota of rice sourdoughs and description of *Lactobacillus spicheri* sp. nov. Syst Appl Microbiol 27:151–159
29. Kleynmans U, Heinzl H, Hammes WP (1989) *Lactobacillus suebicus* sp. nov., an obligately heterofermentative *Lactobacillus* species isolated from fruit mashes. Syst Appl Microbiol 11:267–271
30. Vogel RF, Böcker G, Stolz P, Ehrmann M, Fanta D, Ludwig W, Pot B, Kersters K, Schleifer KH, Hammes WP (1994) Identification of lactobacilli from sourdough and description of *Lactobacillus pontis* sp. nov. Int J Syst Bacteriol 44:223–229
31. Rossi J (1996) The yeasts in sourdough. Adv Food Sci 18:201–211
32. Garofalo C, Silvestri G, Aquilanti L, Clementi F (2008) PCR-DGGE analysis of lactic acid bacteria and yeast dynamics during the production processes of three varieties of Panettone. J Appl Microbiol 105:243–254
33. Gatto V, Torriani S (2004) Microbial population changes during sourdough fermentation monitored by DGGE analysis of 16 S and 26 S rRNA gene fragments. Ann Microbiol 54:31–42
34. Pulvirenti A, Solieri L, Gullo M, De Vero L, Giudici P (2004) Occurence and dominance of yeast species in sourdough. Lett Appl Microbiol 38:113–117
35. Vernocchi P, Valmorri S, Gatto V, Torriani S, Gianotti A, Suzzi G, Guerzoni ME, Gardini F (2004) A survey on yeast microbiota associated with an Italian traditional sweet-leavened baked good fermentation. Food Res Int 37:469–476
36. Iacumin L, Cecchini F, Manzano M, Osualdini M, Boscolo D, Orlic S, Comi G (2009) Description of the microflora of sourdoughs by culture-dependent and culture-independent methods. Food Microbiol 26:128–135
37. Hammes WP, Vogel RF (1995) The genus *Lactobacillus*. In: Woods BJB, Holzapfel WH (eds) The genera of lactic acid bacteria. Blackie Academic and Professional, London, pp 19–54
38. Corsetti A, De Angelis M, Dellaglio F, Paparella A, Fox PF, Settanni L, Gobbetti M (2003) Characterization of sourdough lactic acid bacteria based on genotypic and cell-wall protein analyses. J Appl Microbiol 94:641–654
39. Ricciardi A, Parente E, Piratino P, Paraggio M, Romano P (2005) Phenotypic characterization of lactic acid bacteria from sourdoughs for Altamura bread produced in Apulia (Southern Italy). Int J Food Microbiol 98:63–72
40. De Angelis M, Di Cagno R, Gallo G, Curci M, Siragusa S, Crecchio C, Parente E, Gobbetti M (2007) Molecular and functional characterization of *Lactobacillus sanfranciscensis* strains isolated from sourdoughs. Int J Food Microbiol 114:69–82

41. Torriani S, Felis GE, Dellaglio F (2001) Differentiation of *Lactobacillus plantarum*, *L. pentosus*, and *L. paraplantarum* by *recA* gene sequence analysis and multiplex PCR assay with *recA* gene-derived primers. Appl Environ Microbiol 67:3450–3454
42. Gänzle MG, Vogel RF (2003) Contribution of reutericyclin to the stable persistence of *Lactobacillus reuteri* in an industrial sourdough fermentation. Int J Food Microbiol 80:31–45
43. Zapparoli G, Torriani S, Dellaglio F (1998) Differentiation of *Lactobacillus sanfranciscensis* strains by randomly amplified polymorphic DNA and pulsed-field gel electrophoresis. FEMS Microbiol Lett 166:324–332
44. Barnett JA, Payne RW, Yarrow D (2000) Yeast: characteristics and identification, 3rd edn. Cambridge University Press, Cambrige
45. Fernandez-Espinar MT, Martorell P, de Llanos R, Querol A (2006) Molecular methods to identify and characterize yeasts in foods and beverages. In: Yeasts in food and beverages. Springer, Berlino, pp 55–82
46. Onno B (1994) Les levains. In: Guinet R, Godin B (eds) La panification française. Lavoisier, Paris, cap. 8
47. Gobbetti M, De Angelis M, Corsetti A, Di Cagno R (2005) Biochemistry and physiology of sourdough lactic acid bacteria. Trends Food Sci Tech 16:57–69
48. Brandt MJ (2007) Sourdough products for convenient use in baking. Food Microbiol 24:161–164
49. Carnevali P, Ciati R, Leporati A, Paese M (2007) Liquid sourdough fermentation: industrial application perspectives. Food Microbiol 24:150–154
50. Collar C, Benedito de Barber C, Martínez-Anaya MA (1994) Microbial sourdoughs influence acidification properties and bread-making potential of wheat dough. J Food Sci 59:629–633
51. De Man JC, Rogosa M, Sharpe ME (1960) A medium for the cultivation of lactobacilli. J Appl Bacteriol 23:130–135

Chapter 5
Taxonomy and Biodiversity of Sourdough Yeasts and Lactic Acid Bacteria

Geert Huys, Heide-Marie Daniel, and Luc De Vuyst

5.1 Taxonomy of Sourdough Yeasts and Lactic Acid Bacteria

5.1.1 Taxonomy of Sourdough Yeasts

Yeasts are microscopic fungi that undergo typical vegetative growth by budding or fission resulting in an unicellular appearance and a sexual reproduction without a within or upon which the resulting spores are formed [1]. From the agro-alimentary and scientific point of view, yeasts are among the most important eukaryotes. Yeast species found in sourdough microbial communities share an adaptation to the specific and stressful environment created mainly by a low pH, high carbohydrate concentrations and high cell densities of lactic acid bacteria (LAB). Such adaptations can be found in species located mostly on one branch of the evolutionary tree

G. Huys
BCCM/LMG Bacteria Collection and Laboratory for Microbiology,
Department of Biochemistry and Microbiology, Faculty of Sciences, Ghent University
K.L. Ledeganckstraat 35, 9000 Ghent, Belgium

H.-M. Daniel
Mycothèque de l'Université catholique de Louvain (MUCL),
Earth and Life Institute, Applied Microbiology, Mycology, Université catholique de Louvain,
Croix du Sud 3 bte 6, 1348 Louvain-la-Neuve, Belgium

L. De Vuyst (✉)
Research Group of Industrial Microbiology and Food Biotechnology, Faculty of Sciences
and Bio-engineering Sciences, Vrije Universiteit Brussel,
Pleinlaan 2, 1050 Brussels, Belgium
e-mail: ldvuyst@vub.ac.be

M. Gobbetti and M. Gänzle (eds.), *Handbook on Sourdough Biotechnology*,
DOI 10.1007/978-1-4614-5425-0_5,

of fungi, which accommodates the ascomyceteous yeasts. Within this branch, recognized by the current classification in the phylum Ascomycota, as the subphylum Saccharomycotina, class Saccharomycetes, order Saccharomycetales [2], sourdough yeasts belong to different genera. The major sourdough yeasts belong to genera that are currently placed in the family Saccharomycetaceae, although family assignment of yeast genera is still difficult because of a lack of informative data. Basidiomycetous yeasts and dimorphic ascomycetes, also adapted to growth in liquid environments by unicellular growth forms, lack the fermentative abilities that are common to the Saccharomycetales and that are important for growth under oxygen limitations. The taxonomy of the Saccharomycetales, classically based on morphology and physiology, is in the process of being adapted to the increasing knowledge of evolutionary relationships reconstructed from gene sequences, in other words, a phylogenetic system of classification is being developed [2]. This implies a number of name changes. The new genus names have the advantage of reflecting the common genetic background of related yeast species, hereby providing an informative classification in contrast to the former largely artificial classification.

An overview of recent name changes restricted to species that have been obtained from sourdough is given here. The name changes most relevant to the yeasts found in sourdough concern the genera *Saccharomyces* Meyen ex Reess and *Pichia* E.C. Hansen emend. Kurtzman. The genus *Saccharomyces* has been limited to the group of species known as *Saccharomyces sensu stricto*, including the type species of the genus, *Saccharomyces cerevisiae,* on the basis of multiple gene sequences [3]. The group of species formerly often addressed as *Saccharomyces sensu lato* has been divided into several genera. The new genus *Kazachstania* is accommodating the former *Saccharomyces exiguus*, *Saccharomyces unisporus,* and *Saccharomyces barnettii* as *Kazachstania exigua*, *Kazachstania unispora*, and *Kazachstania barnettii*, respectively. *Saccharomyces kluyveri* has been assigned to the new genus *Lachancea* as *Lachancea kluyveri*. The genus *Pichia* has been restricted to species closely related to the generic type species *Pichia membranifaciens*, including *Pichia fermentans* [4]. The former genus *Issatchenkia* has been integrated into the newly defined genus *Pichia* as its species are located on the same branch as the type species *P. membranifaciens* on the phylogenetic tree based on multiple gene sequences used for the redefinition of genera. While the species epithet of *Issatchenkia occidentalis* has been preserved in its new name *Pichia occidentalis*, a complete name change of *Issatchenkia orientalis* to *Pichia kudriavzevii* was necessary as the combination *Pichia orientalis* had been used for a different species before. The former species *Pichia anomala* and *Pichia subpelliculosa* were found to be only distantly related to the generic type species *P. membranifaciens* and have therefore been assigned to the newly created genus *Wickerhamomyces* as *Wickerhamomyces anomalus* and *Wickerhamomyces subpelliculosus*. A review of the taxonomic considerations including the earlier genus name *Hansenula* has been given by Kurtzman [5, 6]. Other, only occasionally from sourdough isolated species from the former genus *Pichia* have been reassigned to new genera, while preserving their species epithet and include *Kodamaea ohmeri*, *Meyerozyma guilliermondii*, *Millerozyma farinosa*, *Ogataea polymorpha*, *Saturnispora saitoi,* and *Scheffersomyces stipitis*.

To determine the entities, or taxa,[1] that deserve attention in the sourdough context, about 40 original publications were reviewed and the repeatedly reported species are listed in Table 5.1. These publications span the time from the early 1970s until present and it is obvious that not all of them are based on identification techniques that are currently considered as the most accurate. However, most of the six regularly (seven or more reports) encountered species *S. cerevisiae* (syn.[2] *S. fructuum*); *Candida humilis* (syn. *Candida milleri*); *P. kudriavzevii* (syn. *I. orientalis*, anamorph *Candida krusei*); *K. exigua* (syn. *S. exiguus*, anamorph *Candida [Torulopsis] holmii*); *Torulaspora delbrueckii*, anamorph *Candida colliculosa*; and *W. anomalus* (syn. *P. anomala, Hansenula anomala*, anamorph *Candida pelliculosa*) can be distinguished from each other reasonably well by classical methods based on morphology and physiology as used in the reviewed literature up to the late 1990s. However, comparing studies using phenotypic identification techniques (n = 19) with those using DNA-based techniques (n = 23), the incidence of *C. humilis* has increased markedly in the DNA-based studies, probably at the cost of a decreased frequency of detecting *K. exigua*. These two phylogenetically closely related species may be mistaken for each other if using phenotypical identification methods. For example, originally *S. exiguus* was reported from San Francisco sourdough [7], while some strains from this study that were preserved in culture collections, later served for the description of *C. milleri*, currently a synonym of *C. humilis*. The sourdough isolate M14 reported as *S. exiguus* and deposited also as CBS 7901 was suggested to belong to *C. humilis* after DNA-based analyses [29, 50]. The regularity with which *S. cerevisiae, C. humilis, P. kudriavzevii, K. exigua, T. delbrueckii*, and *W. anomalus* are encountered in sourdough is an indication of their common association with this substrate. This is not necessarily an exclusive association with sourdoughs and some of the species have to be considered as generalists, able to thrive in a wide range of environmental conditions, as for example *W. anomalus* [51], while a specific ecological niche of the most frequently encountered species *S. cerevisiae* has been elucidated [52]. Other species are less frequently detected in sourdough, namely *Candida glabrata*; *P. membranifaciens* (anamorph *Candida valida*); *Candida parapsilosis*; *Candida tropicalis*; *Candida stellata* (syn. *Torulopsis stellata*); *K. unispora* (syn. *S. unisporus, Torulopsis unisporus*); *Kluyveromyces marxianus* (anamorph *Candida kefyr*); *M. guilliermondii* (syn. *Pichia guilliermondii*, anamorph *Candida guilliermondii*); and *Saccharomyces pastorianus*. Finally, 14 species have been mentioned only in single reports and may be considered as rather transient or present fortuitously in the sourdough ecosystem (Table 5.2).

[1] A taxon (plural taxa) refers to a group of individuals that are judged to form a single unit. A taxon may or may not be given name or rank (species, genus, family, etc.). Primarily, the term serves communication about taxonomic units without the necessity or the possibility to be more specific about them.

[2] Synonymous names and those of asporogenic forms (asexual or anamorphic forms, not producing ascospores) are only selectively mentioned if used in the sourdough literature. It is preferable to use the name of the sporogenous form (sexual or teleomorphic form) if it exists and if no strong reasons require the explicit referral to the asporogenous form.

Table 5.1 Overview of original studies that analyzed the yeast diversity of sourdoughs. Only species repeatedly obtained from sourdough are listed. Numbers in species columns indicate the number of obtained isolates. These numbers depend on the isolation method (i.e., one or multiple representatives of the same colony morphotype) and are mostly of value to judge species abundance within a given study. An "x" was used if the isolate number was not specified. Multiple x's indicate the relative abundance of a detected species if mentioned. Studies that report on more than one sourdough may include sourdoughs of differing yeast species composition. However, a more detailed assignment of yeast species to single sourdoughs was not possible as not all studies reported the necessary data. Synonyms or asporogenic forms (anamorphs) are listed only if used in the cited references or if resulting from recent taxonomic changes

Reference	Wheat (W), rye (R), sorghum (S), barley (B), maize (M), spelt (Sp), buckwheat (Bw), teff (T)	Number of analyzed sourdoughs	Phenotypic (P) or molecular (M) identification	Country	***Saccharomyces cerevisiae*** syn. *S. fructuum*	***Candida humilis*** *Candida milleri* syn. of *C. humilis*	***Pichia kudriavzevii*** syn. *Issatchenkia orientalis* Anamorph *C. krusei*	***Kazachstania exigua*** syn. *S. exiguus* Anamorph *C. (Torulopsis) holmii*	***Torulaspora delbrueckii*** Anamorph *C. colliculosa*	***Wickerhamomyces anomalus*** syn. *P. anomala, Hansenula anomala* Anamorph *C. pelliculosa*	***Candida glabrata***	***Pichia membranifaciens*** Anamorph *C. valida*	***Candida parapsilosis***	***Candida stellata*** syn. *T. stellata*	***Candida tropicalis***	***Kazachstania unispora*** syn. *S. unisporus, T. unisporus*	***Kluyveromyces marxianus*** Anamorph *C. kefyr*	***Meyerozyma guilliermondii*** syn. *P. guilliermondii* Anamorph *C. guilliermondii*	***Saccharomyces pastorianus***
[7]	W	5	P	USA		148[a]													
[8]	W	3	P	Italy				x											
[9]	R	2	P	Germany	11		27	4											
[10]	W	3	P	Spain	X														
[11]	R	20	P	Finland	11			6		2				1		1			
[12]	W	2	P	Spain	29													6	
[13]	W, R	10	P	Italy	X	x		x						x					
[14]	S	2	P	Sudan	[d]		x												

[15]	W	12	P	France	220				25	25	27						
[16]	B, M	10	P	Morocco	21[d]		23[b]										
[17]	W	24	P	Italy	51			13	12		1						
[18]	M	10	P	Ghana	x[d]			x						x		x	
[19]	W	7	P	Italy	25												
[20]	R, M, W	33	P	Portugal	27			11		13	5		12			1	
[21]	W	10	P	Italy	x[d]				xx								
[22]	R	5	P, M	Finland	3	10[c]											
[23]	R, M	n.s.	P	Portugal	7					1	2						
[24]	W	5	P, M	Greece	14								25				
[25]	R	2	P	Denmark	x[d]												
[26]	W	25	P	Italy	17			1	1								
[27]	n.s.	14	P, M	Italy	33[d]	2	7	8									
[28]	W	1	M	Italy	[d]	403											
[29]	R	3	P, M	Germany	48		36	50				19					
[30]	n.s.	35	P, M	Italy	118	42	7	8	3								
[31]	W	13	P, M	Italy	58[d]			3		5		2	3				
[32]	W	8	P, M	Italy	6[d]		6										1
[33]	W	1	M	Italy	6[d]												
[34]	n.s.	35	P, M	Italy	102		48	8	3								
[35]	W	1	P, M	Italy	45		36										
[36]	W	1	P, M	Italy	13	27											
[37]	n.s.	7	M	Italy	23	36											2
[38]	W	10	P, M	Italy	37		15										
[39]	W	3	M	Italy	x		xx										
[40]	W, R, Sp	33	P, M	Belgium	102[d]					2	177	62			1		
[41]	W	4	M	Italy	60[d]	18	2										

(continued)

Table 5.1 (continued)

Reference	Wheat (W), rye (R), sorghum (S), barley (B), maize (M), spelt (Sp), buckwheat (Bw), teff (T)	Number of analyzed sourdoughs	Phenotypic (P) or molecular (M) identification	Country	***Saccharomyces cerevisiae*** syn. *S. fructuum*	***Candida humilis***	*Candida milleri* syn. of *C. humilis*	***Pichia kudriavzevii*** syn. *Issatchenkia orientalis* Anamorph *C. krusei*	***Kazachstania exigua*** syn. *S. exiguus* Anamorph *C.* (*Torulopsis*) *holmii*	***Torulaspora delbrueckii*** Anamorph *C. colliculosa*	***Wickerhamomyces anomalus*** syn. *P. anomala, Hansenula anomala* Anamorph *C. pelliculosa*	***Candida glabrata***	***Pichia membranifaciens*** Anamorph *C. valida*	***Candida parapsilosis***	***Candida stellata*** syn. *T. stellata*	***Candida tropicalis***	***Kazachstania unispora*** syn. *S. unisporus, T. unisporus*	***Kluyveromyces marxianus*** Anamorph *C. kefyr*	***Meyerozyma guilliermondii*** syn. *P. guilliermondii* Anamorph *C. guilliermondii*	***Saccharomyces pastorianus***
[42]	W	4	P, M	Thailand	4[d]			1						1		1				
[43]	W	9	P, M	Italy	33	3														
[44]	W	15	P	Pakistan	x															
[45]	W	10	P, M	Greece	4[d]					163										
[46]	W	20	P, M	Italy	49[d]		10	1	2	5										

[47]	Bw, T	2	M	Ireland	x[d]						x								
[48]	W	7	M	China	43	22	2	2	6	3			2					1	
Total occurrences					***38***	***19***	***14***	***12***	***8***	***7***	***4***	***3***	***2***	***2***	***2***	***2***	***2***	***2***	***2***

[a]Originally designated as *S. exiguus*, but in culture collections deposited representative strains were reassigned to *C. humilis*. One of them served as the type strain for the description of the currently not taxonomically recognized population *C. milleri* (NRRL Y-7244, NRRL Y-7245, NRRL Y-7246, NRRL Y-7248; [49]). *C. humilis* and *C. milleri* isolates are listed here in partially separated columns as some studies distinguished them despite their taxonomic synonymy
[b]Labeled as *C. milleri* by the authors, ITS sequence of one strain deposited as CBS 7541 shows intermediate similarities to *C. humilis* and *C. milleri*, while its ITS RFLP HaeIII type is similar to *C. humilis*
[c]Retrospective identification based on the presence of type and reference strains [22]
[d]No baker's yeast added as explicitly mentioned by the authors
n.s. Not specified by the original authors

Table 5.2 Yeast species isolated from sourdough mentioned by single reports. None of these reports refer to these species as the sole or dominating yeast found in sourdough. Species considered as rare contaminants (e.g., *Schizosaccharomyces pombe*, *Rhodotorula glutinis*, *R. mucilaginosa*, *Endomycopsis fibulinger*) or reported as unidentified were not included

Species	Synonyms[a]	Reference
Candida boidinii		[12]
Candida parapsilosis		[42]
Hanseniaspora uvarum		[38]
Kodamaea ohmeri	*Pichia ohmeri*	[23]
Lachancea kluyveri	*Saccharomyces kluyveri*	[20]
Meyerozyma guilliermondii	*P. guilliermondii*, anamorph *C. guilliermondii*	[12]
Ogataea polymorpha	*P. polymorpha*	[10]
Pichia fermentans	Anamorph *C. lambica*	[31]
Pichia occidentalis	*Issatchenkia occidentalis*	[20]
Saccharomyces bayanus	*Saccharomyces inusitatus*	[7]
Saturnispora saitoi	*P saitoi*	[9]
Scheffersomyces stipitis	*P. stipitis*	[42]
Wickerhamomyces subpelliculosus	*P. subpelliculosa*, *Hansenula subpelliculosa*	[10]
Yarrowia lipolytica		[24]

[a]Synonyms or asporogenic forms (anamorphs) were listed only if used in the cited references or if resulting from recent taxonomic changes

The species complex *C. humilis/C. milleri* frequently detected in sourdough deserves a detailed elucidation. In terms of classification it was placed in the artificial genus *Candida*, because no sexual reproduction could be observed. Phylogenetically, *C. humilis/C. milleri* belongs to the same group as the genus *Kazachstania* [3]. It has, however, not yet been taxonomically placed in this genus, as the phylogenetic reclassification is treating sexually reproducing taxa with priority. The species *C. humilis* has been described based on a yeast strain associated with South African bantu beer, made from kaffir corn (*Sorghum caffrorum*) or finger millet (*Eleusine coracana*) [53]. The species *C. milleri* has been described to accommodate yeast strains isolated from San Francisco sourdough fermentations and initially assigned to *S. exiguus* [7, 49]. The basis for this reassignment were significantly higher guanine-plus-cytosine contents in selected San Franscisco sourdough strains compared to the type and other reference strains of *S. exiguus* and growth stimulation of *C. milleri* by calcium pantothenate. *C. humilis* and *C. milleri* were indicated to be conspecific based on their identical D1/D2 region large subunit (LSU) ribosomal DNA (rDNA) sequences, a locus that in rare cases may not suffice to resolve closely related species [54, 55]. DNA-DNA reassociation of at least 90% between the *C. milleri* type strain CBS 6897 and strain CBS 2664, the type strain of *T. holmii* var. *acidilactici*, a synonym of *C. milleri*, are of interest in this situation and confirms the conspecificity of these strains [54, 56]. Strain CBS 2664 shows ten substitutions in the internal transcribed spacer (ITS) regions of the rDNA, if compared to the type strain of *C. milleri*, indicating the degree of ITS divergence within this species.

In the sourdough context, *C. humilis* is often distinguished from its synonym *C. milleri* by targeting the ITS rDNA region [27]. This is done by restriction fragment length polymorphisms (RFLPs) of the ITS generated by the restriction enzyme *HaeIII* that is recognizing and cutting the nucleotide sequence GGCC. This site is made unrecognizable by a single nucleotide change from C to T (resulting in GGCT) in a population represented by the *C. milleri* type strain. As a result of this single nucleotide substitution the RFLP analysis shows two fragments, in contrast to three fragments for strains with the intact recognition site. A comparison of relevant publicly available ITS sequences (CBS 5658: AY046174, CBS 6897: AY188851, SY13: DQ104399; CBS 2664 and CBS 7541: yeast database at www.cbs.knaw.nl/) shows the transitional position of some strains between the type strain sequences of *C. humilis* and *C. milleri*, especially CBS 7541, indicative of a continuum of ITS sequence variants between both type strains. Therefore, these two taxa might be best considered as populations of one species. Eventual heterogeneity of the species *C. humilis*, especially of applied value, should be documented and accompanied by the deposition of the isolates in public culture collections for further study.

5.1.2 Taxonomy of Sourdough Lactic Acid Bacteria

LAB comprise a heterogeneous group of Gram-positive, nonsporulating, strictly fermentative lactic acid-producing bacteria that play an important role in the organoleptic, health-promoting, technological, and safety aspects of various fermented food products. As a result of natural contamination through the flour or the environment or by deliberate introduction via dough ingredients, a wide taxonomic range of LAB has also been found in sourdoughs. In sourdough environments, LAB live in association with yeasts and are generally considered to contribute most to the process of dough acidification, while yeasts are primarily responsible for the leavening. Although also obligately homofermentative LAB have been isolated from sourdoughs, obligately or facultatively heterofermentative LAB species have the best potential and competitiveness to survive and grow in this particular food environment [57, 58].

Initially, classification of LAB was based on morphology, ecology, and physiological characteristics [59]. At a later stage, also chemotaxonomical properties such as cellular fatty acid and cell wall composition were included. In LAB as well as in many other bacterial groups, phenotypic characters are often limited in their taxonomic usefulness for discrimination of closely related species and suffer from poor interlaboratory exchangeability. As a result, differentiation of LAB solely based on phenotypic traits is generally only considered reliable at the genus level. The introduction of DNA-based techniques such as genomic mol % GC, DNA-DNA hybridization, and sequencing of ribosomal RNA (rRNA) genes has brought significant changes to LAB taxonomy [60–62]. Especially the use of rRNA gene sequences as evolutionary chronometers has allowed the elucidation of phylogenetic relationships between LAB species. As a result, comparison of 16S rRNA gene sequences with sequences in public online databases has become a standard approach for identification of unknown LAB isolates. However, the low evolutionary rate of ribosomal genes may compromise differentiation between LAB species exhibiting identical or nearly identical 16S rRNA gene sequences [63–66]. Alternatively, the use of multiple housekeeping genes encoding essential cellular functions has been proposed for sequence-based identification of LAB. For instance, classification of *Lactobacillus* species based on sequence analysis of the housekeeping genes *pheS* and *rpoA* proved to be highly congruent with 16S rRNA gene phylogeny [67].

The LAB species diversity associated with sourdoughs has been reviewed by several authors in recent years [57, 58, 68–70]. On the basis of these reviews, an updated overview of the LAB species most commonly found in fermented sourdough is compiled in Table 5.3. As is the case for many food ecosystems, this overview again highlights that lactobacilli are by far the most frequently recovered LAB species from sourdough ecosystems. The taxonomy of the genus *Lactobacillus* is extremely complex; according to the April 2011 update, at least 171 species names have so far been proposed in this genus (www.bacterio.cict.fr/l/lactobacillus.html). However, as several of the proposed species names have meanwhile been synonymized, the actual number of phylogenetically unique species is lower. In sourdoughs, more than 55 *Lactobacillus* species have been identified, of which the large majority are obligately heterofermentative (Table 5.3). Given the taxonomic complexity of this genus, accurate identification of unknown *Lactobacillus* isolates requires specific expertise, for example the use of methods that offer sufficient taxonomic resolution and the correct interpretation of identification results by comparison with complete and up-to-date databases. In most of the older studies, however, identification of lactobacilli mainly or even exclusively relied on phenotypic approaches with limited taxonomic resolution at species level. Therefore, it is safe to assume that some of the *Lactobacillus* species previously reported in sourdough environments may have been incorrectly identified at the species or even at the genus level. A typical example is the taxonomic situation in the *Lactobacillus plantarum* group where discrimination between the ubiquitous sourdough bacterium *Lb. plantarum* and the phylogenetically highly related *Lactobacillus paraplantarum* and *Lactobacillus pentosus* may be problematic when identification methods with insufficient taxonomic resolution are used. In this regard, Torriani and colleagues [71] were among the first to suggest that sequences of housekeeping genes such as *recA* rather than 16S rRNA gene sequences are recommended to distinguish between members of this phylogenetically tight species group. Likewise, several sourdough isolates initially assigned to *Lactobacillus alimentarius* may in fact belong to the later described and closely related *Lactobacillus paralimentarius* due to phenotypic misidentification [72]. Also in this case, it has been shown that molecular fingerprint- or sequence-based methods are required to differentiate between both species [67, 73]. In *Lactobacillus rossiae*, a remarkable intraspecific heterogeneity leading to the identification of several subspecific clusters based on *pheS* gene sequencing may complicate unambiguous identification of *Lb. rossiae* strains [77]. Finally, nomenclatural issues may also be a cause for taxonomic confusion. Corrections of originally misspelled specific epithets, such as "*Lactobacillus*

Table 5.3 LAB species generally associated with sourdough fermentation or found in fermented sourdoughs[a]

Obligately heterofermentative[b]	Facultatively heterofermentative	Obligately homofermentative
Lb. acidifarinae	*Lb. alimentarius*	*E. casseliflavus*
Lb. brevis	*Lb. casei/paracasei*	*E. durans*
Lb. buchneri	*Lb. coleohominis*	*E. faecalis*
Lb. cellobiosus	*Lb. kimchi*	*E. faecium*
Lb. collinoides	*Lb. paralimentarius*	*Lb. acidophilus*
Lb. crustorum	*Lb. pentosus*	*Lb. amylolyticus*
Lb. curvatus	*Lb. perolens*	*Lb. amylovorus*
Lb. fermentum	*Lb. plantarum*	*Lb. crispatus*
Lb. fructivorans	*Lb. sakei*	*Lb. delbrueckii* subsp. *delbrueckii*
Lb. frumenti	*P. acidilactici*	*Lb. farciminis*
Lb. hammesii	*P. dextrinicus*	*Lb. gallinarum*
Lb. hilgardii	*P. pentosaceus*	*Lb. gasseri*
Lb. homohiochii		*Lb. helveticus*
Lb. kefiri		*Lb. johnsonii*
Lb. kunkeei		*Lb. mindensis*
Lb. lindneri		*Lb. nagelii*
Lb. mucosae		*Lb. salivarius*
Lb. namurensis		*Lc. lactis* subsp. *lactis*
Lb. nantensis		*S. constellatus*
Lb. nodensis		*S. equinus*
Lb. oris		*S. suis*
Lb. panis		
Lb. parabuchneri		
Lb. pontis		
Lb. reuteri		
Lb. rossiae		
Lb. sanfranciscensis		
Lb. secaliphilus		
Lb. siliginis		
Lb. spicheri		
Lb. vaginalis		
Lb. zymae		
Le. citreum		
Le. gelidum		
Le. mesenteroides subsp. *cremoris*		
Le. mesenteroides subsp. *dextranicum*		
Le. mesenteroides subsp. *mesenteroides*		
W. cibaria		
W. confusa		

(continued)

Table 5.3 (continued)

Obligately heterofermentative[b]	Facultatively heterofermentative	Obligately homofermentative
W. hellenica		
W . kandleri		
W. paramesenteroides		
W. viridescens		

E. Enterococcus, *Lb. Lactobacillus*, *Lc. Lactococcus*, *Le. Leuconostoc*, *P. Pediococcus*, *S. Streptococcus*, *W. Weissella*

[a]Data compiled from [57, 58, 68, 70]

[b]Classification in glucose fermentation types according to Felis and Dellaglio [59]

sanfrancisco" [75] (now *Lactobacillus sanfranciscensis*) and "*Lactobacillus rossii*" [76] (now *Lb. rossiae*), and the synonymization of species, such as the recognition of *Lactobacillus suntoryeus* as a synonym of *Lactobacillus helveticus* [77], can take a while to be introduced in subsequent taxonomic literature.

Triggered by the introduction of molecular DNA-based taxonomic methods in sourdough microbiology and the growing number of 16S rRNA gene sequences in public databases, in recent years many new *Lactobacillus* species have been described which were first isolated from a sourdough environment. Since 2000, 13 new *Lactobacillus* species originally isolated from sourdoughs have been proposed, i.e., *Lactobacillus frumenti* [78], *Lactobacillus mindensis* [79], *Lactobacillus spicheri* [80], *Lactobacillus acidifarinae* [81], *Lactobacillus zymae* [81], *Lactobacillus hammesii* [82], *Lb. rossiae* [76], *Lactobacillus siliginis* [83], *Lactobacillus nantensis* [84], *Lactobacillus secaliphilus* [85], *Lactobacillus crustorum* [86], *Lactobacillus namurensis* [87], and *Lactobacillus nodensis* [88] (Table 5.3). However, many of these species have only been reported once or are rarely isolated from this type of fermented food, and are represented by only a few strains. Therefore, it is not clear which of these species are really typical for sourdough environments and if so, what their geographical distribution is. In fact, only a few *Lactobacillus* species such as the obligately heterofermentative *Lb. sanfranciscensis* and the facultatively heterofermentative *Lb. paralimentarius* seem to be optimally adapted to this specific environment and are rarely isolated from other sources. Other heterofermentative species such as *Lactobacillus brevis* and the facultatively heterofermentative *Lb. plantarum* are also frequently isolated from fermented sourdoughs, but have also been found in many other food and nonfood environments [69].

Although the LAB microbiota of sourdoughs is clearly dominated by lactobacilli, other less predominant or subdominant LAB species may also be found, including members of the genera *Weissella*, *Pediococcus*, *Leuconostoc*, *Lactococcus*, *Enterococcus* and *Streptococcus* (Table 5.3). Of these, specific species of *Weissella*, *Pediococcus*, and *Leuconostoc* are particularly well adapted to survive and grow in plant-derived materials [57, 89]. The taxonomy of the latter three genera is much less complex than in *Lactobacillus*, and their presence in sourdoughs is restricted to only a few species. Weissellas are obligately heterofermentative LAB of which a number of species produce dextran, the best-documented exopolysaccharide formed by heterofermentative LAB. In sourdoughs, the dextran-producing species *Weissella cibaria* and *Weissella confusa* are most frequently found. Both taxa are positioned together on one of the four phylogenetic branches in this genus based on 16S rRNA gene analyses [90], but can be differentiated using restriction analysis of the amplified 16S rDNA [91], randomly amplified polymorphic DNA-PCR (RAPD-PCR) [92], and PCR targeting the ribosomal ITS [26]. Within the facultatively heterofermentative pediococci, the species *Pediococcus acidilactici* and *Pediococcus pentosaceus* are most commonly found in sourdoughs. Differentiation between these two biochemically and phylogenetically related species can be achieved by fingerprinting methods such as ribotyping [93, 94], restriction analysis of the amplified 16S rDNA (16S-ARDRA) [95], RAPD-PCR [96, 97], and by sequence analyses of the 16S rRNA gene, the ribosomal ITS regions and the heat-shock protein 60 gene [98]. In the obligately heterofermentative genus *Leuconostoc*, the majority of sourdough isolates so far identified belong to *Leuconostoc mesenteroides* and *Leuconostoc citreum.* In the former species, further taxonomic distinction is made at subspecies level between *Le. mesenteroides* subsp. *mesenteroides*, *Le. mesenteroides* subsp. *dextranicum,* and *Le. mesenteroides* subsp. *cremoris* (www.bacterio.cict.fr/l/leuconostoc.html). The three subspecies can be separated by RAPD-PCR fingerprinting [99].

5.2 Microbial Species Diversity of Sourdoughs

5.2.1 Influence of Geography

5.2.1.1 The Origin of Sourdough

Historically, sourdough production started as a *conditio sine qua non* to process cereals for the production of baked goods [100]. Indeed, thousands of years ago the first bread production must have been based on spontaneous wild lactic acid fermentation whether or not associated with yeasts and with little or no leavening. Leavening could not have been very pronounced because of the use of barley (*Hordeum vulgare*) and ancient grains [such as spelt (*Triticum aestivum* subsp. *spelta*), emmer (*Triticum turgidum* subsp. *dicoccum*), and kamut (*T. turgidum* subsp. *turanicum*)] in early times and no addition of yeast for leavening. This form of flat (sour) bread production is still daily practice in many countries of the world, in particular in African countries and the Middle East. Leavening must have been an accidental discovery when yeasts from the air or the flour were allowed to ferment the cereal dough mixture extensively. However, it was only from the late nineteenth century onwards that yeast starter cultures were introduced for bread production from wheat (*T. aestivum* and *Triticum durum*) flour, first by using brewing yeasts as remnants of beer production followed by intentionally produced commercial bakers' yeast, *S. cerevisiae*. Consequently, sourdough bread must have been consumed for a long time. Also afterwards, bread production in countries relying on cereals other than wheat such as rye (*Secale cereale*) had still to be supported by lactic acid fermentation. Rye bread baking requires dough acidification to inhibit the abundant α-amylase in the rye flour and to make rye starch and pentosans more water-retaining to form a good dough texture since not enough gluten is present in rye. Hence, various (rye) breads from Germany, Central European countries, and Scandinavia are based on sourdough. In the USA, sourdough was the main base of bread supply in Northern California during the California Gold Rush, because of the easy way to store it and to keep it active for daily bread production. Today, San Francisco sourdough bread is commercially produced in the San Francisco area and it remains a part of the culture of the San Francisco bay area. In the early 1970s, the responsible sourdough bacterium was identified as *Lb. sanfrancisco*, now *Lb. sanfranciscensis*, named according to the area where it was discovered [75]. Notice that this LAB species is actually identical to *Lb. brevis* var. *lindneri* (now *Lb. sanfranciscensis*), which was found to be responsible for various sourdough breads produced in Europe [17]. Also, it was shown that these sourdough LAB species occur in a stable association with the yeasts *C. humilis* (syn. *C. milleri*) and *K. exigua* (syn. *S. exiguus*), respectively [7, 101]. Nowadays, sourdough is used for its technological (dough processing and bread texture, flavor, and shelf life) and nutritional effects [57, 58, 69, 102, 103]. Moreover, sourdough products are appreciated for their traditional value, gastronomic quality, and natural and healthy status [104].

5.2.1.2 The Origin of Sourdough Variation

Thanks to the fact that craftsmanship has determined bakery practice for a long time, a huge variety of bakery products, in particular those based on sourdough, exists, which may differ considerably from region to region. Most of these products, including breads, cakes, snacks, and pizzas, originate from very old traditions. For instance, in Italy, numerous different types of sourdough breads exist, often called according to the name of the region, such as Altamura bread and Pugliese bread [101]. Also, seasonal varieties exist, which are traditionally produced on the occasion of religious festivities. For instance, Panettone cake in Milan and Pandoro in Verona are made for Christmas, while Colomba is a Milanese cake made for Easter. Alternatively, sourdough-based products such as crackers, French baguettes, and Italian ciabattas are much more common, although the original sourdough recipe has been replaced by the faster growing bakers' yeast that is in only few cases allowed to ferment longer to enable contaminating LAB to develop for flavor formation (pre-doughs or type 0 sourdoughs). Fortunately, numerous bakery sourdoughs have been kept alive for tens of years through backslopping procedures, i.e., repeated cyclic re-inoculation of a new batch of flour and water from a previous one with a so-called "sour" during refreshment of the flour-water mixture by the baker, thereby assuring the quality of the baked goods produced thereof. It turned out that these traditional sourdoughs harbor a mixture of distinctive yeast and LAB strains, which may be held responsible for the typical organoleptic quality of the breads made thereof, as backslopping results in a prevalence of the best-adapted strains [57, 68, 69]. This diversity of natural sourdough starters likely accounts for the variety of artisan sourdoughs produced by bakeries, whether or not typical for a certain geographical region. Alternatively, sourdough starter cultures, comprised of one or more defined strains, are commercially available now. In addition to *Lb. sanfranciscensis* strains, commercially available strains of other LAB species include *Lb. brevis*, *Lactobacillus delbrueckii*, *Lactobacillus fermentum*, and *Lb. plantarum*, albeit that not all strains in use are competitive enough to dominate the sourdough fermentation processes that have to be started up [69]. These starter cultures are used for rapid acidification of the raw materials and flavor formation upon fermentation. Also, industrial manufacturers produce dried sourdough powders that are used as nonliving flavor ingredients in industrial bread production.

5.2.1.3 Region-Specific Sourdoughs and Their Associated Microbiota

Whereas it was initially thought that a relationship could be seen between the presence of certain LAB species and the geographical origin of a particular sourdough, it turned out through systematic and detailed taxonomic investigations that the species diversity of both LAB and yeasts of local sourdoughs has nothing to do with the geography of the sourdough production process (Tables 5.1 and 5.4; 57, 68, 69). For instance, the typical sourdough bacterium of San Francisco sourdough bread of the San Francisco bay area, *Lb. sanfranciscensis*, has been found in various wheat sourdoughs throughout Europe, and hence its (unique) presence should be ascribed to other factors, which are mainly based on the fermentation technology and practical conditions applied [121, 152–154]. In various countries, such as in Italy, several studies have been focused on region-specific sourdoughs (Table 5.4). However, no clear-cut relationship could be shown between for the region typical sourdoughs and their associated microbiota. In contrast, Italian sourdoughs harbor simple to very complex communities of LAB species depending on the final products examined, among which *Lb. brevis, Lb. (par) alimentarius, Lb. plantarum, Lb. sanfranciscensis*, *Lb. fermentum, P. pentosaceus*, and *W. confusa* are widespread. Similarly, Belgian bakery sourdoughs have been analyzed extensively and are characterized by LAB consortia of *Lb. brevis, Lb. hammesii, Lb. nantensis, Lb. paralimentarius*, *Lb. plantarum, Lb. pontis, Lb. sanfranciscensis*, and/or *P. pentosaceus* [105, 106, 155]. Sourdoughs with both stable large and stable restricted species diversities may occur [106]. Also, *Lb. rossiae* seems to have a wide distribution in sourdoughs, as has been shown through its isolation from sourdoughs in Central and Southern Italy, Belgium, and elsewhere [74, 105, 106, 156–158]. LAB species are responsible for the acidification of the dough and contribute to flavor formation. Besides LAB species, a large variety of yeast species are found in sourdough ecosystems (Tables 5.1 and 5.2). *S. cerevisiae* has been found in almost every sourdough study (38 out of 42 reviewed) regardless as to whether or not bakers' yeast is added (14 studies mention *S. cerevisiae*, although they indicate that no bakers' yeast was added). A single sourdough usually harbors only one or two yeast species at a given time, among which *C. humilis* (and *K. exigua*) and *P. kudriavzevii* occur most frequently [40]. Yeasts are responsible for the leavening of the dough and also contribute to flavor formation.

Table 5.4 Nonexhaustive overview of the species diversity of LAB in sourdoughs made from various flours and of different geographical origins. The method of identification of the LAB species is indicated

Country	Cereal flour type Method of identification	LAB species reported	Reference(s)
Belgium	Wheat Molecular	*Lb. acidifarinae, Lb. brevis, Lb. buchneri, Lb. crustorum, Lb. hammesii, Lb. helveticus, Lb. namurensis, Lb. nantensis, Lb. parabuchneri, Lb. paracasei, Lb. paralimentarius, Lb. plantarum, Lb. pontis, Lb. rossiae, Lb. sakei, Lb. sanfranciscensis, Lb. spicheri, Leuc. mesenteroides, P. pentosaceus, W. cibaria, W. confusa*	[86, 87, 105, 106]
	Rye Molecular	*Lb. brevis, Lb. hammesii, Lb. nantensis, Lb. paralimentarius, Lb. plantarum*	[105, 106]
	Wheat-rye Molecular	*E. mundtii, Lb. brevis, Lb. curvatus, Lb. fermentum, Lb. paracasei, Lb. paralimentarius, Lb. plantarum, Lb. pontis, Lb. rossiae, Lb. sanfranciscensis, P. acidilactici, P. pentosaceus, W. confusa*	[105, 106]
	Spelt Molecular	*Lb. brevis, Lb. curvatus, Lb. hammesii, Lb. nantensis, Lb. paralimentarius, Lb. plantarum, Lb. pontis, Lb. rossiae, Lb. sakei, Lb. sanfranciscensis, P. pentosaceus, W. cibaria*	[105, 106]
	Wheat (laboratory) Molecular	*Lb. fermentum, Lb. plantarum*	[107]
	Spelt (laboratory) Molecular	*Lb. brevis, Lb. paraplantarum, Lb. plantarum, Lb. rossiae*	[107]
	Wheat (laboratory) Molecular + microarray	*Lb. curvatus, Lb. fermentum, Lb. plantarum, P. pentosaceus*	[108]
	Spelt (laboratory) Molecular + microarray	*Lb. fermentum, Lb. plantarum, P. pentosaceus, W. confusa*	[108]
	Rye (laboratory) Molecular	*Lb. brevis, Lb. fermentum, Lb. plantarum, P. pentosaceus*	[109]

China	Wheat Molecular	*Lb. brevis, Lb. crustorum, curvatus, Lb. fermentum, Lb. guizhouensis, Lb. helveticus, Lb. mindensis, Lb. paralimentarius, Lb. plantarum, Lb. rossiae, Lb. sanfranciscensis, Lb. zeae, Lc. lactis, Leuc. citreum, W. cibaria, W. confusa*	[48]
Denmark	Rye Phenotypical	*Lb. amylovorus, Lb. panis, Lb. reuteri*	[25]
Ethiopia	Teff (enjera) Phenotypical	*E. faecalis, Lb. brevis, Lb. fermentum, Lb. plantarum, Leuc. mesenteroides, P. cerevisiae*	[110]
Finland	Rye Phenotypical	*Lb. acidophilus, Lb. casei, Lb. plantarum*	[11]
France	Wheat Phenotypical	*Lb. acidophilus, Lb. brevis, Lb. casei, Lb. curvatus, Lb. delbrueckii* subsp. *delbrueckii, Lb. plantarum, Leuc. mesenteroides* subsp. *dextranicum, Leuc. mesenteroides* subsp. *mesenteroides, P. pentosaceus*	[111]
	Wheat Phenotypical	*Lb. brevis, Lb. curvatus, Lb. plantarum, Lb. rhamnosus, Leuc. mesenteroides, P. pentosaceus*	[112]
	Wheat Molecular	*Lb. frumenti, Lb. hammesii, Lb. nantensis, Lb. panis, Lb. paralimentarius, Lb. pontis, Lb. sanfranciscensis, Lb. spicheri, Leuc. mesenteroides* subsp. *mesenteroides*	[113]
	Wheat Molecular	*Lb. frumenti, Lb. hammesii, Lb. nantensis, Lb.panis, Lb. paralimentarius, Lb. sanfranciscensis, Lb. spicheri, Leuc. mensenteroides* subsp. *mensenteroides*	[114]
	Wheat Molecular	*E. hirae, Lb. brevis, Lb. curvatus, Lb. paracasei, Lb. paraplantarum, Lb. pentosus, Lb. plantarum, Lb. sakei, Lb. sanfranciscensis, Lc. lactis, Leuc. citreum, Leuc. mesenteroides, P. pentosaceus, W. cibaria, W. confusa*	[115]
Germany	Wheat Phenotypical	*Lb. brevis, Lb. buchneri, Lb. casei, Lb. delbrueckii, Lb. fermentum, Lb. plantarums*	[116]

(continued)

Table 5.4 (continued)

Country	Cereal flour type / Method of identification	LAB species reported	Reference(s)
	Rye Phenotypical	*Lb. acidophilus, Lb. alimentarius, Lb. brevis, Lb. buchneri, Lb.casei, Lb. farciminis, Lb. fermentum, Lb. fructivorans, Lb. plantarum, Lb. sanfranciscensis*	[9, 117]
	Rye – use of a starter Phenotypical	*Lb. acidophilus, Lb. alimentarius, Lb. brevis, Lb. buchneri, Lb. casei, Lb. farciminis, Lb. fermentum, Lb. fructivorans, Lb. plantarum, Lb. sanfrancis-censis, Pediococcus* spp.	[118]
	Wheat (Panettone, wheat bread from Germany, Italy, Sweden, and Switzerland) Phenotypical	*Lb. brevis, Lb. casei, Lb. farciminis, Lb. hilgardii, Lb. homohiochii, Lb. plantarum* (spontaneous) *Lb. brevis, Lb. hilgardii, Lb. sanfranciscensis, W. viridescens* (masa madre)	[119]
	Rye Molecular	*Lb. amylovorus, Lb. frumenti, Lb. pontis/panis, Lb. reuteri* (type II)	[120]
	Rye (bran) – use of a starter Molecular	*Lb. mindensis, Lb. sanfranciscensis* (rye, type I) *Lb. crispatus, Lb. fermentum, Lb. frumenti, Lb. panis, Lb. pontis* (rye, type II) *Lb. johnsonii, Lb. reuteri* (rye bran, type II)	[29]
	Rice Phenotypic + molecular	*Lb. paracasei, Lb. paralimentarius, Lb. spicheri*	[80]
	Wheat Molecular	*E. faecium, Lb. casei, Lb. fermentum, Lb. plantarum, Lb. sanfranciscensis, P. pentosaceus, W. confusa*	[121]
	Rye Molecular	*Lb. pontis, Lb. sanfranciscensis*	[121]

	Amaranth (from India, Peru, Mexico, Germany) Molecular	*Lb. plantarum, Lb. sakei, P. pentosaceus*	[122]
	Cereals and pseudocereals Molecular	*Lb. fermentum, Lb. helveticus, Lb. paralimentarius, Lb. plantarum, Lb. pontis, Lb. spicheri*	[123]
Greece	Wheat Phenotypical + molecular	*Lb. brevis, Lb. paralimentarius, Lb. sanfranciscensis, Lb. zymae, W. cibaria*	[72]
Italy	Wheat (panettone) Phenotypical	*Lb. brevis, Lb. plantarum*	[8]
	Wheat (panettone, brioche) Phenotypical	*Lb. fermentum, Lb. plantarum, Lb. sanfranciscensis, Leuc. mesenteroides, Pediococcus* spp.	[13]
	Wheat (Umbria) Phenotypical	*Lb. farciminis, Lb. plantarum, Lb. sanfranciscensis*	[17]
	Pizza from Naples Phenotypical	*Lb. plantarum, Lb. sakei, Leuc. gelidum, Leuc. mesenteroides*	[124]
	Wheat (Molise, Umbria, Venise) Molecular	*Lb. sanfranciscensis*	[125, 126]
	Mother sponges (Lombardy) Molecular	*Lb. sanfranciscensis*	[21]
	Wheat (Apulia) Molecular	*Lb. acidophilus, Lb. alimentarius, Lb. brevis, Lb. delbrueckii* subsp. *delbrueckii, Lb. fermentum, Lb. plantarum, Lb. sanfranciscensis, Lc. lactis* subsp. *lactis, Leuc. citreum, W. confusa*	[26]
	Wheat (Southern Italy) Molecular	*Lb. alimentarius, Lb. brevis, Lb. farciminis, Lb. fermentum, Lb. fructivorans, Lb. planta rum, Lb. sanfranciscensis, W. confusa*	[127]

(continued)

Table 5.4 (continued)

Country	Cereal flour type Method of identification	LAB species reported	Reference(s)
	Wheat (Molise) Molecular	*Lb. arizonensis/plantarum, Lb. brevis, Lb. kimchii/paralimentarius, Lb. paraplantarum, Lb. sanfranciscensis*	[33]
	Wheat (Sicily) Molecular	*Lb. casei, Lb. kimchii/alimentarius, Lb. plantarum, Lb. sanfranciscensis*	[128]
	Wheat (Molise) Molecular	*Lb. brevis, Lb. fermentum, Lb. plantarum, Lb. sanfranciscensis*	[129]
	Wheat (Altamura, Apulia) Phenotypical + molecular	*Lb. brevis, Lb. casei, Lb. paracasei, Lb. plantarum, Leuc. mesenteroides*	[130]
	Wheat (Sardinia) Molecular	*Lb. brevis, Lb. pentosus, Lb. plantarum, Lb. sanfranciscensis, W. confusa*	[131]
	Wheat (Abruzzo) Physiological + molecular	*Lb. alimentarius, Lb. brevis, Lb. farciminis, Lb. fermentum, Lb. fructivorans, Lb. frumenti, Lb. hilgardii, Lb. mindensis, Lb. panis, Lb. paralimentarius, Lb. paraplantarum, Lb. pentosus, Lb. plantarum, Lb. pontis, Lb. rossiae, Lb. sanfranciscensis*	[133]
	Wheat (Abruzzo) Molecular	*E. faecium, P. pentosaceus*	[133]
	Wheat (Panettone from Central Italy) Molecular	*Lb. brevis, Lb. farciminis, Lb. plantarum, Lb. sanfranciscensis, Leuc. mesenteroides*	[39]
	Wheat (Cornetto, Basilicata) Phenotypic + molecular	*Lb. curvatus, Lb. paraplantarum, Lb. pentosus, Lb. plantarum, Leuc. mensenteroides, W. cibaria*	[134]
	Wheat – use of a starter Phenotypic + molecular	*Lb. brevis, Lb. paralimentarius, Lb. plantarum, Lb. sanfranciscensis, Lc. lactis, Leuc. durionis, Leuc. fructosus, P. pentosaceus, W. cibaria*	[41]

	Wheat (Marche) Phenotypic + molecular	*Lb. brevis, Lb. casei, Lb. curvatus, Lb. fermentum, Lb. paracasei, Lb. paralimentarius, Lb. plantarum, Lb. rhamnosus, Lb. sakei, Lb. sanfranciscensis, Leuc. citreum, Leuc. pseudomesenteroides, W. cibaria, W. confusa*	[43]
	Wheat – use of a starter Molecular	*Lb. paralimentarius, Lb. plantarum, Lb. rossiae, Lb. sanfranciscensis, Lc. lactis* subsp. *lactis, P. pentosaceus, W. cibaria, W. cibaria/confusa*	[135]
	Barley – use of a starter Molecular	*Lb. alimentarius, Lb. brevis, Lb. paralimentarius, Lb. plantarum*	[136]
	Emmer Molecular	*Lb. plantarum, Lb. rossiae, W. confusa*	[137]
	Spelt Molecular	*Lb. brevis, Lb. curvatus, Lb. plantarum, Lb. sanfranciscensis, P. pentosaceus/acidilactici, W. confusa*	[137]
	Wheat Molecular	*Lb. plantarum, Lb. rossiae, P. pentosaceus*	[138]
	Wheat germ Molecular	*Lb. plantarum, Lb. rossiae, P. pentosaceus, W. confusa*	[139]
Iran	Sangak Phenotypical	*Lb. brevis, Lb. plantarum, Leuc. mesenteroides, P. cerevisiae*	[140]
Ireland	Oat Molecular	*Lb. coryniformis, Leuc. argentinum, P. pentosaceus, W. cibaria*	[141]
	Buckwheat Molecular	*Lb. crispatus, Lb. fermentum, Lb. gallinarum, Lb. graminis, Lb. plantarum, Lb. sakei, Lb. vaginalis, Leuc. holzapfelii, P. pentosaceus, W. cibaria*	[47]
	Buckwheat – use of a starter Molecular	*Lb. amylovorus, Lb. brevis, Lb. fermentum, Lb. frumenti, Lb. paralimentarius, Lb. plantarum, Leuc. argentinum, W. cibaria*	[142]
	Teff Molecular	*Lb. fermentum, Lb. gallinarum, Lb. pontis, Lb. vaginalis, Leuc. holzapfelii, P. pentosaceus*	[47]

(continued)

Table 5.4 (continued)

Country	Cereal flour type / Method of identification	LAB species reported	Reference(s)
	Teff – use of a starter Molecular	*Lb. amylovorus, Lb. brevis, Lb. fermentum, Lb. frumenti, Lb. paralimentarius, Lb. plantarum, Lb. pontis, Lb. reuteri, Lb. sanfranciscensis, P. acidilactici*	[142]
Mexico	Maize (pozol) Molecular	*Lb. alimentarius, Lb. casei, Lb. delbrueckii, Lb. lactis, Lb. plantarum, S. suis*	[143]
Morocco	Sourdough ferments, traditional starter sponges Phenotypical	*Lb. brevis, Lb. buchneri, Lb. casei, Lb. plantarum, Leuc. mesenteroides, Pediococcus* spp.	[16]
	Soft wheat flour Phenotypical	*Lb. buchneri, Lb. casei, Lb.delbrueckii, Lb. plantarum, Lb. sanfranciscensis, Leuc. mesenteroides, P. pentosaceus*	[144]
Nigeria	Maize	*Lb. brevis, Lb. casei, Lb. fermentum, Lb. plantarum, Leuc. mesenteroides, P. acidilactici*	[145]
Portugal	Maize (broa, Northern Portugal) Phenotypical	*E. casseiliflavus, E. durans, E. faecium, Lb. brevis, Lb. curvatus, Lc. lactis subsp. lactis, Leuconostoc* spp., *S. constellantus, S. equines*	[21]
Russia	Rye Phenotypical	*Lb. brevis, Lb. fermentum, Lb. plantarum*	[146]
Spain	Wheat Phenotypical	*Lb. brevis, Lb. plantarum*	[10]
	Wheat Phenotypical	*Lb. brevis, Lb. cellobiosus, Lb. plantarum, Leuc. mesenteroides*	[12, 147]
Sudan	Sorghum (kisra) Phenotypical	*Lb. amylovorus, Lb. fermentum, Lb. reuteri*	[14]
	Kisra Molecular	*E. faecalis, Lb. fermentum, Lb. helveticus, Lb. reuteri, Lb. vaginalis, Lc. lactis*	[148]

Sweden	Rye-wheat Phenotypical	*Lb. acidophilus, Lb. brevis, Lb. delbrueckii, Lb. farciminis, Lb. fermentum, Lb. plantarum, Lb. rhamnosus, Lb. sanfranciscensis, W. viridescens*	[149]
	Rye Phenotypical	*Lactobacillus* sp., *P. pentosaceus*	[150]
Thailand	Wheat Phenotypic + molecular	*Lb. casei, Lb. plantarum*	[42]
Turkey	Wheat Phenotypic	*C. divergens, Lb. acetotolerans, Lb. amylophilus, Lb. brevis, Lb. plantarum, Lb. sakei, P. acidilactici, P. pentosaceus, T. halophilus*	[151]
USA	Wheat Phenotypic	*Lb. sanfranciscensis*	[75]
	Wheat Molecular	*Lb. paralimentarius, Lb. plantarum, Lb. sanfranciscensis*	[121]

C. Carnobacterium, E. Enterococcus, Lb. Lactobacillus, Lc. Lactococcus, Le. Leuconostoc, P. Pediococcus, S. Streptococcus, W. Weissella

Single isolations of yeast and LAB species have caused former misinterpretations of their association with certain sourdough-producing regions, not only because of the random isolation itself but also regarding the single habitat (sourdough) explored. Thus, the association of, for instance, *Lb. spicheri* with German rice sourdoughs [80] might represent accidental discoveries [57, 68, 69]. Instead, the dedicated use of basic raw materials as well as the technological procedures applied rather determines the stability and persistence of the yeast and LAB communities involved in the sourdough fermentation process. Indeed, the presence of *Lb. sanfranciscensis* in wheat sourdoughs, which was for a long time the sole habitat wherein this LAB species could be found [now, it has been detected in rye sourdoughs (Table 5.4) and the insect gut as well [159]], can be ascribed to its selection by the type of technology applied, i.e., backslopping practices, temperature of incubation of the dough, pH of the dough, and/or microbial interactions [113, 121, 135, 160, 161]. However, the use of certain raw materials, encompassing cereal types and other ingredients such as adjunct carbohydrates, salt, yoghurt, herbs, etc., and operational practices, such as dough yield and refreshment times, may be linked to local traditions, may favor particular microorganisms as a result of trophic and metabolic relationships and interactions (both cooperation and antibiosis), and hence may associate specific LAB and/or yeast species with specific geographical regions. For instance, the use of rye may select for amylase-positive homofermentative *Lactobacillus amylovorus*, although higher temperatures cause a shift toward the predominance of heterofermentative lactobacilli [120, 162]. The dominance of mainly heterofermentative LAB species in traditional sourdoughs is caused by a highly adapted carbohydrate metabolism, a dedicated amino acid assimilation, and environmental stress responses [57, 58, 68, 163–166]. In particular, maltose, as the most abundant fermentable energy source in dough, is metabolized via the maltose phosphorylase pathway and the pentose phosphate shunt by strictly heterofermentative LAB species such as *Lb. sanfranciscensis*, *Lactobacillus reuteri*, and *Lb. fermentum*. This efficient maltose metabolism coupled to the use of external alternative electron acceptors such as fructose, together with specific pathways such as the arginine deiminase (ADI) pathway, and various environmental stress responses such as response to acidic conditions, increases its competitiveness in the harsh sourdough environment [167–172]. Moreover, maltose-positive LAB species such as *Lb. sanfranciscensis* often form a stable association with maltose-negative yeast species such as *C. humilis*, thereby preventing competition for the same carbohydrate sources [58, 164, 165]. Also, the production of specific inhibitory compounds, maintained through backslopping, such as the antibiotic reutericyclin produced by *Lb. reuteri*, may favor the dominance of this LAB species, as is the case for certain German type II sourdoughs [173].

5.2.2 *Influence of Cereals and Other Raw Materials*

Cereal flours are not sterile. Their microbiological stability is related to their low water activity. As yeasts and LAB naturally occur on plant materials, cereal flours carry both groups of microorganisms and competitive yeast and LAB species reach numbers above those of the adventitious microbiota upon fermentation of a flour-water mixture [57]. However, whether the type of flour mainly directs the growth of sourdough LAB species remains controversial [25, 26, 47, 89, 105, 107, 121, 123, 135]. For instance, whereas it was assumed that rice sourdough fermentation selects for *Lb. spicheri* [80], this LAB species cannot always be found in rice sourdoughs [123]. This has been ascribed to competitiveness of the microorganisms that are present in the sourdough ecosystem. Indeed, microbial interactions between the spontaneous microbiota and an added sourdough starter culture may lead to the dominance of autochthonous LAB species and/or strains. Among other mechanisms, competitiveness may explain the apparent prevalence of LAB species in specific sourdough preparations, such as evidenced by single reports on *Lb. amylovorus* in rye sourdoughs [120], *Lactobacillus sakei* in amaranth sourdoughs [122], and *Lb. pontis* in teff sourdoughs [47]. Yet, spontaneous sourdough fermentations carried out in the laboratory with flour as the sole nonsterile ingredient indicate that the type and quality (microbiological and nutritional) of the cereal flour used is indeed an important source of autochthonous LAB and yeasts occurring in the ripe sourdoughs [40, 107]. Hence, the flour plays a key role in establishing stable microbial consortia within a short time. In this context, it has been shown that laboratory sourdoughs based on wheat, rye, or spelt, backslopped daily for 10 days at 30 °C, whether or not initiated with a *Lb. sanfranciscensis* starter culture [as tested in the case of wheat sourdough fermentations in the study of Siragusa *et al.* [135]], reach an equilibrium of LAB species through a three-step fermentation process: (1) prevalence of sourdough-atypical LAB species (e.g., *Enterococcus* spp. and *Lc. lactis* subsp. *lactis*); (2) prevalence of sourdough-typical LAB species (e.g., species of *Lactobacillus*, *Leuconostoc*, *Pediococcus*, and *Weissella*); and (3) prevalence of highly adapted sourdough-typical LAB species (e.g., *Lb. fermentum* and *Lb. plantarum*) [107–109, 135]. Indeed, it has been shown that the LAB species *Lb. fermentum* (strictly heterofermentative) and *Lb. plantarum* (facultatively heterofermentative) dominate several sourdough fermentation processes, irrespective of the type of flour or the addition of starter cultures that are not robust enough [47, 108, 123, 136–138, 142, 172, 174, 175]. Concerning yeasts, *C. glabrata* and *W. anomalus* prevail during laboratory sourdough fermentations [40]. Further, it has been shown that previous introduction of flour into the bakery environment helps to build up a so-called house microbiota that serves as an important inoculum for subsequent bakery sourdough fermentations [155]. Indeed, LAB strains adapted to the sourdough and bakery environment (apparatus, air, etc.), which have been shown to be genetically indistinguishable, may be repetitively introduced in consecutive sourdough batches during backslopping. The widespread use of bakers' yeast may be responsible for the prevalence of *S. cerevisiae* in bakery sourdoughs [26, 29, 31, 34–36, 39, 40]. However, there are also indications through reliable molecular data of a large strain diversity of *S. cerevisiae* in single sourdoughs, that suggest an autochthonous wheat flour origin of this yeast species in sourdough too [27, 30, 34]. Supportive of an autochthonous origin of *S. cerevisiae* is also the presence of this species in rye flour [29]. Yet, during laboratory fermentations with flour as the sole nonsterile ingredient and without added bakers' yeast, other species such as *C. glabrata* and *W. anomalus* emerge [40]. Anyway, the direct environment is another important source of (accidental) contamination of the flour by LAB and yeasts. Consequently, hygienic conditions in the sourdough and bakery environments will play a role as well. Finally, microorganisms occurring on cereals and subsequently in sourdoughs may be of intestinal origin, due to fertilization practices on the grain fields, mouse feces or insects in the flour mills, or fecal contamination of the sourdough production environment [70, 175–178]. It may explain the opportunistic presence of *Lactobacillus acidophilus*, *Lactobacillus johnsonii*, *Lb. reuteri*, and *Lb. rossiae*, which are common gastrointestinal inhabitants.

5.2.3 *Influence of Technology*

Besides the cereals and other dough ingredients, which are mainly responsible as the source of metabolic activity in the form of flour enzymes and endogenous microorganisms, specific technological process parameters determine the species diversity, number, and metabolic activity of the microorganisms (whether or not added) present in the stable, ripe sourdough. These process parameters include chemical composition and coarseness of the flour, leavening and storage temperature, fermentation time, pH, redox potential, dough yield, refreshment time and number of propagation steps, and interactions between the microorganisms [57, 102, 164, 165, 179].

Different types of sourdough exist, on the basis of the processing conditions and/or technology used for production, with a specific microbiota occurring in each type [57, 180]. Type I or traditional sourdoughs are manufactured by continuous, (daily) backslopping, at ambient temperature (<30 °C), to keep the microorganisms in an active state. Therefore, mother doughs are used as an inoculum for subsequent doughs by addition of the desired amount of dough to a fresh flour-water batch according to defined cycles of preparation. These small-scale sourdough productions are used in traditional (home-made) sourdough bread making. Natural sourdoughs frequently harbor *Lb. sanfranciscensis* and *C. humilis/K. exigua* as prevalent LAB and yeast species, respectively. Type II or industrial sourdoughs are produced through one-step propagation processes of long duration (typically 2–5 days) at a fermentation temperature above 30 °C and with high water content. These large-scale sourdough productions result in semifluid preparations, which are used as dough acidifiers or flavor ingredients. *Lb. amylovorus, Lb. fermentum*, *Lb. pontis*, and *Lb. reuteri* are commonly found in type II wheat and rye sourdoughs. Type III sourdoughs are prepared in dried form to be used as nonliving acidifier supplement and flavor carriers for (sourdough) bread production. In contrast to type I doughs, doughs of types II and III require the addition of baker's yeast for leavening.

Commercially available bulk starter cultures to prepare type II and III sourdoughs aim at standardizing the end products through acidification of and flavor formation in the dough [69, 181, 182]. New trends tend to develop starter cultures that lead to improved functional properties other than acidification and flavor formation, such as texture improvement, antibacterial and antifungal activities, and health-promoting effects [69, 166]. In this context, strain robustness and fitness towards microbial competitors and environmental conditions should be the driving force in the selection of useful starters for sourdough fermentation processes, as it has been shown that autochthonous strains often emerge [69, 135, 138, 142]. However, studies on the industrial exploitation of sourdough starter cultures are scarce [183, 184]. Recently, the use of starter cultures in type I propagated sourdoughs has been investigated [89, 123, 135, 185]. It is of course well known that the fermentation temperature affects the ratio of lactic acid to acetic acid [185, 187]. In general, homofermentative LAB starter cultures are used at high temperature and for short fermentation times (e.g., 37 °C for 36 h) and heterofermentative LAB starter cultures are used at low temperature and for long fermentation times (e.g., 25 °C for 48 h), resulting in sourdoughs with mainly lactic acid and acetic acid, respectively. However, it would be of great value to know the circumstances for the expression of other functional properties that are of added value to sourdoughs [182].

The fermentation temperature, one of the criteria to distinguish type I and II sourdoughs, is essential for the community dynamics and stability of a sourdough microbiota [29, 160, 175, 179, 185, 188, 189]. For instance, spontaneous wheat sourdough backslopping fermentations (type I) carried out at 23 °C for 10 days select for *Le. citreum* instead of *Lb. fermentum* that prevails at 30 °C and 37 °C [175]. Similarly, rye fermentations initiated with commercial sourdough starter cultures maintain the presence of *Lb. mindensis* and *Lb. sanfranciscensis* at 25 °C (type I), but select for *Lactobacillus crispatus* and *Lb. pontis* at 30 °C and for *Lb. crispatus*, *Lb. frumenti*, and *Lb. panis* at 40 °C (both type II) [184]. Whereas *Lb. sanfranciscensis* prefers long fermentation times at relatively low temperature, conditions that often prevail during type I sourdough preparations, this species grows optimally at 32 °C [159, 189]. However, whereas *C. humilis* grows optimally at 27–28 °C but does not grow above 35 °C [160, 189], the association of *Lb. sanfranciscensis*-*C. humilis* grows optimally at 25 °C and 30 °C and may explain its stability between 20 °C and 30 °C [160, 190]. The abundance of *Lb. sanfranciscensis* in wheat sourdoughs made at ambient temperature indicates a low competitiveness of other LAB species such as *Lb. fermentum* that prefers higher temperatures for optimal growth. Similarly, temperature may be responsible for a selection toward *Lb. helveticus* during Sudanese sorghum sourdough fermentations, which are carried out at 37 °C [148].

For the growth of sourdough LAB, also the pH plays an important role [160, 179, 188, 189]. For instance, *Lb. sanfranciscensis* cannot grow below pH 3.8–4.0 [160, 189], whereas *C. humilis* is not influenced by the pH [190]. An optimal pH for growth of around 5.0 has been found for *Lb. sanfranciscensis*. This pH value corresponds approximately to that observed during the first stage of dough fermentation. However, the growth of lactobacilli is favored over yeast growth at pH values above 4.5 [160]. Hence, the rate of acidification of the dough may determine the level of *Lb. sanfranciscensis* in the dough. Natural sourdough fermentations displaying higher pH values are often dominated by a different microbiota, encompassing *Enterococcus*, *Lactococcus*, *Leuconostoc*, *Pediococcus*, *Streptococcus*, and *Weissella*, which are commonly present in the cereal flour [57, 133] or during the early fermentation process but die off when a significant pH decrease occurs upon fermentation [89, 108].

Although sourdough fermentation proceeds anaerobically, the presence of oxygen in the beginning of the fermentation and when small amounts of dough (high ratio of surface to volume) are used may favor certain LAB and yeast species [179, 190, 192]. For instance, mild aeration has a positive influence on the competitiveness of *Lb. amylovorus* DCE 471 [179]. Similarly, *P. kudriavzevii* can only grow when enough oxygen is available during fermentation [190]. Further, the ionic strength and salt concentration of the dough affects microbial growth [160, 179, 193, 194]. Similarly, the presence of organic acids in and the buffering capacity of the flour influence growth of both yeasts and LAB [160, 179]. In general, sourdough LAB are acid-tolerant and their growth is favored in the presence of salt, as is the case for, for instance, *Lb. amylovorus* DCE 471 [188, 193]. Alternatively, the growth of *C. humilis* and *S. exiguus* is completely inhibited by 4% NaCl; also, the growth of these yeasts is strongly inhibited in the presence of acetic acid and to a much lesser extent by lactic acid [160, 179].

Whereas backslopping practices select for mainly heterofermentative LAB, the amounts of dough used for backslopping and the frequency of the refreshments determine the community dynamics and stability of the sourdough microbiota as well. The amount of backslopping dough defines the initial pH and in this way influences the growth and acidification rates of the LAB species involved [107, 189, 190]. Also, the amount of backslopping dough determines the dough yield and hence the availability of water (water activity of the dough). Short refreshment times may select for rapidly growing LAB species, which in turn depends on the fermentation temperature and influences the acidification rate. In this regard, *Lb. fermentum* is most competitive at 30 °C and 37 °C with backsloppings every 24 h, while a mixture of *Lb. fermentum* and *Lb. plantarum* prevails at 30 °C with backsloppings every 48 h [175]. This may explain why *Lb. sanfranciscensis* is sometimes missed during laboratory-scale fermentation processes [136, 175]. Also, a short refreshment time seems to favor *C. humilis* during sourdough fermentation compared to *S. cerevisiae* [190].

Finally, interactions between LAB and yeasts are an important aspect for the community dynamics and stability of the sourdough microbiota [50, 57, 165, 189, 195, 196]. Interactions encompass both cooperative and antagonistic ones. During some sourdough fermentation processes yeasts cannot develop at all, perhaps because of inhibition of yeast growth by nutritional competition or the presence of inhibitory compounds [47, 123, 175]. In other processes, mutualistic interactions lead to stable associations, not only between LAB species and yeasts (besides *Lb. sanfranciscensis*/*C. humilis*, also *Lb. sanfranciscensis*/*K. barnettii*, *Lb. plantarum*/*S. cerevisiae*, and *Lb. brevis*/*Candida* spp.) but also among LAB species (e.g., between *Lb. sanfranciscensis* and *Lb. plantarum* or *Lb. paralimentarius*) [40, 41, 58, 69, 142]. Nevertheless, the competitiveness of LAB and yeasts in sourdough seems to be strain-specific and not species-specific, as has been shown for *Lb. sanfranciscensis* [135] and *Lb. plantarum* strains [138] in wheat sourdoughs recently.

5.3 Isolation of Sourdough Yeasts and Lactic Acid Bacteria

5.3.1 Isolation of Sourdough Yeasts

Yeast isolation from mature sourdoughs is relatively uncomplicated as a stable sourdough usually harbors a homogeneous yeast population as part of the resident microbiota. However, the follow-up of different developmental stages of a sourdough or the search for minor components requires a strategy that is optimized towards the detection of subdominant components. Detailed information on food yeast isolation was provided by Deak [197]. Here, only a brief discussion of the currently practiced yeast isolation methods from sourdough is given, including some references to more complete methodological resources. After sample homogenization and dilution, yeast growth is suitably effected on solid media. This sequence of manipulations should be performed with minimal delay to avoid the settling of yeast cells and cell death. Diluents, usually distilled water, peptone water, saline, or Ringer solution, may influence the resulting cell counts, with peptone water having resulted in the highest cell recovery [198]. Overviews of classical growth media and isolation techniques for yeasts and foodborne yeasts are given by Yarrow [199] and Beuchat [200], respectively. The cultivation media used in sourdough analyses are rich media containing complex compounds such as peptone (e.g., Sabouraud agar), tryptone (e.g., Wallerstein Laboratory nutrient agar), yeast extract (e.g., yeast extract peptone dextrose agar), malt (e.g., yeast and malt extract agar, wort agar), and potato infusion (e.g., potato dextrose agar), together with an additional component to inhibit bacterial growth. Most often chloramphenicol is used as an antibiotic that can be added to the medium before sterilization without losing its activity. Acidification of the medium is sometimes used to restrict bacterial growth, while this is known to affect growth of some yeasts (namely of the genus *Schizosaccharomyces*). However, the acidity of the sourdough lets it appear unlikely for acid-sensitive strains to be present in a ripe sourdough. In general, most yeasts show good vegetative growth at room temperature, although some may grow at subzero temperatures and others up to 45 °C [201]. Cardinal growth temperatures of yeasts are species- and strain-specific. The most suitable incubation temperature for sourdough yeasts would be the temperature at which the sourdough is in its most active state. The most frequently applied temperatures are 25–30 °C. First yeast growth can under such conditions usually be observed after 2–3 days, while daily inspection of the plates for 5 up to 10 days is recommended to allow full differentiation of colony morphology and detection of more slowly growing components.

Other than the consideration of growth conditions, the selection of yeast colonies for further characterization and identification is the most important factor influencing the completeness of a diversity survey. Even though sourdough samples often present a homogeneous yeast population, one needs to bear in mind that the colony morphology of different yeast species is often very similar. Each observed morphotype should therefore be sampled more than once. A logical strategy would be to recover a number of colonies of each type that represents a reasonable percentage of the total number of colonies of that particular morphotype. Each morphoptype's percentage provides important species abundance data in the sourdough if more than one yeast species is isolated. The generation of pure cultures is crucial before characterization and identification of the isolates. The purity should be tested microscopically and on antibiotic-free medium to exclude any carry-over of the accompanying bacterial microbiota.

5.3.2 Isolation of Sourdough LAB

Isolation of LAB from sourdough environments is challenging for three main reasons. First, sourdoughs are complex ecosystems not only in terms of their microbial composition but also in terms of the interactive effects among types of breadmaking processes and ingredients. The utilization of soluble carbohydrates by LAB and, thus, their energy yield are greatly influenced by the associated yeasts and vary according to the type of carbohydrates [195]. However, as many media for selective isolation of LAB incorporate yeast-inhibiting agents such as cycloheximide, pimaricin, and amphotericin B, the trophic interaction between LAB and yeasts is in these cases disturbed, which may affect the recovery potential of LAB strains that strongly rely on this association. Secondly, sourdough fermentation is a dynamic process in which fast-acidifying LAB initially dominate the ecosystem and are then gradually replaced by typical sourdough LAB that largely contribute to the organoleptic and textural properties of the end product. Depending on whether the early subdominant LAB and/or the final dominant LAB are the target of the isolation approach, it may thus be necessary to include multiple samples taken at different time points. Finally, the LAB communities in sourdoughs may consist of metabolically very diverse groups, including obligately homofermentative and facultatively or obligately heterofermentative species. As some of these species have specific growth requirements in terms of the incubation medium and conditions (e.g., temperature, pH, atmosphere, etc.), it seems inevitable that different medium formulations and/or sets of incubation parameters are required to cover the entire metabolic LAB spectrum present in a sourdough sample.

Initially, sourdough LAB were mostly isolated on de Man-Rogosa-Sharpe (MRS) medium [202], which is the general medium used for the isolation and enumeration of lactobacilli from fermented food products. The MRS medium contains glucose as the main carbohydrate source. Triggered by growing insights in the species diversity of sourdough-associated LAB, a number of more specialized media have been developed for the selective isolation of typical sourdough species. For the specific detection of *Lb. sanfranciscensis*, Kline and Sugihara [75] proposed the SourDough Bacteria medium which contains maltose as the carbohydrate source in addition to freshly prepared yeast extract (FYE) to further enhance growth. The Sanfrancisco medium was developed for the isolation and description of *Lb. pontis* and *Lb. mindensis* [79, 203]. This medium contains three carbohydrates (maltose, fructose, and glucose), FYE, cysteine and rye or wheat bran. In parallel to the design of new media, several authors also described variations of the original MRS medium formulation for isolation of sourdough LAB. Vogel and co-workers [203] proposed a modified MRS medium, referred to as MRS "Vogel", with higher pH value (6.3), whereas the MRS5 medium [185] contains the three major carbohydrates present in the sourdough ecosystem (i.e., maltose, fructose, and glucose) in addition to cystein and a vitamin mixture. In subsequent studies, the MRS5 medium has been successfully used for the isolation of several novel *Lactobacillus* species from sourdough such as *Lb. spicheri* [80], *Lb. namurensis* [87], and *Lb. crustorum* [86]. From these recent descriptions, it thus appears that the use of a modified MRS formulation with a lowered pH (<6.0) and supplemented with an additional carbon source such as maltose and/or fructose as well as with amino acids and vitamins under anaerobic conditions is one of the most successful strategies for the isolation of (new) sourdough LAB species. In a recent study, the qualitative and quantitative performance of 11 elective and selective culture media was compared for isolation of lactobacilli from type I sourdoughs [204]. On the basis of the identification results obtained with protein profiling, the largest species diversity was recovered on maltose-containing MRS medium. However, the fact that MRS5 medium allowed the isolation of a specific (but unidentified) subpopulation only found on this medium indicates that there is no single efficient medium for the recovery of all lactobacilli from type I sourdoughs.

5.4 Identification of Sourdough Yeasts and Lactic Acid Bacteria

Traditionally, identification of sourdough microorganisms relied on (selective) culturing, selection and purification of a limited number of isolates, and identification of purified isolates with phenotypic and/or genotypic methods. Although this approach has significantly contributed to our current knowledge of the sourdough-associated yeast and LAB species diversity, the use of culture media holds a number of intrinsic limitations. In a culture-based approach, species with very specific nutrient and growth conditions may only sporadically or even not be recovered which leads to an underestimation of the actual microbial species diversity present in the complex sourdough ecosystem [68]. In contrast, culture-independent techniques that are based on phylogenetic dissection of the metagenomic DNA extracted directly from the sample allow one to unravel the species diversity and dynamics of sourdough yeasts and LAB without the need to isolate and culture its single components. However, also these DNA-based methods have a number of limitations, including poor detection capacity of subdominant species and inadequate taxonomic resolution between phylogenetically closely related species. Depending on the aim of the study, conventional culturing and molecular methods are therefore often combined to obtain a more complete picture of the microbial species diversity of sourdough ecosystems [41, 106].

5.4.1 Culture-Dependent Approaches

5.4.1.1 Yeasts

Identification is the localization of individuals in a classification scheme by means of diagnostic characteristics resulting in the assignment of names. The diagnostic characteristics to be used are provided in the species descriptions and, in the case of yeasts, are collected in the monograph "The yeasts: a taxonomic study," currently in its fourth edition [1], with the fifth edition about to be released [204]. While the fourth edition still included instructions for the phenotypic identification of yeasts [199], the fifth edition reformulates them as phenotypic "characterization" instead [205]. This is a consequence of the need to use DNA-based methods to recognize the since 1998 twofold increased number of yeast species. Nevertheless, the accurate description of fermentation and assimilation abilities as well as other phenotypic characters of yeasts continues to be of interest in technological, ecological and taxonomic frameworks.

Among the DNA-based methods currently applied to yeast identification the partial sequencing of the large subunit (LSU) ribosomal ribonucleic acid genes occupies a key position [206]. The DNA sequences of the variable regions D1 and D2 located at the 5' end of the LSU of virtually all known yeast species are documented in the public databases of the International Nucleotide Sequence Database Collaboration (INSDC, including GenBank, the European Molecular Biology Laboratory, and the DNA Data Bank of Japan). The entries of the three submission hubs are bundled, daily updated, and made available for searches by the NCBI (www.ncbi.nlm.nih.gov/nucleotide/). Few distinct yeast species show no or low sequence divergence in the D1/D2 LSU rDNA, can therefore not reliably be distinguished, and require complementary analyses such as *Saccharomyces bayanus* and *S. pastorianus* [206], *Hanseniaspora meyeri* and *Hanseniaspora clermontia; Hanseniaspora guilliermondii* and *Hanseniaspora opuntiae* [207], *Meyerozyma guilliermondii* and *Meyerozyma caribbica; Trichomonascus ciferrii* and *Candida mucifera; K. marxianus* and *Kluyveromyces lactis* [55], *Debaryomyces hansenii, Debaryomyces fabryi* and *Debaryomyces subglobosus* [208]. Genetic regions that show in most cases larger divergence than the D1/D2 LSU rDNA and for which substantial sequence records have been accumulated include the ITS region of the ribosomal gene cluster. This region is favored by mycologists as the barcoding locus of fungi, although no common threshold value to distinguish intraspecific from interspecific variation can be defined [209, 210]. No systematic evaluation of ITS sequence variation to answer this question for ascomycetous yeast species exists to date. In comparison to sequencing, RFLP analysis of ITS sequences offers simplified access to partial DNA sequence information. The accessed information is determined by the recognition sites of the applied restriction enzymes and typically includes only a few nucleotides, necessitating a range of restriction enzymes to reliably distinguish the species in question.

Single-copy protein-coding gene sequences also accumulate in the public databases and may be used to complement those of D1/D2 LSU and ITS sequences in cases where the ribosomal genes do not allow a conclusive species identification. Such protein-coding genes include the actin gene (*ACT1*) [55], translation elongation factor 1 alpha (*TEF1*), the mitochondrial cytochrome oxidase 2 gene (*COX2*) [211], and the largest and second largest subunits of the RNA polymerase II gene (*RPB1, RPB2*) [212]. In contrast to the multicopy ribosomal DNA (e.g., D1/D2 LSU, ITS), single-copy nuclear genes may be more difficult to amplify, as the design of universal primers effective in phylogenetically distant species is not always possible and as only single primer binding sites are available in a haploid genome in contrast to hundreds of binding sites for ribosomal genes. Highly variable mitochondrial genes (e.g., *COX2*) bear the possibility of having been subject to a different evolutionary path than the nuclear genome, in other words, are prone to potential horizontal gene transfers across species borders. As databases for complementary gene sequences are far more restricted than for the D1/D2 LSU, one needs to assure the existence of reference sequences. Databases that allow searching the available sequences for specific strains, such as the yeast database of the Centraalbureau voor Schimmelcultures, The Netherlands (www.cbs.knaw.nl/yeast/BioloMICS.aspx) and the StrainInfo portal (www.straininfo.net/), Ghent University, Belgium, may be used for the selection of complementary sequencing targets. The comparison of a query sequence with reliable type strain sequences is essential, as type strains are the only valid taxonomic reference for a species. Sequence alignments outside the commonly consulted BLAST (Basic Local Alignment and Search Tool) tabulated results are helpful, because type strain sequences are often not recognizable from the sequence entry title line. It is recommended to include type strain sequences of the phylogenetically most closely related species in these alignments to confirm the differentiation of the species in question by the given DNA region.

While the so far discussed methods use genetic information of few or single loci, the techniques commonly known as DNA fingerprinting exploit genetic information that is distributed throughout the genome. A large variety of protocols exists and the more reproducible among them are based on the specific binding of PCR primers to mini- or microsatellite sequences in contrast to arbitrary binding realised in RAPD-PCR. The primer most frequently applied to sourdough yeasts was derived from a ubiquitous minisatellite sequence found in the protein II gene of the bacteriophage M13 [213]. The primer referred to as M13 results from a consensus sequence of 12 partially incomplete repeats. After its use in a PCR assay as the single primer and visualization of the PCR reaction products by agarose gel electrophoresis, usually species-specific banding profiles based on the different lengths of amplifiable sequences enclosed or flanked by M13 minisatellites are observed. The important influence of experimental factors such as the DNA extraction method and PCR and electrophoresis parameters on the resulting profiles implies the need to include type strains for ideally side-by-side-comparisons if a complete identification is to be performed. However, PCR-fingerprinting without type strains may be used to group larger numbers of isolates and to select those that are representative of each group for identification by DNA sequencing [40].

The species *S. cerevisiae* has been observed to show extensive pheno- and genotypic intraspecies diversity (reviewed in [214]). Part of such strain diversity has been traced by molecular analyses to genomic variability associated with Ty-element insertion sites [215]. Ty elements belong to a group of eukaryotic transposable elements that are also called retrotransposons because of some similarities with retroviruses. Ty-elements are flanked by long terminal repeat (LTR) sequences, in turn formed by repeated delta elements. Delta elements are also found in larger numbers independently from LTR sequences and have been used to design and optimize specific PCR primers to amplify the sequences between two delta elements in *S. cerevisiae*, resulting in an effective strain-typing method for this species [216, 217].

5.4.1.2 LAB

Although largely abandoned and replaced by molecular tools, characterization and identification of sourdough LAB species by phenotypic methods is in some cases still useful, or even mandatory when it concerns new species descriptions. The conventional phenotypic approaches for identification of sourdough LAB species may include physiological and chemotaxonomic tests and determination of major fermentation pathways, carbohydrate utilization patterns, lactic acid configuration, and peptidoglycan types. To determine carbohydrate patterns and enzymatic properties in a faster and more reproducible way, miniaturized biochemical test systems such as the API system (Biomérieux, France) can be used for phenotypic characterization of sourdough LAB species [23]. However, it should be stressed that the identifications obtained by comparison with commercial databases such as those linked to API are only tentative and need verification with other taxonomic methods. A more advanced phenotypic identification approach is offered by chemotaxonomic methods, which are based on the use of analytical methods to detect and characterize one or several chemical cell components. Protein profiling by sodium dodecyl sulfate-polyacrylamide gel electrophoresis (SDS-PAGE) has been used for identification of LAB isolates recovered from Italian [127] and Greek [72] sourdoughs. SDS-PAGE of cellular proteins generally offers sufficient discrimination of LAB isolates at species level but may fail to discriminate between species in the *Lb. acidophilus* group [218] and the *Lb. plantarum* group [71], both of which are prominent members of sourdough ecosystems. Although not yet evaluated for sourdough LAB, the use of mass spectrometry (MS) methods is probably the most powerful phenotypic approach currently available for classification and identification of bacteria [219]. One of these methods, matrix-assisted laser desorption/ionization-time-of-flight (MALDI-TOF) MS, allows one to measure peptides and other compounds in the presence of salts and to analyze complex peptide mixtures, which makes it an ideal method for measuring nonpurified extracts and intact bacterial cells. The resulting MALDI-TOF MS spectra can be used to generate identification libraries for simple and high-throughput identification of unknown bacterial isolates. De Bruyne and co-workers [220] constructed such identification libraries for the LAB genera *Leuconostoc*, *Fructobacillus*, and *Lactococcus*, and reported that 84% of the leuconostocs and fructobacilli and 94% of the lactococci were correctly identified at species or subspecies level. Identification accuracies for the former two groups further increased to 94–98% when machine learning was applied, which indicates the important role played by advanced techniques for the analysis of complex MALDI-TOF MS profiles.

Essentially, molecular approaches for identification of sourdough LAB are either DNA fingerprint- or sequence-based. DNA fingerprinting methods rely on the use of restriction enzyme analysis, the use of specific or random PCR primers, or a combination thereof. Ribotyping, one of the first DNA fingerprinting techniques used in bacterial taxonomy [221], relies on a combination of restriction analysis of total genomic DNA and Southern hybridization to visualize a subset of restriction fragments with labeled rDNA probes targeting conserved domains of ribosomal 16S and 23S rRNA encoding genes. Although ribotyping generally provides high discriminatory power at species to subspecies level, it has been used only sporadically as an identification technique for LAB species. In specific cases, ribotyping has proven particularly useful in the classification and identification of sourdough LAB, for example for discrimination between the genomically and phenotypically highly similar sourdough LAB *W. cibaria* and *W. confusa* [63] and for intraspecific differentiation of *Lb. sanfranciscensis* strains from different sourdoughs [121]. Despite its high resolution at species as well as at strain level, amplified fragment length polymorphism (AFLP) fingerprinting was also only used sporadically for LAB identification purposes. Essentially, AFLP combines the power of restriction fragment length polymorphism with the flexibility of PCR-based methods by ligating primer-recognition sequences (adaptors) to the digested DNA [222]. This whole-genome fingerprinting technique has proved useful to support the description of the sourdough species *Lb. hammesii* [82], *Lb. crustorum* [86] and *Lb. namurensis* [87] and for molecular source tracking of *Lb. spicheri*, *Lb. plantarum* and *Lb. sanfranciscensis* in the production environment of artisan sourdough bakeries [155]. In contrast to the aforementioned methods, RAPD-PCR and repetitive DNA element (rep)-PCR are technically less demanding DNA fingerprinting techniques based on a single PCR step. Both methods are fast, relatively inexpensive, and exhibit a high discriminatory power ranging from genus to intraspecific level, which explains their wide application range for the identification and classification of LAB. RAPD-PCR has been used in multiple studies on sourdough LAB for species identification [129], strain differentiation [80, 89, 120, 126, 127, 131, 148, 158] and strain monitoring purposes [70, 122, 135, 138, 223]. The reproducibility of RAPD-PCR is highly influenced by various factors, such as DNA purity and concentration and minimal differences in the PCR temperature programme [224], for which reason this method is less suitable for interlaboratory comparisons. Because of the use of longer PCR primers complementary to bacterial interspersed repetitive DNA elements such as ERIC, BOX, REP or $(GTG)_5$ and higher annealing temperatures, rep-PCR protocols are more robust and display a higher level of reproducibility [225]. rep-PCR using the $(GTG)_5$ primer, i.e., $(GTG)_5$-PCR, has been found particularly useful for differentiation of sourdough LAB at the (sub)species up to the strain level [74, 105, 106, 155, 226, 227]. For high-resolution differentiation of individual sourdough LAB strains by DNA fingerprinting, however, AFLP fingerprinting and pulsed-field gel electrophoresis (PFGE) are the most powerful. Essentially, PFGE involves the electrophoretic separation of genomic macrorestriction fragments obtained by digestion with rare-cutting enzymes in an alternating electric field. In sourdough studies, PFGE has been used as a typing method to differentiate among strains within *Lb. plantarum* and *Lb. sanfranciscensis* [126, 135, 174].

Sequence-based analysis approaches for identification of sourdough LAB have long relied on the use of 16S rRNA genes, and this has become a standard approach to obtain a first preliminary view of the taxonomic diversity among a set of unknown isolates recovered from a sourdough ecosystem [72, 80, 148]. In many of these studies, only partial 16S rRNA gene sequences are determined and used in comparisons with public sequence databases. In many cases, the use of partial sequences will only allow a tentative identification, of which the reliability is likely to improve when the entire 16S rRNA gene is sequenced [228]. Despite its established use as a standard method for identification of LAB species, 16S rRNA gene sequencing does not allow differentiation of phylogenetically closely related species [63–66, 71]. The growing availability of whole-genome sequences has triggered the search for alternative genes that offer a higher taxonomic resolution than the 16S rRNA gene. The use of protein-encoding genes or so-called housekeeping genes essentially combines the technological advantages of 16S rRNA gene sequencing and the taxonomic resolution offered by a number of fingerprinting methods. Sequencing of one or preferably multiple of these genes as taxonomic markers is a crucial step forward in the development of standardized and globally accessible methods for the identification of LAB. Housekeeping genes such as *pheS* (encoding the phenylalanyl-tRNA synthase) and *rpoA* (encoding the DNA-dependent RNA polymerase alpha-subunit) display higher divergence rates than the 16S rRNA gene, and allow discrimination between closely related LAB species with almost identical 16S rRNA gene sequences [67, 229]. Several studies on sourdough LAB species diversity have used such protein-encoding genes as phylogenetic markers in a single-locus sequence approach. In conjunction with $(GTG)_5$-PCR fingerprinting, the *pheS* gene has been successfully used for the identification of LAB species from sourdough fermentations at laboratory scale [107] and from Belgian artisan bakery sourdoughs and their environment [105, 106, 155], as well as for unraveling the intraspecific diversity in the sourdough species *Lb. rossiae* [74]. Settanni and coworkers [156] used the *recA* gene, encoding a protein essential for repair and maintenance of DNA, in a multiplex PCR assay to discriminate between the phylogenetically highly related *Lb. plantarum*, *Lb. pentosus* and *Lb. paraplantarum* in sourdough ecosystems. The *recA* gene has also been used in combination with the 16S rRNA gene to unravel the identity of LAB isolates recovered during wheat flour sourdough type I propagation [135]. Sequences derived from the *tuf* gene, which encodes the elongation factor Tu, have revealed a higher discriminatory power compared to 16S rRNA gene sequences and have been used to support the delineation of the new sourdough species *Lb. secaliphilus* [85]. Although single-locus sequence analysis approaches are now commonly used within specific LAB groups, it has been argued that the phylogenetic information obtained from only a single gene may be influenced by lateral gene transfer (LGT) and may lead to incorrect identifications. To compensate for possible LGT events, it has been suggested that multilocus sequence analysis (MLSA) of at least five housekeeping genes from diverse chromosomal loci and with wide distribution among taxa is required to reliably distinguish a species from related taxa [230]. After a more thorough evaluation, however, Konstantinidis and co-workers [231] concluded that three genes are sufficient to anticipate the possible effects of LGT in MLSA-based identification schemes. For LAB, MLSA based on the combined sequence analysis of the genes *atpA, pheS,* and *rpoA* has been successfully explored for species identification of enterococci [229], lactobacilli [67], leuconostocs [232], and pediococci [233]. For sequence-based differentiation of LAB at strain level, multilocus schemes typically include six or seven housekeeping genes. The resulting multilocus sequence typing (MLST) approach has so far mainly been applied to study community structure, evolution and phylogeography of bacterial pathogens [234]. A few MLST schemes have been specifically developed for *Lactobacillus* species, including *Lb. casei* [235, 236], *Lb. plantarum* [237] and *Lb. salivarius* [238].

5.4.2 Culture-Independent Approaches

The first approaches used to identify sourdough microorganisms independent of culturing relied on the use of oligonucleotide probes targeting ribosomal gene sequences specific for individual species or groups of species. The majority of these probe-based methods made use of partial 16S rRNA gene sequences that were identified as molecular signatures unique to specific LAB species [203, 239]. Gradually, the relatively laborious probe hybridizations were replaced by faster community PCR assays using species-specific oligonucleotide primers. The success of both approaches strongly depended on rigorous *in silico* probe or primer design and required in vitro and in vivo validation using taxonomically well-characterized type and reference strains and spiked sourdough samples, respectively. Species-specific PCR primers complementary to signature sequences in the 16S or 23S rRNA gene or in the 16S–23S rRNA intergenic spacer region have been applied for the culture-independent identification of sourdough LAB species [26, 156, 161, 240–242]. By combining multiple sets of primers, several typical sourdough LAB species can be simultaneously detected. In this way, Settanni and co-workers [156] developed a two-step multiplex community PCR assay that enabled rapid identification of up to 16 *Lactobacillus* species in sourdough samples. The introduction of real-time PCR technology has allowed one to further increase the sensitivity of PCR-based identification assays and enables the simultaneous detection and quantification of food microorganisms [243]. For this purpose, SYBR Green-based real-time PCR assays based on the detection of the *pheS* gene have been used for source tracking of *Lb. plantarum* and *Lb. sanfranciscensis* in traditional sourdoughs and their production environments [155].

Whereas probe- and primer-based identification approaches offer the specificity and selectivity required to detect and monitor specific LAB in sourdough samples, they were not designed to offer a complete picture of the predominant LAB species diversity or to reveal new or unknown LAB species diversity in sourdough ecosystems. In contrast, community fingerprinting methods such as denaturing gradient gel electrophoresis (DGGE) and temporal temperature gradient gel electrophoresis (TTGE) do not require prior knowledge of the ecosystem's diversity and are universally applicable to study the species diversity and dynamics of complex bacterial communities in food environments [244]. The universal use of DGGE fingerprinting is based on the sequence-dependent separation of a mixture of equally sized PCR amplicons generated from a common taxonomic marker such as the 16S rRNA gene. For the design of PCR primers, the V1, V3, and V6-V8 hypervariable regions of the 16S rRNA gene are most commonly used. Taxonomic information on individual members of the sample community can be obtained by band position analysis provided that an identification database is available, clone library analysis, sequencing of excised and purified DGGE bands or hybridization using species-specific probes. Major drawbacks of DGGE fingerprinting include its inability to detect subdominant (i.e., <1%) community members and the fact that a single strain or species may be represented by multiple bands in the DGGE profile due to heterogeneous rRNA operons and/or heteroduplex molecules. Either using universal or group-specific 16S rRNA gene primers, DGGE has been widely applied to inventorize LAB communities in sourdoughs [41, 47, 106, 128, 245, 246] and to investigate the dynamics, adaptation, and source of predominant sourdough LAB communities [34, 39, 80, 107, 123, 142, 155, 185]. Likewise, primers targeting the 26S LSU rDNA have been used for DGGE fingerprinting analysis of sourdough yeast communities [29, 46]. To maximally cover the microbial species diversity present in a sourdough ecosystem, a number of DGGE studies have combined the use of 16S and 23S rRNA gene primers to determine in parallel the predominant LAB and yeast composition of sourdough samples [33, 39, 41, 47, 123, 142, 245]. Compared to DGGE, TTGE has been used to a much lesser extent for culture-independent analysis of the sourdough microbiota [112]. In many of the cited studies, the sequence heterogeneity of the multicopy 16S rRNA gene is mentioned as an important limitation in DGGE, as this may lead to an overestimation of the LAB species diversity. The degree of overestimation can be estimated by scoring individual bands by position analysis with a reference database and/or by band sequencing. Alternatively, single-copy genes that do not exhibit this heterogeneity such as *rpoB* have been evaluated for DGGE fingerprinting of LAB species during food fermentations [247].

Microarray technology represents one of the most recent culture-independent approaches to study the diversity and identify individual members of the sourdough microbiota. Phylogenetic microarrays, containing partial 16S rRNA gene sequences as targets, are ideally suited for this purpose but are currently not available for sourdough microbiota. Alternatively, a functional gene microarray can be used when the original annotation information allows one to link the responding oligonucleotides to the original species. Weckx and co-workers [108, 109] used a LAB functional gene microarray to analyze time-related RNA samples that represented the metatranscriptome of sourdough fermentations maintained by daily backslopping. The resulting set of hybridization data allowed one to monitor the LAB community dynamics in the sourdough ecosystem and to identify the major LAB species that contributed to the establishment of a stable ecosystem through its three successive phases.

References

1. Kurtzman CP, Fell JW (eds) (1998) The yeasts: a taxonomic study, 4th edn. Elsevier, Amsterdam, pp 111–947
2. Suh S-O, Blackwell M, Kurtzman CP, Lachance M-A (2006) Phylogenetics of Saccharomycetales, the ascomycetous yeasts. Mycologia 98:1006–1017
3. Kurtzman CP (2003) Phylogenetic circumscription of *Saccharomyces, Kluyveromyces* and other members of the Saccharomycetaceae, and the proposal of the new genera *Lachancea, Nakaseomyces, Naumovia, Vanderwaltozyma* and *Zygotorulaspora*. FEMS Yeast Res 4:233–245
4. Kurtzman CP, Robnett CJ, Basehoar-Powers E (2008) Phylogenetic relationships among species of *Pichia, Issatchenkia* and *Williopsis* determined from multigene sequence analysis, and the proposal of *Barnettozyma* gen. nov., *Lindnera* gen. nov. and *Wickerhamomyces* gen. nov. FEMS Yeast Res 8:939–954
5. Kurtzman CP (2011) Phylogeny of the ascomycetous yeasts and the renaming of *Pichia anomala* to *Wickerhamomyces anomalus*. Antonie van Leeuwenhoek 99:13–23
6. Daniel HM, Redhead SA, Schürer J, Naumov GI, Kurtzman CP (2012) Proposals to conserve the name Wickerhamomyces against Hansenula and to reject the name Saccharomyces sphaericus (Ascomycota: Saccharomycotina). Taxon 61 (2):459–461.
7. Sugihara TF, Kline L, Miller MW (1971) Microorganisms of the San Francisco sour dough bread process. Appl Microbiol 21:456–458
8. Galli A, Ottogalli G (1973) The microflora of the sour dough of "Panettone" cake. Ann Microbiol 23:39–49
8. Spicher G, Schröder R, Schöllhammer K (1979) Die Mikroflora des Sauerteiges. Z Lebensm Unters Forsch 169:77–81
10. Barber S, Báguena R, Martínez-Anaya MA, Torner MJ (1983) Microflora of the sour dough of wheat flour bread. I. Identification and functional properties of microorganisms of industrial sour doughs. Rev Agroquím Tecnol Aliment 23:552–562
11. Salovaara H, Savolainen J (1984) Yeast type isolated from Finnish sour rye dough starters. Act Aliment Pol 10:241–246
12. Barber S, Báguena R (1988) Microflora of the sour dough of wheat flour bread. V. Isolation, identification and evaluation of functional properties of sourdough's microorganisms. Rev Agroquím Tecnol Aliment 28:67–78
13. Galli A, Franzetti L, Fortina MG (1988) Isolation and identification of sour dough microflora. Microbiol Aliment Nutr 6:345–351
14. Hamad SH, Böcker G, Vogel RF, Hammes WP (1992) Microbiological and chemical analysis of fermented sorghum dough for Kisra production. Appl Microbiol Biotechnol 37:728–731
15. Infantes M, Schmidt JL (1992) Characterization of the yeast flora of natural sourdoughs located in various French areas. Sci Alim 12:271–287
16. Boraam F, Faid M, Larpent JP, Breton A (1993) Lactic acid bacteria and yeasts associated with traditional Moroccan sour-dough bread fermentation. Sci Alim 13:501–509
17. Gobbetti M, Corsetti A, Rossi J, La Rosa F, De Vincenzi S (1994) Identification and clustering of lactic acid bacteria and yeasts from wheat sourdoughs of Central Italy. Ital J Food Sci 6:85–94

18. Obiri-Danso K (1994) Microbiological studies on corn dough fermentation. Cer Chem 71:186–188
19. Iorizzo M, Coppola R, Sorrentino E, Grazia L (1995) Micobiological characterization of sourdough from Molise. Industrie Alimentari XXXIV:1290–1294
20. Almeida MJ, Pais CS (1996) Characterization of the yeast population from traditional corn and rye bread doughs. Lett Appl Micobiol 23:154–158
21. Foschino R, Terraneo R, Mora D, Galli A (1999) Microbial characterization of sourdoughs for sweet baked products. Ital J Food Sci 11:19–28
22. Mäntynen VH, Korhola M, Gudmundsson H, Turakainen H, Alfredsson GA, Salovaara H, Lindström K (1999) A polyphasic study on the taxonomic position of industrial sour dough yeasts. Syst Appl Microbiol 22:87–96
23. Rocha JM, Malcata FX (1999) On the microbiological profile of traditional Portuguese sourdough. J Food Prot 62:1416–1429
24. Paramithiotis S, Muller MRA, Ehrmann MA, Tsakalidou E, Seiler H, Vogel R, Kalantzopoulos G (2000) Polyphasic identification of wild yeast strains isolated from Greek sourdoughs. Syst Appl Microbiol 23:156–164
25. Rosenquist H, Hansen A (2000) The microbial stability of two bakery sourdoughs made from conventionally and organically grown rye. Food Microbiol 17:241–250
26. Corsetti A, Lavermicocca P, Morea M, Baruzzi F, Tosti N, Gobbetti M (2001) Phenotypic and molecular identification and clustering of lactic acid bacteria and yeasts from wheat (species *Triticum durum* and *Triticum aestivum*) sourdoughs of Southern Italy. Int J Food Microbiol 64:95–104
27. Pulvirenti A, Caggia C, Restuccia C, Gullo M, Giudici P (2001) DNA fingerprinting methods used for identification of yeasts isolated from Sicilian sourdoughs. Ann Microbiol 51:107–120
28. Gullo M, Romano AD, Pulvirenti A, Giudici P (2003) *Candida humilis* – dominant species in sourdoughs for the production of durum wheat bran flour bread. Int J Food Microbiol 80:55–59
29. Meroth CB, Hammes WP, Hertel C (2003) Identification and population dynamics of yeasts in sourdough fermentation processes by PCR-denaturing gradient gel electrophoresis. Appl Environ Microbiol 69:7453–7461
30. Solieri L, De Vero L, Pulvirenti A, Gullo M (2003) A phenotypical and molecular study of yeast species in home-made sourdoughs. Industr Aliment XLII:971–978
31. Succi M, Reale A, Andrighetto C, Lombardi A, Sorrentino E, Coppola R (2003) Presence of yeasts in southern Italian sourdoughs from *Triticum aestivum* flour. FEMS Microbiol Lett 225:143–148
32. Foschino R, Gallina S, Andrighetto C, Rossetti L, Galli A (2004) Comparison of cultural methods for the identification and molecular investigation of yeasts from sourdoughs for Italian sweet baked products. FEMS Yeast Res 4:609–618
33. Gatto V, Torriani S (2004) Microbial population changes during sourdough fermentation monitored by DGGE analysis of 16S and 26S rRNA gene fragments. Ann Microbiol 54:31–42
34. Pulvirenti A, Solieri L, Gullo M, De Vero L, Giudici P (2004) Occurrence and dominance of yeast species in sourdough. Lett Appl Microbiol 38:113–117
35. Vernocchi P, Valmorri S, Gatto V, Torriani S, Gianotti A, Suzzi G, Guerzoni ME, Gardini F (2004) A survey on yeast microbiota associated with an Italian traditional sweet-leavened baked good fermentation. Food Res Int 37:469–476
36. Vernocchi P, Valmorri S, Dalai I, Torriani S, Gianotti A, Suzzi G, Guerzoni ME, Mastrocola D, Gardini F (2004) Characterization of the yeast population involved in the production of a typical Italian bread. J Food Sci 69:M182–M186
37. Lombardi A, Zilio F, Andrighetto C, Zampese L, Loddo A (2007) Micobiological characterization of sourdoughs of Veneto region. Industr Aliment XLVI:147–151
38. Restuccia C, Randazzo C, Pitino I, Caggia C, Fiasconaro N (2007) Phenotypic and genotypic characterization of lactic acid bacteria and yeasts from sourdough for Panettone production. Industr Aliment XLVI:1231–1236
39. Garofalo C, Silvestri G, Aquilanti L, Clementi F (2008) PCR-DGGE analysis of lactic acid bacteria and yeast dynamics during the production processes of three varieties of Panettone. J Appl Microbiol 105:243–254

40. Vrancken G, De Vuyst L, Van der Meulen R, Huys G, Vandamme P, Daniel HM (2010) Yeast species composition differs between artisan bakery and spontaneous laboratory sourdoughs. FEMS Yeast Res 10:471–481
41. Iacumin L, Cecchini F, Manzano M, Osualdini M, Boscolo D, Orlic S, Comi G (2009) Description of the microflora of sourdoughs by culture-dependent and culture-independent methods. Food Microbiol 26:128–135
42. Luangsakul N, Keeratipibul S, Jindamorakot S, Tanasupawat S (2009) Lactic acid bacteria and yeasts isolated from the starter doughs for Chinese steamed buns in Thailand. LWT- Food Sci Technol 42:1404–1412
43. Osimani A, Zannini E, Aquilanti L, Mannazzu I, Comitini F, Clementi F (2009) Lactic acid bacteria and yeasts from wheat sourdoughs of the Marche region. Ital J Food Sci 21:269–286
44. Saeed M, Anjum FM, Zahoor T, Nawaz H, Rehman SU (2009) Isolation and characterization of starter culture from spontaneous fermentation of sourdough. Int J Agric Biol 11:329–332
45. Paramithiotis S, Tsiasiotou S, Drosinos E (2010) Comparative study of spontaneously fermented sourdoughs originating from two regions of Greece: Peloponnesus and Thessaly. Eur Food Res Technol 231:883–890
46. Valmorri S, Tofalo R, Settanni L, Corsetti A, Suzzi G (2010) Yeast microbiota associated with spontaneous sourdough fermentations in the production of traditional wheat sourdough breads of the Abruzzo region (Italy). Antonie van Leeuwenhoek 97:119–129
47. Moroni AV, Arendt EK, Dal Bello F (2011) Biodiversity of lactic acid bacteria and yeasts in spontaneously-fermented buckwheat and teff sourdoughs. Food Microbiol 28:497–502
48. Zhang J, Liu W, Sun Z, Bao Q, Wang F, Yu J, Chen W, Zhang H (2011) Diversity of lactic acid bacteria and yeasts in traditional sourdoughs collected from western region in Inner Mongolia of China. Food Contr 22:767–774
49. Yarrow D (1978) *Candida milleri* sp. nov. Int J Syst Bacteriol 28:608–610
50. Gobbetti M, Corsetti A, Rossi J (1994) The sourdough microflora. Interactions between lactic acid bacteria and yeasts: metabolism of carbohydrates. Appl Microbiol Biotechnol 41:456–460
51. Daniel HM, Moons MC, Huret S, Vrancken G, De Vuyst L (2011) *Wickerhamomyces anomalus* in the sourdough microbial ecosystem. Antonie van Leeuwenhoek 99:63–73
52. Sampaio JP, Gonçalves P (2008) Natural populations of *Saccharomyces kudriavzevii* in Portugal are associated with oak bark and are sympatric with *S. cerevisiae* and *S. paradoxus*. Appl Environ Microbiol 74:2144–2152
53. Nel EE, van der Walt JP (1968) *Torulopsis humilis*, sp. n. Mycopathol Mycol Appl 36:94–96
54. Meyer SA, Payne RW, Yarrow D (1998) *Candida* Berkout. In: Kurtzman CP, Fell JW (eds) The yeasts: a taxonomic study, 4th edn. Elsevier, Amsterdam, pp 454–573
55. Daniel HM, Meyer W (2003) Evaluation of ribosomal RNA and actin gene sequences for the identification of ascomycetous yeasts. Int J Food Microbiol 86:61–78
56. Vaughan-Martini A (1995) *Saccharomyces barnetti* and *Saccharomyces spencerorum*: two new species of *Saccharomyces sensu lato* (van der Walt). Antonie van Leeuwenhoek 68:111–118
57. De Vuyst L, Neysens P (2005) The sourdough microflora: biodiversity and metabolic interactions. Trends Food Sci Technol 16:43–56
58. Corsetti A, Settanni L (2007) Lactobacilli in sourdough fermentation. Food Res Int 40:539–558
59. Orla-Jensen S (1919) The lactic acid bacteria. Fred Host and Son, Copenhagen
60. Felis GE, Dellaglio F (2007) Taxonomy of lactobacilli and bifidobacteria. Curr Iss Intest Microbiol 8:44–61
61. Schleifer KH, Ludwig W (1995) Phylogeny of the genus *Lactobacillus* and related genera. Syst Appl Microbiol 18:461–467
62. Stiles ME, Holzapfel WH (1997) Lactic acid bacteria of foods and their current taxonomy. Int J Food Microbiol 36:1–29
63. Björkroth KJ, Schillinger U, Geisen R, Weiss N, Hoste B, Holzapfel WH, Korkeala HJ, Vandamme P (2002) Taxonomic study of *Weissella confusa* and description of *Weissella cibaria* sp. nov., detected in food and clinical samples. Int J Syst Evol Microbiol 52:141–148
64. Cachat E, Priest FG (2005) *Lactobacillus suntoryeus* sp. nov., isolated from malt whisky distilleries. Int J Syst Evol Microbiol 55:31–34

65. Leisner JJ, Vancanneyt M, Lefebvre K, Vandemeulebroecke K, Hoste B, Vilalta NE, Rusul G, Swings J (2002) *Lactobacillus durianis* sp. nov., isolated from an acid-fermented condiment (tempoyak) in Malaysia. Int J Syst Evol Microbiol 52:927–931
66. Yoon JH, Kang SS, Mheen TI, Ahn JS, Lee HJ, Kim TK, Park CS, Kho YH, Kang KH, Park YH (2000) *Lactobacillus kimchii* sp. nov., a new species from kimchi. Int J Syst Evol Microbiol 50:1789–1795
67. Naser SM, Dawyndt PSR, Hoste B, Gevers D, Vandemeulebroecke K, Cleenwerck I, Vancanneyt M, Swings J (2007) Identification of lactobacilli by *pheS* and *rpoA* gene sequence analyses. Int J Syst Evol Microbiol 57:2777–2789
68. De Vuyst L, Vancanneyt M (2007) Biodiversity and identification of sourdough lactic acid bacteria. Food Microbiol 24:120–127
69. De Vuyst L, Vrancken G, Ravyts F, Rimaux T, Weckx S (2009) Biodiversity, ecological determinants, and metabolic exploitation of sourdough microbiota. Food Microbiol 26:666–675
70. Ehrmann MA, Vogel R (2005) Molecular taxonomy and genetics of sourdough lactic acid bacteria. Trends Food Sci Technol 16:31–42
71. Torriani S, Felis GE, Dellaglio F (2001) Differentiation of *Lactobacillus plantarum, L. pentosus*, and *L. paraplantarum* by *recA* gene sequence analysis and multiplex PCR assay with *recA* gene-derived primers. Appl Environ Microbiol 67:3450–3454
72. De Vuyst L, Schrijvers V, Paramithiotis S, Hoste B, Vancanneyt M, Swings J, Kalatzopoulos G, Tsakalidou E, Messens W (2002) The biodiversity of lactic acid bacteria in Greek traditional sourdoughs is reflected in both composition and metabolite formation. Appl Environ Microbiol 68:6059–6069
73. Gevers D, Huys G, Swings J (2001) Applicability of rep-PCR fingerprinting for identification of *Lactobacillus* species. FEMS Microbiol Lett 205:31–36
74. Scheirlinck I, Van der Meulen R, Vrancken G, De Vuyst L, Settanni L, Vandamme P, Huys G (2009) Polyphasic taxonomic characterization of *Lactobacillus rossiae* isolates from Belgian and Italian sourdoughs reveals intraspecific heterogeneity. Syst Appl Microbiol 32:151–156
75. Kline L, Sugihara TF (1971) Microorganisms of the San Francisco sourdough bread process. II. Isolation and characterization of undescribed bacterial species responsible for the souring activity. Appl Microbiol 21:459–465
76. Corsetti A, Settanni L, van Sinderen D, Felis GE, Dellaglio F, Gobbetti M (2005) *Lactobacillus rossii* sp. nov., isolated from wheat sourdough. Int J Syst Evol Microbiol 55:35–40
77. Naser SM, Hagen KE, Vancanneyt M, Cleenwerck I, Swings J, Tompkins TA et al (1980) *Lactobacillus suntoryeus* Cachat and Priest 2005 is a later synonym of *Lactobacillus helveticus* (Orla-Jensen 1919) Bergey *et al.* 1925 (Approved Lists 1980). J Syst Evol Microbiol 2006(56):355–360
78. Müller MRA, Ehrmann MA, Vogel RF (2000) *Lactobacillus frumenti* sp. nov., a new lactic acid bacterium isolated from rye-bran fermentations with a long fermentation period. Int J Syst Evol Microbiol 50:2127–2133
79. Ehrmann MA, Müller MRA, Vogel RF (2003) Molecular analysis of sourdough reveals *Lactobacillus mindensis* sp. nov. *Int.* J Syst Evol Microbiol 53:7–13
80. Meroth CB, Hammes WP, Hertel C (2004) Characterisation of the microbiota of rice sourdoughs and description of *Lactobacillus spicheri*, sp. nov. Syst Appl Microbiol 27:151–159
81. Vancanneyt M, Neysens P, De Wachter M, Engelbeen K, Snauwaert C, Cleenwerck I, Van der Meulen R, Hoste B, Tsakalidou E, De Vuyst L, Swings J (2005) *Lactobacillus acidifarinae* sp. nov. and *Lactobacillus zymae* sp. nov., from wheat sourdoughs. Int J Syst Evol Microbiol 55:615–620
82. Valcheva R, Korakli M, Onno B, Prévost H, Ivanova I, Ehrmann MA, Dousset X, Gänzle MG, Vogel RF (2005) *Lactobacillus hammessi* sp. nov., isolated from French sourdough. Int J Syst Evol Microbiol 55:763–767
83. Aslam Z, Im WT, Ten LN, Lee MJ, Kim KH, Lee ST (2006) *Lactobacillus siliginis* sp. nov., isolated from wheat sourdough in South Korea. Int J Syst Evol Microbiol 56:2209–2213
84. Valcheva R, Ferchichi MF, Korakli M, Ivanova I, Gänzle MG, Vogel RF, Prévost H, Onno B, Dousset X (2006) *Lactobacillus nantensis* sp. nov., isolated from French wheat sourdough. Int J Syst Evol Microbiol 56:587–591

85. Ehrmann MA, Brandt M, Stolz P, Vogel RF, Korakli M (2007) *Lactobacillus secaliphilus* sp. nov, isolated from type II sourdough fermentation. Int J Syst Evol Microbiol 57:745–750
86. Scheirlinck I, Van der Meulen R, Van Schoor A, Huys G, Vandamme P, De Vuyst L, Vancanneyt M (2007) *Lactobacillus crustorum* sp. nov., isolated from two traditional Belgian wheat sourdoughs. Int J Syst Evol Microbiol 57:1461–1467
87. Scheirlinck I, Van der Meulen R, Van Schoor A, Cleenwerck I, Huys G, Vandamme P, De Vuyst L, Vancanneyt M (2007) *Lactobacillus namurensis* sp. nov., isolated from a traditional Belgian sourdough. Int J Syst Evol Microbiol 57:223–227
88. Kashiwagi T, Suzuki T, Kamakura T (2009) *Lactobacillus nodensis* sp. nov., isolated from rice bran. Int J Syst Evol Microbiol 59:83–86
89. Corsetti A, Settanni L, Valmorri S, Mastrangelo M, Suzzi G (2007) Identification of subdominant sourdough lactic acid bacteria and their evolution during laboratory-scale fermentations. Food Microbiol 24:592–600
90. Björkroth J, Dicks LMT, Holzapfel WH (2009) Genus III. *Weissella*. In: De Vos P, Garrity GM, Jones D, Krieg NR, Ludwig W, Rainey FA, Schleifer K-H, Whitman WB (eds) Bergey's manual of systematic bacteriology, vol 3, The Firmicutes. Springer, Dordrecht/New York, pp 643–654
91. Villani F, Moschetti G, Blaiotta G, Coppola S (1997) Characterization of strains of *Leuconostoc mesenteroides* by analysis of soluble whole-cell protein pattern, DNA fingerprinting and restriction of ribosomal DNA. J Appl Microbiol 82:578–588
92. Nigatu A (2000) Evaluation of numerical analyses of RAPD and API 50 CH patterns to differentiate *Lactobacillus plantarum*, *Lact. fermentum*, *Lact. rhamnosus*, *Lact. sake*, *Lact. parabuchneri*, *Lact. gallinarum*, *Lact. casei*, *Weissella minor* and related taxa isolated from *kocho* and *tef*. J Appl Microbiol 89:969–978
93. Satokari R, Mattila-Sandholm T, Suihko ML (2000) Identification of pediococci by ribotyping. J Appl Microbiol 88:260–265
94. Barney M, Volgyi A, Navarro A, Ryder D (2001) Riboprinting and 16S rRNA sequencing for identification of brewery *Pediococcus* isolates. Appl Environ Microbiol 67:553–560
95. Rodas AM, Ferrer S, Pardo I (2003) 16S-ARDRA, a tool for identification of lactic acid bacteria isolated from grape must and wine. Syst Appl Microbiol 26:412–422
96. Nigatu A, Ahrne S, Gashe BA, Molin G (1998) Randomly amplified polymorphic DNA (RAPD) for discrimination of *Pediococcus pentosaceus* and *Ped. acidilactici* and rapid grouping of *Pediococcus* isolates. Lett Appl Microbiol 26:412–416
97. Mora D, Fortina MG, Parini C, Daffonchio D, Manachini PL (2000) Genomic subpopulations within the species *Pediococcus acidilactici* detected by multilocus typing analysis: relationships between pediocin AcH/PA-producing and non-producing strains. Microbiology 146:2027–2038
98. Dobson CM, Deneer H, Lee S, Hemmingsen S, Glaze S, Ziola B (2002) Phylogenetic analysis of the genus *Pediococcus: Pediococcus claussenii* sp. nov., a novel lactic acid bacterium isolated from beer. Int J Syst Evol Microbiol 52:2003–2010
99. Pérez G, Cardell E, Zárate V (2002) Random amplified polymorphic DNA analysis for differentiation of *Leuconostoc mesenteroides* subspecies isolated from Tenerife cheese. Lett Appl Microbiol 34:82–85
100. Hutkins RW (ed) (2006) Microbiology and technology of fermented foods. Blackwell Publishing, Iowa, pp 1–473
101. Ottogalli G, Galli A, Foschino R (1996) Italian bakery products obtained with sour dough: characterization of the typical microflora. Adv Food Sci 18:131–144
102. Hammes WP, Brandt MJ, Francis KL, Rosenheim J, Seitter MFH, Vogelmann A (2005) Microbial ecology of cereal fermentations. Trends Food Sci Technol 16:4–11
103. Arendt EK, Ryan LAM, Dal Bello F (2007) Impact of sourdough on the texture of bread. Food Microbiol 24:165–174
104. Leroy F, De Vuyst L (2004) Lactic acid bacteria as functional starter cultures for the food fermentation industry. Trends Food Sci Technol 15:67–78
105. Scheirlinck I, Van der Meulen R, Van Schoor A, Vancanneyt M, De Vuyst L, Vandamme P, Huys G (2007) Influence of geographical origin and flour type on diversity of lactic acid bacteria in traditional Belgian sourdoughs. Appl Environ Microbiol 73:6262–6269

106. Scheirlinck I, Van der Meulen R, Van Schoor A, Vancanneyt M, De Vuyst L, Vandamme P, Huys G (2008) Taxonomic structure and stability of the bacterial community in Belgian sourdough ecosystems as assessed by culture and population fingerprinting. Appl Environ Microbiol 74:2414–2423
107. Van der Meulen R, Scheirlinck I, Van Schoor A, Huys G, Vancanneyt M, Vandamme P, De Vuyst L (2007) Population dynamics and metabolite target analysis of lactic acid bacteria during laboratory fermentations of wheat and spelt sourdoughs. Appl Environ Microbiol 73:4741–4750
108. Weckx S, Van der Meulen R, Allemeersch J, Huys G, Vandamme P, Van Hummelen P, De Vuyst L (2010) Community dynamics of sourdough fermentations as revealed by their metatranscriptome. Appl Environ Microbiol 76:5402–5408
109. Weckx S, Van der Meulen R, Maes D, Scheirlinck I, Huys G, Vandamme P, De Vuyst L (2010) Lactic acid bacteria community dynamics and metabolite production of rye sourdough fermentations share characteristics of wheat and spelt sourdough fermentations. Food Microbiol 27:1000–1008
110. Gashe BA (1985) Involvement of lactic acid bacteria in the fermentation of tef (*Eragrostis tef*), an Ethiopian fermented food. J Food Sci 50:800–801
111. Infantes L, Tourneur C (1991) Survey on the lactic flora of natural sourdoughs located in various French areas. Sci Alim 11:527–545
112. Gabriel V, Lefebvre D, Vayssier Y, Faucher C (1999) Characterization of microflora from natural sourdoughs. Microbiol Alim Nutr 17:171–179
113. Ferchichi M, Valcheva R, Prevost H, Onno B, Dousset X (2007) Molecular identification of the microbiota of French sourdough using temporal temperature gradient gel electrophoresis. Food Microbiol 24:678–686
114. Ferchichi M, Valcheva R, Oheix N, Kabadjova P, Prevost H, Onno B, Dousset X (2008) Rapid investigation of French sourdough microbiota by restriction fragment length polymorphism of the 16S-23S rRNA gene intergenic spacer region. World J Microbiol Biotechnol 24:2425–2434
115. Robert H, Gabriel V, Fontagné-Faucher C (2009) Biodiversity of lactic acid bacteria in French wheat sourdough as determined by molecular characterization using species-specific PCR. Int J Food Microbiol 135:53–59
116. Spicher G (1959) Die Mikroflora des Sauerteiges. I. Mitteilung: Untersuchungen über die Art der in Sauerteigen anzutreffenden stäbchenförmigen Milchsäurebakterien (Genus *Lactobacillus* Beijerinck). Zeitblatt fur Bakteriologie II Abt 113:80–106
117. Spicher G, Schröder R (1978) Die Mikroflora des Sauerteiges. IV. Mitteilung: Untersuchungen über die Art der in "Reinzuchtsauern" anzutreffenden stäbchenförmigen Milchsäurebakterien (Genus *Lactobacillus* Beijerinck). Z Lebensm Unters Forsch 167:342–354
118. Spicher G (1984) Weitere Unterschungen über die Zusammensetzung und die Variabilität der Mikroflora handelsüblicher Sauerteig-Starter. Z Lebensm Unters Forsch 178:106–109
119. Spicher G (1987) Die Mikroflora des Sauerteiges. XXII. Mitteilung: Die in Weizensauerteigen vorkommenden Lactobacillen. Z Lebensm Unters Forsch 184:300–303
120. Müller MRA, Wolfrum G, Stolz P, Ehrmann MA, Vogel RF (2001) Monitoring the growth of *Lactobacillus* species in a sourdough fermentation. Food Microbiol 18:217–227
121. Kitahara M, Sakata S, Benno Y (2005) Biodiversity of *Lactobacillus sanfranciscensis* strains isolated from five sourdoughs. Lett Appl Microbiol 40:353–357
122. Sterr Y, Weiss A, Schmidt H (2009) Evaluation of lactic acid bacteria for sourdough fermentation of amaranth. Int J Food Microbiol 136:75–82
123. Vogelmann SA, Seitter M, Singer U, Brandt MJ, Hertel C (2009) Adaptability of lactic acid bacteria and yeasts to sourdoughs prepared from cereals, pseudocereals and cassava and use of competitive strains as starters. Int J Food Microbiol 130:205–212
124. Coppola S, Pepe O, Masi P, Sepe M (1996) Characterization of leavened doughs for pizza in Naples. Adv Food Sci 18:160–162
125. Zapparoli G, De Benedictis P, Salardi C, Veneri G, Torriani S, Dellaglio F (1996) Lactobacilli of sourdoughs from Verona bakery: a preliminary investigation. Adv Food Sci 18:163–166

126. Zapparoli G, Torriani S, Dellaglio F (1998) Differentiation of *Lactobacillus sanfranciscensis* strains by randomly amplified polymorphic DNA and pulsed-field gel electrophoresis. FEMS Microbiol Lett 166:325–332
127. Corsetti A, De Angelis M, Dellaglio F, Paparella A, Fox PF, Settanni L, Gobbetti M (2003) Characterization of sourdough lactic acid bacteria based on genotypic and cell-wall protein analyses. J Appl Microbiol 94:641–654
128. Randazzo CL, Heilig H, Restuccia C, Giudici P, Caggia C (2005) Bacterial population in traditional sourdough evaluated by molecular methods. J Appl Microbiol 99:251–258
129. Reale A, Tremonte P, Succi M, Sorrentino E, Coppola R (2005) Exploration of lactic acid bacteria ecosystem of sourdoughs from the Molise region. Ann Microbiol 55:17–22
130. Ricciardi A, Parente E, Piraino P, Paraggio M, Romano P (2005) Phenotypic characterization of lactic acid bacteria from sourdoughs for Altamura bread produced in Apulia (Southern Italy). Int J Food Microbiol 98:63–72
131. Catzeddu P, Mura E, Parente E, Sanna M, Farris GA (2006) Molecular characterization of lactic acid bacteria from sourdough breads produced in Sardinia (Italy) and multivariate statistical analyses of results. Syst Appl Microbiol 29:138–144
132. Valmorri S, Settanni L, Suzzi G, Gardini F, Vernocchi P, Corsetti A (2006) Application of a novel polyphasic approach to study the lactobacilli composition of sourdoughs from the Abruzzo region (Central Italy). Let Appl Microbiol 43:343–349
133. Corsetti A, Settanni L, Lopez CC, Felis GE, Mastrangelo M, Suzzi G (2007) A taxonomic survey of lactic acid bacteria isolated from wheat (*Triticum durum*) kernels and non-conventional flours. Syst Appl Microbiol 30:561–571
134. Zotta T, Piraino P, Parente E, Salzano G, Ricciardi A (2008) Characterization of lactic acid bacteria isolated from sourdoughs for Cornetto, a traditional bread produced in Basilicata (Southern Italy). World J Microbiol Biotechnol 24:1785–1795
135. Siragusa S, Di Cagno R, Ercolini D, Minervini F, Gobbetti M, De Angelis M (2009) Taxonomic structure and monitoring of the dominant population of lactic acid bacteria during wheat flour sourdough type I propagation using *Lactobacillus sanfranciscensis* starters. Appl Environ Microbiol 75:1099–1109
136. Zannini E, Garofalo C, Aquilanti L, Santarelli S, Silvestri G, Clementi F (2009) Microbiological and technological characterization of sourdoughs destined for bread-making with barley flour. Food Microbiol 26:744–753
137. Coda R, Nionelli L, Rizzello CG, De Angelis M, Tossut P, Gobbetti M (2010) Spelt and emmer flours: characterization of the lactic acid bacteria microbiota and selection of mixed starters for bread making. J Appl Microbiol 108:925–935
138. Minervini F, De Angelis M, Di Cagno R, Pinto D, Siragusa S, Rizzello CG, Gobbetti M (2010) Robustness of *Lactobacillus plantarum* starters during daily propagation of wheat flour sourdough type I. Food Microbiol 27:897–908
139. Rizzello CG, Nionelli L, Coda R, De Angelis M, Gobbetti M (2010) Effect of sourdough fermentation on stabilisation, and chemical and nutritional characteristics of wheat germ. Food Chem 119:1079–1089
140. Azar M, Ter-Sarkissian N, Ghavifek H, Ferguson T, Ghassemi H (1977) Microbiological aspects of Sangak bread. J Food Sci Technol 14:251–254
141. Hüttner EK, Dal Bello F, Arendt EK (2010) Identification of lactic acid bacteria isolated from oat sourdoughs and investigation into their potential for the improvement of oat bread quality. Eur Food Res Technol 230:849–857
142. Moroni AV, Arendt EK, Morrissey JP, Dal Bello F (2010) Development of buckwheat and teff sourdoughs with the use of commercial starters. Int J Food Microbiol 142:142–148
143. Escalante A, Wacher C, Farres A (2001) Lactic acid bacterial diversity in the traditional Mexican fermented dough pozol as determined by 16S rDNA sequence analysis. Int J Food Microbiol 64:21–31
144. Faid M, Boraam F, Zyani I, Larpent JP (1994) Characterization of sourdough bread ferments made in the laboratory by traditional methods. Z Lebensm Unters Forsch 198:287–291
145. Edema MO, Sanni AI (2006) Micro-population of fermenting maize meal for sour maize bread production in Nigeria. Nig J Microbiol 20:937–946

146. Kazanskaya LN, Afanasyeva OV, Patt VA (1983) Microflora of rye sours and some specific features of its accumulation in bread baking plants of the USSR. In: Holas J, Kratochvil F (eds) Developments in food science, vol 5B, Progress in cereal chemistry and technology. Elsevier, London, pp 759–763
147. Barber S, Báguena R (1989) Microflora de la masa madre panaria. XI. Evolución de la microflora de masas madre durante el proceso de elaboración por el sistema de 'refrescos' sucesivos y de sus correspondientes masa panarias. Revista de Agroquímica y Tecnologia de Alimentos 29:478–491
148. Hamad SH, Dieng MC, Ehrmann MA, Vogel RF (1997) Characterization of the bacterial flora of Sudanese sorghum flour and sorghum sourdough. J Appl Microbiol 83:764–770
149. Spicher G, Lönner C (1985) Die Mikroflora des Sauerteiges. XXI. Mitteilung: Die in Sauerteigen Schwedischer Bäckereien vorkommenden Lactobacillen. Z Lebensm Unters Forsch 181:9–13
150. Lönner C, Welander T, Molin N, Dostálek M, Blickstad E (1986) The microflora in a sour dough started spontaneously on typical Swedish rye meal. Food Microbiol 3:3–12
151. Gül H, Özçelik S, Sağdıç O, Certel M (2005) Sourdough bread production with lactobacilli and *S. cerevisiae* isolated from sourdoughs. Proc Biochem 40:691–697
152. Gobbetti M, Corsetti A, Rossi J (1996) *Lactobacillus sanfrancisco,* a key sourdough lactic acid bacterium: physiology, genetic and biotechnology. Adv Food Sci 18:167–175
153. Foschino R, Arrigoni C, Picozzi C, Mora D, Galli A (2001) Phenotypic and genotypic aspects of *Lactobacillus sanfranciscensis* strains isolated from sourdoughs in Italy. Food Microbiol 18:277–285
154. Picozzi C, Bonacina G, Vigentini I, Foschino R (2001) Genetic diversity in Italian *Lactobacillus sanfranciscensis* strains assessed by multilocus sequence typing and pulsed-field gel electrophoresis analyses. Microbiology 156:2035–2045
155. Scheirlinck I, Van der Meulen R, De Vuyst L, Vandamme P, Huys G (2009) Molecular source tracking of predominant lactic acid bacteria in traditional Belgian sourdoughs and their production environments. J Appl Microbiol 106:1081–1092
156. Settanni L, Van Sinderen D, Rossi J, Corsetti A (2005) Rapid differentiation and *in situ* detection of 16 sourdough *Lactobacillus* species by multiplex PCR. Appl Environ Microbiol 71:3049–3059
157. Settanni L, Valmorri S, van Sinderen D, Suzzi G, Paparella A, Corsetti A (2006) Combination of multiplex PCR and PCR-denaturing gradient gel electrophoresis for monitoring common sourdough-associated *Lactobacillus* species. Appl Environ Microbiol 72:3793–3796
158. Di Cagno R, De Angelis M, Gallo G, Settanni L, Berloco MG, Siragusa S, Parente E, Corsetti A, Gobbetti M (2007) Genotypic and phenotypic diversity of *Lactobacillus rossiae* strains isolated from sourdough. J Appl Microbiol 103:821–835
159. Groenewald WH, Van Reenen CA, Todorov SD, Du Toit M, Witthuhn R, Holzapfel WH, Dicks LMT (2006) Identification of lactic acid bacteria from vinegar flies based on phenotypic and genotypic characteristics. Am J Enol Viticult 57:519–525
160. Gänzle MG, Ehmann M, Hammes WP (1998) Modeling of growth of *Lactobacillus sanfranciscensis* and *Candida milleri* in response to process parameters of sourdough fermentation. Appl Environ Microbiol 64:2616–2623
161. Picozzi C, D'Anchise F, Foschino R (2006) PCR detection of *Lactobacillus sanfranciscensis* in sourdough and Panettone baked product. Eur Food Res Technol 222:330–335
162. Leroy F, De Winter T, Foulquié Moreno MR, De Vuyst L (2007) The bacteriocin producer *Lactobacillus amylovorus* DCE 471 is a competitive starter culture for type II sourdough fermentations. J Sci Food Agricult 87:1726–1736
163. Hammes WP, Stolz P, Gänzle M (1996) Metabolism of lactobacilli in traditional sourdoughs. Adv Food Sci 18:176–184
164. Gobbetti M, De Angelis M, Corsetti A, Di Cagno R (2005) Biochemistry and physiology of sourdough lactic acid bacteria. Trends Food Sci Technol 16:57–69
165. Gänzle MG, Vermeulen N, Vogel RF (2007) Carbohydrate, peptide and lipid metabolism of lactic acid bacteria in sourdough. Food Microbiol 24:128–138
166. Gänzle MG (2009) From gene to function: metabolic traits of starter cultures for improved quality of cereal foods. Int J Food Microbiol 134:29–36

167. Dal Bello F, Walter J, Roos S, Jonsson H, Hertel C (2005) Inducible gene expression in *Lactobacillus reuteri* LTH5531 during type II sourdough fermentation. Appl Environ Microbiol 71:5873–5878
168. Hüfner E, Britton RA, Roos S, Jonsson H, Hertel C (2008) Global transcriptional response of *Lactobacillus reuteri* to the sourdough environment. Syst Appl Microbiol 31:323–338
169. Vrancken G, Rimaux T, De Vuyst L, Leroy F (2008) Kinetic analysis of growth and sugar consumption by *Lactobacillus fermentum* IMDO 130101 reveals adaptation to the acidic sourdough ecosystem. Int J Food Microbiol 128:55–66
170. Vrancken G, Rimaux T, Weckx S, De Vuyst L, Leroy F (2009) Environmental pH determines citrulline and ornithine release through the arginine deiminase pathway in *Lactobacillus fermentum* IMDO 130101. Int J Food Microbiol 135:216–222
171. Vrancken G, Rimaux T, Wouters D, Leroy F, De Vuyst L (2009) The arginine deiminase pathway of *Lactobacillus fermentum* IMDO 130101 responds to growth under stress conditions of both temperature and salt. Food Microbiol 26:720–727
172. Weckx S, Allemeersch J, Van der Meulen R, Vrancken G, Huys G, Vandamme P, Van Hummelen P, De Vuyst L (2011) Metatranscriptome analysis for insight into whole-ecosystem gene expression during spontaneous wheat and spelt sourdough fermentations. Appl Environ Microbiol 77:618–626
173. Gänzle MG, Vogel R (2003) Contribution of reutericyclin production to the stable persistence of *Lactobacillus reuteri* in an industrial sourdough fermentation. Int J Food Microbiol 80:31–45
174. Pepe O, Blaiotta G, Anatasio M, Moschetti G, Ercolini D, Villani F (2004) Technological and molecular diversity of *Lactobacillus plantarum* strains isolated from naturally fermented sourdoughs. Syst Appl Microbiol 27:443–453
175. Vrancken G, Rimaux T, Weckx S, Leroy F, De Vuyst L (2011) Influence of temperature and backslopping time on the microbiota of a type I propagated laboratory wheat sourdough fermentation. Appl Environ Microbiol 77:2716–2726
176. Du Toit M, Dicks LMT, Holzapfel WH (2003) Identification of heterofermentative lactobacilli isolated from pig faeces by numerical analysis of total soluble cell protein and RAPD patterns. Lett Appl Microbiol 37:12–16
177. Park SH, Itoh K (2005) Species-specific oligonucleotide probes for the detection and identification of *Lactobacillus* isolated from mouse faeces. J Appl Microbiol 99:51–57
178. De Angelis M, Siragusa S, Berloco M, Caputo L, Settanni L, Alfonsi G, Amerio M, Grandi A, Ragni A, Gobbetti M (2006) Selection of potential probiotic lactobacilli from pig feces to be used as additives in pelleted feeding. Res Microbiol 157:792–801
179. Neysens P, De Vuyst L (2005) Kinetics and modelling of sourdough lactic acid bacteria. Trends Food Sci Technol 16:95–103
180. Böcker G, Stolz P, Hammes WP (1995) Neue Erkentnisse zum Ökosystem Sauerteig und zur Physiologie der Sauerteig-typischen Stämme *Lactobacillus sanfranciscensis* und *Lactobacillus pontis*. Getreide Mehl und Broth 49:370–374
181. Vogel RF, Ehrmann MA, Gänzle MG (2002) Development and potential of starter lactobacilli resulting from exploration of the sourdough ecosystem. Antonie van Leeuwenhoek 81:631–638
182. Ravyts F, De Vuyst L (2011) Prevalence and impact of single-strain starter cultures of lactic acid bacteria on metabolite formation in sourdough. Food Microbiol 28:1129–1139
183. Brandt MJ (2007) Sourdough products for convenient use in baking. Food Microbiol 24:161–164
184. Carnevali P, Ciati R, Leporati A, Paese M (2007) Liquid sourdough fermentation: industrial application perspectives. Food Microbiol 24:150–154
185. Meroth CB, Walter J, Hertel C, Brandt M, Hammes WP (2003) Monitoring the bacterial population dynamics in sourdough fermentation processes by using PCR-denaturing gradient gel electrophoresis. Appl Environ Microbiol 69:475–482
186. Decock P, Cappelle S (2005) Bread technology and sourdough technology. Trends Food Sci Technol 16:113–120
187. Katina K, Heinio R-L, Autio K, Poutanen K (2006) Optimization of sourdough process for improved sensory profile and texture of wheat bread. Food Sci Technol 39:1189–1202

188. Messens W, Neysens P, Vansieleghem W, Vanderhoeven J, De Vuyst L (2002) Modeling growth and bacteriocin production by *Lactobacillus amylovorus* DCE 471 in response to temperature and pH values used for sourdough fermentations. Appl Environ Microbiol 68:1431–1435
189. Brandt MJ, Hammes WP, Gänzle MG (2004) Effects of process parameters on growth and metabolism of *Lactobacillus sanfranciscensis* and *Candida humilis* during rye sourdough fermentation. Eur Food Res Technol 218:333–338
190. Vogelmann SA, Hertel C (2011) Impact of ecological factors on the stability of microbial associations in sourdough fermentation. Food Microbiol 28:583–589
191. Valmorri S, Mortensen HD, Jesperen L, Corsetti A, Gardini F, Suzzi G, Arneborg N (2008) Variations of internal pH in typical Italian sourdough yeasts during co-fermentation with lactobacilli. Food Sci Technol 41:1610–1615
192. Martínez-Anaya MA, Llin ML, Macias MP, Collar C (1994) Regulation of acetic acid production by homo- and heterofermentative lactobacilli in whole-wheat sourdoughs. Z Lebensm Unters Forsch 199:186–190
193. Neysens P, Messens W, De Vuyst L (2003) Effect of sodium chloride on growth and bacteriocin production by *Lactobacillus amylovorus* DCE 471. Int J Food Microbiol 88:29–39
194. Simonson L, Salovaara H, Korhola M (2003) Response of wheat sourdough parameters to temperature, NaCl and sucrose variations. Food Microbiol 20:193–199
195. Gobbetti M (1998) The sourdough microflora: interactions of lactic acid bacteria and yeasts. Trends Food Sci Technol 9:267–274
196. Gobbetti M, Corsetti A, Rossi J (1994) The sourdough microflora – interactions between lactic acid bacteria and yeasts: metabolism of amino acids. World J Microbiol Biotechnol 10:275–279
197. Deak T (2003) Detection, enumeration and isolation of yeasts. In: Boekhout T, Robert V (eds) Yeasts in food, 1st edn. Behr's Verlag, Hamburg, pp 39–68
198. Mian MA, Fleet GH, Hocking AD (1997) Effect of diluent type on viability of yeasts enumerated from foods or pure culture. Int J Food Microbiol 35:103–107
199. Yarrow D (1998) Methods for the isolation, maintenance and identification of yeasts. In: Kurtzman CP, Fell JW (eds) The yeasts: a taxonomic study, 4th edn. Elsevier, Amsterdam, pp 77–100
200. Beuchat LR (1993) Selective media for detecting and enumerating foodborne yeasts. Int J Food Microbiol 19:1–14
201. Arthur H, Watson K (1976) Thermal adaptation in yeasts: growth temperatures, membrane lipid, and cytochrome composition of psychrophilic, mesophilic, and thermophilic yeasts. J Bacteriol 128:56–68
202. de Man JC, Rogosa M, Sharpe ME (1960) A medium for the cultivation of lactobacilli. J Appl Bacteriol 23:130–135
203. Vogel RF, Böcker G, Stolz P, Ehrmann M, Fanta D, Ludwig W, Pot B, Kersters K, Schleifer KH, Hammes WP (1994) Identification of lactobacilli from sourdough and description of *Lactobacillus pontis* sp. nov. Int J Food Microbiol 44:223–229
204. Vera A, Rigobello V, Demarigny Y (2009) Comparative study of culture media used for sourdough lactobacilli. Food Microbiol 26:728–733
205. Kurtzman CP, Fell JW, Boekhout T (2011) The yeasts: a taxonomic study, vol 1-3, 5th edn. Elsevier, Amsterdam
206. Kurtzman CP, Robnett CJ (1998) Identification and phylogeny of ascomycetous yeasts from analysis of nuclear large subunit (26S) ribosomal DNA partial sequences. Antonie van Leeuwenhoek 73:331–371
207. Cadez N, Poot GA, Raspor P, Smith MT (2003) *Hanseniaspora meyeri* sp. nov., *Hanseniaspora clermontiae* sp. nov., *Hanseniaspora lachancei* sp. nov. and *Hanseniaspora opuntiae* sp. nov., novel apiculate yeast species. Int J Syst Evol Microbiol 53:1671–1680
208. Groenewald M, Daniel HM, Robert V, Poot GA, Smith MT (2008) Polyphasic re-examination of *Debaryomyces hansenii* strains and reinstatement of *D. hansenii*, *D. fabryi* and *D. subglobosus*. Persoonia 21:17–27

209. Bergerow D, Nilsson H, Unterseher M, Maier W (2010) Current state and perspectives of fungal barcoding and rapid identification procedures. Appl Microbiol Biotechnol 87:99–108
210. Nilsson RH, Kristiansson E, Ryberg M, Hallenberg N, Larsson K-H (2008) Intraspecies ITS variability in the kingdom Fungi as expressed in the international sequence databases and its implications for molecular species identification. Evol Bioinf 4:193–201
211. Kurtzman CP, Robnett CJ (2003) Phylogenetic relationships among yeasts of the '*Saccharomyces* complex' determined from multigene sequence analyses. FEMS Yeast Res 3:417–432
212. Tsui CKM, Daniel HM, Robert V, Meyer W (2008) Re-examining the phylogeny of clinically relevant *Candida* species and allied genera based on multigene analyses. FEMS Yeast Res 8:651–659
213. Vassart G, Georges M, Monsieur R, Brocas H, Lequarre AS, Christophe D (1987) A sequence in M13 phage detects hypervariable minisatellites in human and animal DNA. Science 235:683–684
214. Landry CR, Townsend JP, Hartl DL, Cavalieri D (2006) Ecological and evolutionary genomics of *Saccharomyces cerevisiae*. Mol Ecol 15:575–591
215. Carreto L, Eiriz MF, Gomes AC, Pereira PM, Schuller D, Santos MAS (2008) Comparative genomics of wild type yeast strains unveils important genome diversity. BMC Genomics 9:524
216. Ness F, Lavallée F, Dubourdieu D, Aigle M, Dulau L (1993) Identification of yeast strains using the polymerase chain reaction. J Sci Food Agric 62:89–94
217. Legras J-L, Karst F (2003) Optimisation of interdelta analysis for *Saccharomyces cerevisiae* strain characterisation. FEMS Microbiol Lett 221:249–255
218. Gancheva A, Pot B, Vanhonacker K, Hoste B, Kersters K (1999) A polyphasic approach towards the identification of strains belonging to *Lactobacillus acidophilus* and related species. Syst Appl Microbiol 22:573–585
219. Sauer S, Kliem M (2010) Mass spectrometry tools for the classification and identification of bacteria. Nature Rev Microbiol 8:74–82
220. De Bruyne K, Slabbinck B, Waegeman W, Vauterin P, De Baets B, Vandamme P (2011) Bacterial species identification from MALDI-TOF mass spectra through data analysis and machine learning. Syst Appl Microbiol 34:20–29
221. Grimont F, Grimont PAD (1986) Ribosomal ribonucleic acid gene restriction patterns as potential taxonomic tools. Ann Inst Pasteur Microbiol 137B:165–175
222. Janssen P, Coopman R, Huys G, Swings J, Bleeker M, Vos P, Zabeau M, Kersters K (1996) Evaluation of the DNA fingerprinting method AFLP as a new tool in bacterial taxonomy. Microbiology 142:1881–1893
223. Settanni L, Massitti O, Van Sinderen D, Corsetti A (2005) In situ activity of a bacteriocin-producing *Lactococcus lactis* strain. Influence on the interactions between lactic acid bacteria during sourdough fermentation. J Appl Microbiol 99:670–681
224. Olive DM, Bean P (1999) Principles and applications of methods for DNA-beased typing of microbial organisms. J Clin Microbiol 37:1661–1669
225. Versalovic J, Schneider M, De Bruijn FJ, Lupski JR (1994) Genomic fingerprinting of bacteria using repetitive sequence-based polymerase chain reaction. Methods Mol Cell Biol 5:25–40
226. Bounaix M-S, Robert H, Gabriel V, Morel S, Remaud-Siméon M, Gabriel B, Fontagné-Faucher C (2010) Characterization of dextran-producing *Weissella* strains isolated from sourdoughs and evidence of constitutive dextran sucrase expression. FEMS Microbiol Lett 311:18–26
227. Bounaix M-S, Gabriel V, Robert H, Morel S, Remaud-Siméon M, Gabriel B, Fontagné-Faucher C (2010) Characterization of glucan-producing *Leuconostoc* strains isolated from sourdough. Int J Food Microbiol 144:1–9
228. Stackebrandt E, Goebel BM (1994) A place for DNA-DNA reassociation and 16S ribosomal RNA sequence analysis in the present species definition in bacteriology. Int J Syst Bacteriol 44:846–849
229. Naser SM, Thompson FL, Hoste B, Gevers D, Dawyndt P, Vancanneyt M, Swings J (2005) Application of multilocus sequence analysis (MLSA) for rapid identification of *Enterococcus* species based on *rpoA* and *pheS* genes. Microbiology 151:2141–2150

230. Stackebrandt E, Frederiksen W, Garrity GM, Grimont PAD, Kampfer P, Maiden MCJ, Nesme X, Rossello-Mora R, Swings J, Truper HG, Vauterin L, Ward AC, Whitman WB (2002) Report of the ad hoc committee for the re-evaluation of the species definition in bacteriology. Int J Syst Evol Microbiol 52:1043–1047
231. Konstantinidis KT, Ramette A, Tiedje JM (2006) Toward a more robust assessment of intraspecies diversity, using fewer genetic markers. Appl Environ Microbiol 72:7286–7293
232. De Bruyne K, Schillinger U, Caroline L, Boehringer B, Cleenwerck I, Vancanneyt M, De Vuyst L, Franz CMAP, Vandamme P (2007) *Leuconostoc holzapfelii* sp. nov., isolated from Ethiopian coffee fermentation and assessment of sequence analysis of housekeeping genes for delineation of *Leuconostoc* species. Int J Syst Evol Microbiol 57:2952–2959
233. De Bruyne K, Franz CMAP, Vancanneyt M, Schillinger U, Mozzi F, de Valdez GF, De Vuyst L, Vandamme P (2008) *Pediococcus argentinicus* sp. nov. from Argentinean fermented wheat flour and identification of *Pediococcus* species by *pheS*, *rpoA* and *atpA* sequence analysis. Int J Syst Evol Microbiol 58:2909–2916
234. Achtman M (2008) Evolution, population structure, and phylogeography of genetically monomorphic bacterial pathogens. Ann Rev Microbiol 62:53–70
235. Cai H, Rodriguez BT, Zhang W, Broadbent JR, Steele JL (2007) Genotypic and phenotypic characterization of *Lactobacillus casei* strains isolated from different ecological niches suggests frequent recombination and niche specificity. Microbiology 153:2655–2665
236. Diancourt L, Passet V, Chervaux C, Garault P, Smokvina T, Brisse S (2007) Multilocus sequence typing of *Lactobacillus casei* reveals a clonal population structure with low levels of homologous recombination. Appl Environ Microbiol 73:6601–6611
237. de las Rivas B, Marcobal A, Muñoz R (2006) Development of a multilocus sequence typing method for analysis of *Lactobacillus plantarum* strains. Microbiology 152:85–93
238. Raftis EJ, Salvetti E, Torriani S, Felis GE, O'Toole PW (2011) Genomic diversity of *Lactobacillus salivarius*. Appl Environ Microbiol 77:954–965
239. Ehrmann M, Ludwig W, Schleifer KH (1994) Reverse dot blot hybridization: a useful method for the direct identification of lactic acid bacteria in fermented food. FEMS Microbiol Lett 117:143–150
240. Müller MRA, Ehrmann MA, Vogel RF (2000) Multiplex PCR for the detection of *Lactobacillus pontis* and two related species in a sourdough fermentation. Appl Environ Microbiol 66:2113–2116
241. Valcheva R, Kabadjova P, Rachman C, Ivanova I, Onno B, Prevost H, Dousset X (2007) A rapid PCR procedure for the specific identification of *Lactobacillus sanfranciscensis*, based on the 16S-23S intergenic spacer regions. J Appl Microbiol 102:290–302
242. Zapparoli G, Torriani S (1997) Rapid identification and detection of *Lactobacillus sanfrancisco* in sourdough by species-specific PCR with 16S rRNA-targeted primers. Syst Appl Microbiol 20:640–644
243. Levin RE (2004) The application of real-time PCR to food and agricultural systems: a review. Food Biotechnol 18:97–133
244. Ercolini D (2004) PCR-DGGE fingerprinting: novel strategies for detection of microbes in food. J Microbiol Meth 56:297–314
245. Palomba S, Blaiotta G, Ventorino V, Saccone A, Pepe O (2010) Microbial characterization of sourdough for sweet baked products in the Campania region (southern Italy) by a polyphasic approach. Ann Microbiol. doi:10.1007/s13213-010-0140-2
246. Reale A, Di Renzo T, Succi M, Tremonte P, Coppola R, Sorrentino E (2011) Identification of lactobacilli isolated in traditional ripe wheat sourdoughs by using molecular methods. World J Microbiol Biotechnol 27:237–244
247. Rantsiou K, Comi G, Cocolin L (2004) The *rpoB* gene as a target for PCR-DGGE analysis to follow lactic acid bacterial population dynamics during food fermentations. Food Microbiol 21:481–487

Chapter 6
Physiology and Biochemistry of Sourdough Yeasts

M. Elisabetta Guerzoni, Diana I. Serrazanetti, Pamela Vernocchi, and Andrea Gianotti

6.1 Introduction

Although baker's yeast is used worldwide as a leavening agent for the manufacture of baked products, an overwhelming multitude of baked goods are also produced with the aid of sourdough. These include above all breads from wheat, rye and mixtures thereof as well as the well-known Italian products such as Panettone, Colomba, Pandoro and different types of brioches.

Traditional cultivation methods in combination with phenotypic and genotypic identification adopted to characterise the yeasts of ripe dough revealed the presence of some species belonging, especially, to the genera *Saccharomyces* and *Candida* [1–6]. Their taxonomic features and diversity are described in Chap. 5. Based on studies focused on the yeast population of sourdough, the most frequent species were *Saccharomyces cerevisiae* and *Candida milleri* [7, 8]. In some cases, *Candida milleri* accounted for 65.8% of the isolates in artisanal sourdough produced without the addition of baker's yeast [9].

This chapter is mainly focused on the description of: (1) yeast responses to physico-chemical conditions and their fluctuation during sourdough fermentation; (2) implications of such responses for fermentation products and flavour compounds, including quorum-sensing molecules; and (3) the production of baker's yeast and its use in the bread-making industry.

M.E. Guerzoni (✉) • D.I. Serrazanetti • P. Vernocchi • A. Gianotti
Food Science Department (DISA), University of Bologna,
Via Fanin, 46, Bologna, Italy
e-mail: elisabetta.guerzoni@unibo.it; diana.serrazanetti3@unibo.it;
pamela.vernocchi@unibo.it; andrea.gianotti@unibo.it

M. Gobbetti and M. Gänzle (eds.), *Handbook on Sourdough Biotechnology*, 155
DOI 10.1007/978-1-4614-5425-0_6, © Springer Science+Business Media New York 2013

6.2 Effects of Physico-chemical Conditions on Yeast Growth and Physiology

Cereal dough is a dynamic system that is characterised by continuous changes in nutrient availability during processing. This system is subjected to simultaneous release, by endogenous or added amylases, of fermentable carbohydrates such as glucose, maltose and fructose, and their uptake and consumption by yeasts and lactic acid bacteria.

The availability of specific nutrients maintains yeast fermentation and, in industrial fermentations, nutrient-related stresses have several practical implications [10]. In particular, the concentration of fermentable carbohydrates, as well as the dynamics of their release by amylases depends on many factors including cereal type, environmental and cultural factors, grain storage and milling conditions. In turn, the activity of amylases is conditioned by temperature and, especially, water activity (A_w) [11, 12]. During the first phase of bread making, the readily fermentable sugar concentration is probably limiting. Figure 6.1 describes the effect of commercial amylase supplementation on the rate of volume increase of doughs with different concentrations of NaCl (Fig. 6.1).

Depending on the type of flour and bread-making technology, starvation conditions can also be envisaged. The imbalance between yeast consumption and starch hydrolysis might lead to rapid depletion of soluble carbohydrates. Gene expression response to nutrient starvation involves changes in gene expression that persist until the starvation is alleviated [13, 14]. Saldanha et al. [15] studied the physiological response to limitation by diverse nutrients in batch and steady-state (chemostat) cultures of *S. cerevisiae*. The yeast does not respond to the changing environment with a small adjustment in key control points, but with the coherent transcriptional regulation of a large set of genes. Cells adjust their growth rate to nutrient availability and maintain homeostasis in the same way either in batch or in steady state conditions (continuous culture). Therefore, the patterns of gene expression in the steady state closely resemble those of the corresponding batch cultures just before the yeasts exhaust the nutrients.

Yeasts can use a diverse array of compounds as nitrogen sources and, upon demand, are capable of expressing catabolic enzymes of many different pathways. Extensive studies on nitrogen metabolism and its regulation have been carried out with *S. cerevisiae*. Nitrogen compounds such as ammonia, glutamine and glutamate are used by yeasts, but asparagine is the preferred nitrogen source. When these primary nitrogen sources are not available or are present only in low concentrations that limit growth, many other nitrogen sources are used, for example purines, amides and most of the amino acids [16]. Yeasts may also grow on ammonium ions as the sole nitrogen source, since they possess the entire repertoire of genes that encode for the enzymes responsible for biosynthesis of all amino acids. Ammonium is directly assimilated into amino acids, notably glutamate and glutamine, which serve as donors of the amino group for the other amino acids. The main route for assimilation of ammonium is the reaction of the NADPH-dependent glutamate dehydrogenase

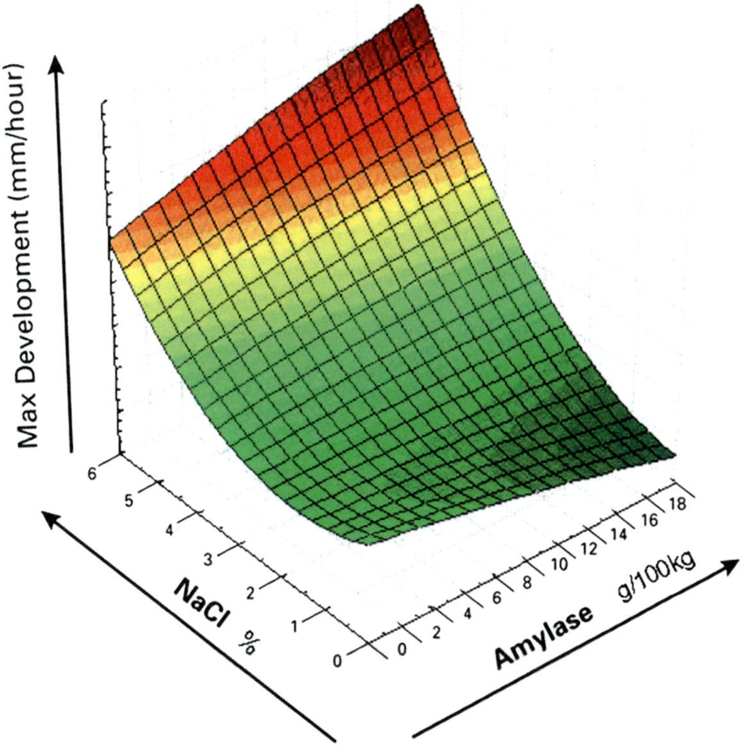

Fig. 6.1 Combined effect of NaCl and α-amylase supplementation on the rate of volume increase of doughs inoculated with *S. cerevisiae* and incubated at 30°C for 3 h (Gianotti et al. unpublished)

(GDH), which forms glutamate from α-ketoglutarate and ammonium. Whenever the concentration of ammonium ions is low, glutamine synthase is activated. This forms glutamine from α-ketoglutarate and ammonium in an ATP-dependent reaction. Glutamine is absolutely required as the prominent precursor of several biosynthetic pathways for asparagine, tryptophane, histidine, arginine, carbamoyl-phosphate, CTP, AMP, GMP, glucosamine and NAD. *Saccharomyces cerevisiae* does not use nitrate as a nitrogen source but very few species of yeasts have that capacity. Nitrate assimilation occurs via the activity of the NADPH-dependent nitrate reductase, which forms nitrite. Further, nitrite is reduced to ammonium by the NADPH-dependent nitrite reductase. The type of nutrients available defines the internal metabolic flow, while their abundance often limits the rate of biomass production and the energy available for growth [17].

The use of the various nitrogen sources is highly regulated. It requires the synthesis of specific catabolic enzymes and permeases, which are subject to nitrogen catabolite repression. The de novo synthesis of permeases and catabolic enzymes is controlled at the level of transcription, and often requires two distinct positive signals.

A global signal indicates nitrogen derepression, and a second, pathway-specific signal indicates the presence of a substrate or of an intermediate of the pathway. This control permits the selective expression of enzymes for a specific catabolic pathway from many potential candidates within the nitrogen regulatory circuit. However, some systems are controlled only by nitrogen metabolite repression and not by induction [16].

The cell yield of yeasts seemed to be unaffected by the presence of lactic acid bacteria in sourdough. *Saccharomyces cerevisiae* grown as a monoculture in dough attained 8 log CFU/g [18]. Almost the same value of cell count was found under co-culture with lactic acid bacteria. Conversely, the maximum specific growth rate was negatively affected. According to Meroth et al. [6], cell counts of endogenous yeasts of ca. 7.5 log CFU/g were reached during the first fermentation period at 30°C, with a dough yield of 200, and remained stable until the end of fermentation. During fermentation of traditional sourdough products (e.g. Panettone or Colomba), the total yeast counts reached the level of 8–8.5 log CFU/g, after addition of baker's yeast [19]. Yeast cells may encounter different environmental states. Maintaining optimal functionality in the presence of such external variability is a central evolutionary constraint. Feedback mechanisms directly link gene expression with internal need. Physiological variables such as the rate of biomass production, the cellular pools of nutrients or the energy feedbacks, to properly tune gene expression with the corresponding functional needs, are part of these strategies [17].

Yeast cells starved for particular amino acids or treated with an inhibitor of protein synthesis (cycloheximide) may exhibit a phenomenon referred to as the stringent response. In *S. cerevisiae*, the stringent response results in a rapid inhibition of the synthesis of rRNA (but not of tRNA or mRNA) and it is thought to be signalled when cells sense a rapid decrease of the overall rate of protein biosynthesis. Nitrogen compounds and, especially, amino acids are not the limiting factors for yeasts. *Saccharomyces cerevisiae* and the other species, which naturally occur in sourdoughs, are prototrophic for amino acids. Figure 6.2 shows the concentration of various free amino acids during fermentation of kamut flour with baker's yeast and sourdough. Compared to the dough leavened with baker's yeast, the fermentation by sourdough lactic acid bacteria caused an increase of some amino acids. Fermentation with baker's yeast resulted in the decrease of free amino acids due to yeast metabolism (see below). Because of the high value of pH during fermentation with yeasts alone, the activation of flour endogenous proteinases is poor, and the concentration of free amino acid increases only after the yeasts have reached the stationary phase of growth [20].

Co-fermentation with lactic acid bacteria and yeasts determines environmental fluctuations in terms of availability of nutrients, synthesis of organic acids, pH decrease and changes of the rheological properties of sourdough. Brandt et al. [21] evaluated the effect of process parameters on the growth and metabolism of *L. sanfranciscensis* and *C. milleri* during rye sourdough fermentation. pH did not affect the growth of the yeast in the range 3.5–5.5, whereas the growth of *L. sanfranciscensis* was inhibited at pH 4.0. The concentration of NaCl of 4% (wt/wt of flour) inhibited the growth of *L. sanfranciscensis* but not that of *C. milleri*.

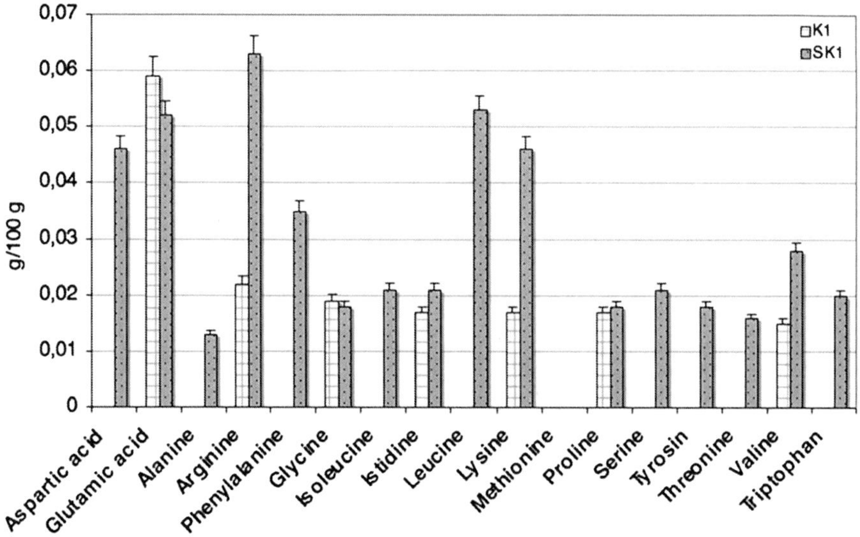

Fig. 6.2 Difference in the content (g/100 g) of various amino acids in ripe dough (K1) and sourdough (SK1) of kamut flour

The effect of such fluctuations on the synthesis of lactate, acetate, ethanol and CO_2 agreed with that on growth. Gänzle et al. [7] used a predictive model to describe the effect of temperature, pH, organic acids and NaCl on the growth of *L. sanfranciscensis* and *C. milleri*. A basic metabolic activity was also found under conditions where growth was inhibited.

6.2.1 The Yeast Carbohydrate Metabolism

Yeasts are classified as facultative anaerobes, i.e. they are capable of both fermentation and respiration. When oxygen is unavailable yeasts carry out fermentation. Under respiration, the hydrogen (electrons) from $NADH+H^+$ is passed to oxygen in the electron transport chain yielding approximately 2.5 ATPs per $NADH+H^+$. During, fermentation the hydrogen (electrons) is passed on acetaldehyde to form ethanol yielding no ATPs per $NADH+H^+$.

Yeasts accumulate energy in the form of ATP from different organic compounds by using different metabolic pathways. Most yeasts use carbohydrates as their main carbon and energy sources, but there are also yeasts that may use non-conventional carbon sources. The preferred energy source for yeasts is glucose, and glycolysis is the general pathway for conversion of glucose to pyruvate, whereby the synthesis ATP is coupled to the generation of intermediates and reducing power in the form of NADH for the biosynthetic pathways. Yeasts respire when oxygen is present and the repression absent; pyruvate enters into mitochondria where it is oxidatively

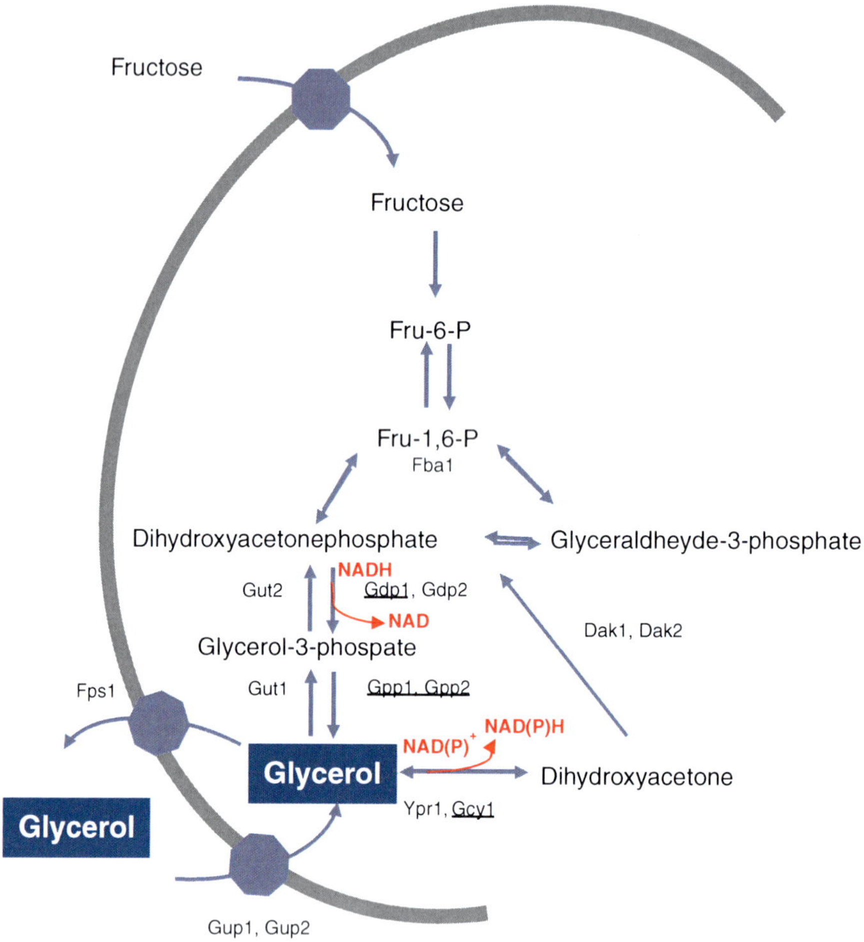

Fig. 6.3 Part of the glycolytic pathway and pathways for production of glycerol. Expression of the genes encoding the proteins that are underlined is stimulated after osmotic shock

decarboxylated to acetyl CoA by means the pyruvate dehydrogenase multi-enzyme complex. This reaction links glycolysis to the citric acid cycle, where acetyl CoA is completely oxidised to give two moles of CO_2 and reductive equivalents in the form of NADH and $FADH_2$. The citric acid cycle is an amphibolic pathway, as it combines both catabolic and anabolic functions. This latter function is the consequence of the production of intermediates for the synthesis of amino acids and nucleotides.

During alcoholic fermentation, yeasts re-oxidise NADH to NAD in a two-step reaction starting from pyruvate. The first reaction is via decarboxylation through pyruvate decarboxylase, followed by the reduction of the acetaldehyde to ethanol, catalysed by the alcohol dehydrogenase (ADH). Simultaneously, glycerol is generated from dihydroxyacetone phosphate to ensure, with an alternative pathway, the regeneration of NAD according to the mechanism described in Fig. 6.3.

Generally, the level of glycerol produced is ca. 0.03 mmol/g of sourdough. Paramithios et al. [18] reported that the synthesis of glycerol is positively affected by the co-cultivation of *S. cerevisiae* with *L. sanfranciscenis* and *L. brevis*. Exposure of baker's yeast to salt stress also increases glycerol production [22].

The glycolytic pathway and related enzymes were conserved during evolution, even though the mechanisms of controlling the carbon and energy metabolism have been adapted to the needs of each species. *Saccharomyces cerevisiae* switches to a mixed respiro-fermentative metabolism, resulting in ethanol production, as soon as the external glucose concentration exceeds 0.8 mmol/kg [23]. Hence, *S. cerevisiae* controls fermentation versus respiration primarily in response to the concentration of glucose. The aerobic synthesis of ethanol by *S. cerevisiae* depends on the relative capacities of the fermentative and respiratory pathways. High glucose levels result in the glycolytic rate exceeding that of the pyruvate dehydrogenase (Pdh) reaction, thereby generating an overflow towards pyruvate decarboxylase (Pdc) and, hence, ethanol production. At low external glucose levels and in the presence of oxygen, *S. cerevisiae* does not synthesise ethanol [24]. The uptake of glucose into *S. cerevisiae* is controlled by multiple hexose transporters (Hxts) [25], which have different substrate specificity and affinity, and are expressed under different, overlapping conditions [26]. When *S. cerevisiae* is exponentially growing under aerobic conditions with glucose or fructose as carbon sources, glucose degradation proceeds mainly via aerobic fermentation. When yeast is growing under aerobic conditions on mannose or galactose, degradation proceeds simultaneously via respiration and fermentation.

Control of carbohydrate metabolism in yeasts is of both fundamental and practical significance. It is regulated by different mechanisms depending on the genus and species as well as on environmental conditions. These mechanisms have been called Pasteur, Kluyver, Custers and Crabtree effects, glucose or catabolite repression, and genera or catabolite inactivation [27]. Table 6.1 describes these regulatory phenomena in *S. cerevisiae* and other genera, including also the species occurring in sourdough. The term Crabtree effect defines the inhibition of the synthesis of respiratory enzymes in the presence of oxygen and high concentration of glucose. *Candida* and *Pichia*, which are frequently associated with sourdough, do not show the Crabtree effect. Depending on environmental conditions, especially, the type of carbon sources, yeasts adjust the energy metabolism according to a process referred to as carbon catabolite repression. Two extreme cases are exponential growth on glucose or ethanol, which lead to the almost exclusive fermentation of the former with extensive secretion of ethanol or to exclusive respiration to carbon dioxide, respectively. Both, fully respiratory and fermentative metabolism are mediated by differentially active transcription factors. During fermentation, the transcription factor complex of Tup1p, Ssn6p and Mig1p mainly represses the expression of respiratory, gluconeogenic and alternative carbon source utilisation genes [28]. The minimal activity of the citric acid cycle for biosynthetic purposes is ensured mainly by the Rtg transcriptional activators [29]. During respiratory growth on non-fermentable carbon sources, the respiratory genes of the citric acid cycle and the respiratory chain are highly induced. This is triggered by the Hap transcription factors, a global activator complex of respiratory genes [29]. Activation of genes for gluconeogenesis during growth on non-fermentable carbon sources is achieved by the transcriptional

Table 6.1 Regulatory phenomena (Adapted from Barnett and Entian [27])

Name	What happens	Yeast species and genera	Underlying factors
Pasteur effect	Sugar used faster anaerobically than aerobically	*S. cerevisiae*	Oxidised cytochrome inactivates 6-phosphofructokinase
Custers effect	D-glucose is fermented to ethanol and CO_2 faster in aerobic than in anaerobic conditions	*Brettanomyces* and *Dekkera* spp.	Much acetic acid is produced via an NAD^+-aldehyde dehydrogenase. Consequently, anaerobically, the high $NADH:NAD^+$ ratio inhibits glycolysis
Kluyver effect	Ability to use oligosaccharide or galactose aerobically, but not anaerobically, although glucose is fermented	*Candida* spp., *Pichia* spp.	Probably caused mainly by slower uptake of sugar anaerobically
Crabtree effect	Adding glucose to tumour cells lowers the respiration rate		Decrease of ADP concentration in mitochondria

activators Cat8p and Sip4p [28–31]. These are the key elements of the yeast transcriptional regulatory network for the extreme cases of fermentative and fully respiratory growth [32].

Saccharomyces cerevisiae carries out respiration and fermentation simultaneously at high growth rates, even under fully aerobic conditions. This phenomenon occurs under chemostat culture conditions by the shift from biomass formation towards fermentative products at increased dilution rate. The Crabtree effect has important consequences for industrial processes aiming at producing yeast biomass, where the formation of fermentative by-products is undesired. Goddard [33] emphasised the importance of the Crabtree effect on the competitive advantage of *S. cerevisiae* during growth in ecosystems characterised by elevated concentrations of carbohydrates. The competitive advantage is lacking, when fermentable carbohydrates are limiting as occurs in the sourdough ecosystem.

An alternative route for glucose oxidation is the hexose phosphate pathway, also known as the pentose phosphate cycle, which provides the cell with the pentose sugars and cytosolic NADPH necessary for biosynthetic reactions. The second step of this pathway is the dehydrogenation of glucose-6-phosphate to 6-phosphogluconolactone and the generation of one mole of NADPH (by glucose-6-phosphate dehydrogenase). Subsequently, 6-phosphogluconate is decarboxylated via the activity of phosphogluconate dehydrogenase to ribulose-5-phosphate and a second mole of NADPH is generated. Besides generating NADPH, the other major function of this pathway is the production of ribose sugars, which serve for the biosynthesis of nucleic acid precursors and nucleotide coenzymes. The reduced carriers, NADH and $FADH_2$, are reoxidised in the respiratory chain located in the inner mitochondrial membrane. The energy released during the transfer of electrons is coupled to the process of oxidative phosphorylation via ATP synthase, an enzyme complex also located in the inner mitochondrial membrane and designed to synthesise ATP from ADP and inorganic phosphate.

For the majority of industrial fermentations with bakers' yeast, the elevated capacity to ferment available carbohydrates is an important characteristic, especially when the biomass is exposed to high sugar concentrations and/or absence of oxygen.

6.2.2 *Stress Response in Sourdough Yeasts*

Yeasts are exposed to constant fluctuations of their growth conditions. Consequently, they have to develop sophisticated responses to adapt to and survive under a variety of conditions. Yeasts, as well as other organisms, employ a concerted response to external stress [34]. One mechanism that yeast cells use to protect the cell from the effects of environmental variation is to initiate a common gene expression program. This program includes about 900 genes, whose expression is stereotypically altered when yeast cells are shifted to stressful environments. The genes that participate in this response amount to almost 14% of the currently predicted genes in the yeast genome [35].

The "general stress" transcription factor could be identified in the zinc-finger transcription factor (Msn2) [14]. Normally, Msn2 is exported from the nucleus, and a cyclin-dependent kinase (Srb10) is concomitantly repressed. Under stress, Msn2 re-localises to the nucleus and, with the relief of Srb10 repression, activates transcription. The stress response is rapid, but quickly attenuated. Bose et al. [36] showed that this attenuation is caused by a nuclear-dependent degradation of Msn2. Msn2 rapidly disappeared from cells after heat or osmotic shock.

Process parameters, including temperature, dough yield, oxygen, pH, as well as the composition of starter cultures, determine the quality and handling properties of sourdough [37] and the metabolic response of microorganisms responsible for the fermentation process [38]. The exposure of microbial cells to stressful and fluctuating conditions during fermentation involves a broad transcriptional response with many induced or repressed genes. The selective pressure exerted by environmental conditions encountered by yeast cells during sourdough fermentation, accounts for the consolidated dominance of selected yeast species. Nutrient availability likely modulates the microbial ecology of sourdough. However, within the sourdough ecosystem there are numerous mechanisms whereby one species may influence the growth of another [38]. Although autochthonous bacteria and yeasts are adapted and competitive in their respective environment, the dough environment can be described as a stressful environment for microorganisms [39].

The conditions of the sourdough microenvironment that principally affect yeast responses and growth rate are: nutrient availability (starvation), pH (acid stress), dough yield and presence of sugars, salts and polysaccharides (osmotic stress), oxygen (oxidative stress), temperature fluctuations (heat shock and cold stress) and interactions between lactic acid bacteria and yeasts (e.g. *S. cerevisiae, C. milleri* and *L. sanfranciscensis*), and between yeasts (e.g. *S. cerevisiae* and *C. milleri*).

6.2.2.1 Low Temperature

In bakery practice, the temperature of the sourdough is an important parameter to control the growth of lactobacilli and yeasts [21]. Häggman and Salovaara [40] studied the effect of process parameters, including low temperature, on the leavening of rye dough. Under the experimental conditions, the endogenous *C. milleri* was responsible for leavening, especially when the temperature was set at 22°C. The low temperature of fermentation also slowed the acidification rate, thus favouring an extended production of CO_2 by *C. milleri*. An appropriate modulation of the temperature and the use of temperatures ranging from 4°C to 8°C is a practical approach to control the fermentation rate and to program the working times.

The effects of the low temperature and other environmental parameters on yeast physiology are conditioned by the exposure dynamics [41]. Transcriptional responses during adaptation to suboptimal temperatures permitting growth (10–20°C) [42, 43] differed from those found after exposure to temperatures below 10°C, where growth ceases [44, 45]. Two different mechanisms of response are distinguished: (1) the early cold response (ECR), occurring within 12 h; and (2) the late cold response (LCR), occurring later than 12 h [43].

Tai et al. [41] used chemostat cultivation to compare different culture conditions and/or microbial strains at fixed specific growth rate. They observed 15% of the genes that showed a consistent transcriptional response in previous batch-culture studies on cold adaptation [42, 43, 45] were also identified under chemostat conditions at 30°C [46]. In batch cultures, the exposure of *S. cerevisiae* to low temperatures invariably induces an increased synthesis of storage carbohydrates (especially trehalose) and transcriptional up-regulation of genes involved in storage carbohydrate metabolism [47]. Transcriptional induction of the trehalose-biosynthesis genes TPS1 and TPS2 is consistently observed in cold-shock studies, and after exposure to near-freezing conditions. Several other genes such as HSP12, HSP26, HSP42, HSP104, YRO2 and SSE2, and those encoding the three cell-wall mannoproteins (Tip1p, Tir1p, and Tir2p), fatty-acid desaturase (Ole1p), which influences the membrane fluidity, and Nsr1p, a nucleolar protein required for pre-rRNA processing and ribosome biogenesis, were consistently associated with cold shock [41, 43, 45]. The stress response element (STRE) binding factors Msn/Msn4 are implicated in the coordinate regulation of low-temperature-responsive genes [43, 47]. Consistently, many genes induced upon a temperature downshift were also induced under a variety of other stress conditions [43]. Contrarily to batch cultures, where the low temperature adaptation is accompanied by marked ESR (Environmental Stress Response) (29% of the differentially expressed genes in batch cultures at low temperature respond to ESR as well), only three ESR-induced genes (YCP1, VPS73,

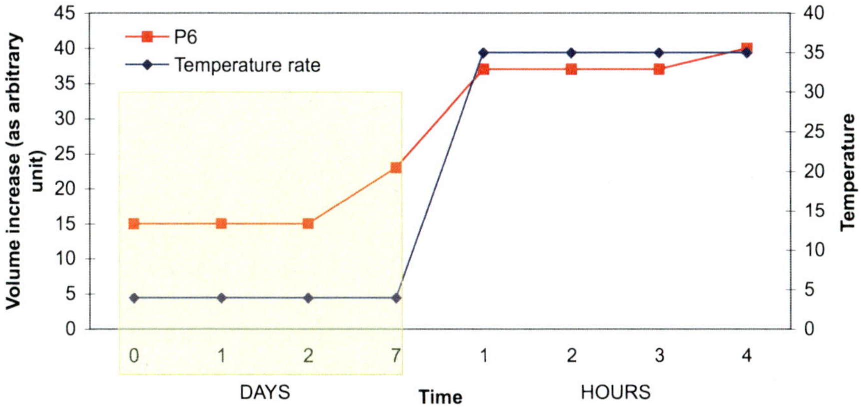

Fig. 6.4 Yeast recovery ability in terms of volume increase of a dough inoculated with the selected strain (P6). The dough was initially stored for 7 days at 4°C (*yellow square*), and then incubated at 35°C

and EMI1) showed higher transcription at 12°C under chemostat cultures. Conversely, 88 ESR-induced genes showed a consistently lower transcript level at 12°C [41]. Studies on cold adaptation in batch cultures of *S. cerevisiae* revealed a clear transcriptional up-regulation at low temperature of chaperone-encoding genes such as HSP26 and HSP42 [43, 44]. The proteins encoded by these genes prevent the aggregation of cytosolic proteins during heat shock [41].

In bakery practice, refrigeration of dough or sourdough is used to control fermentation. Under the refrigerated conditions, yeasts have to maintain and then to recover their fermentative capacity in a very short period of time (15–30 min). At 4–8°C, many yeast strains continue to ferment at a slow rate and induce a slight increase of the dough volume. The fermentation stops at 4°C. Only selected strains recover their leavening capacity when the dough temperature is raised to 28–35°C (Gottardi et al. unpublished data) (Fig. 6.4). As shown in Fig. 6.4, these strains increased the dough volume within 15 min also when the dough had been kept for 7 days at 4°C.

6.2.2.2 Acidity

The acidity of the sourdough depends on lactic and acetic acid production by lactic acid bacteria [40]. Usually, low temperatures of sourdough fermentation delay the lactic acidification and decrease the time of yeast exposure to high acidity. A low fermentation temperature was suggested as a means to improving the synthesis of CO_2 by yeasts [40].

Because only the non-dissociated acids diffuse into the cell, the type of acid more than the pH determines yeast inhibition [49]. Mainly the level of dissociation of acetic acids affects the leavening capacity [40]. The non-dissociated form of the acetic acid inhibits yeasts by acidifying the cytoplasm, which causes physiological stress or suppresses metabolic activity [48]. This especially happens when the pH of the sourdough is below the pK_a of acetic acid, 4.76. Non-dissociated acetic acid decreased the leavening capacity of *C. milleri* when grown together with the hetero-fermentative *L. brevis* compared to homofermentative species. *Candida milleri* adapts to a wide range of pH (3.5–6.0) and has a good inherent acid tolerance, the leavening activity is obviously affected by the fermentation process [7].

The leavening power of baker's yeast is strongly influenced by the environmental conditions of storage [49]. Pre-treatment at 30°C with organic acids (malic, succinic and citric acids), under a wide range of pH values, was assayed before use [49]. The treatment with organic acids variously increased the fermentative activity. When the pH of baker's yeast, containing citric acid, was raised from 3.5 to 7.5, both the fermentative and maltase activities increased. Glycerol-3-phosphate dehydrogenase activity and the levels of internal glycerol also increased in the presence of citrate. On the contrary, baker's yeast containing succinic acid at pH 7.5, showed a decreased viability during storage, despite the maintenance of high fermentative activity.

6.2.2.3 Osmotic Stress

A_w is defined as the chemical potential of the free water in solution. Low and intermediate values of A_w limit the growth of yeasts. The A_w of the cytosol of yeast cells is lower than that of the surrounding medium, corresponding to a higher osmotic pressure (turgor pressure). This turgor pressure drives water into the cell based on the concentration gradient. Turgor pressure is counteracted by the limited ability for expansion of the cell wall and thus determines the shape of the cells [50]. The ability to survive under rapid changes of A_w is an intrinsic characteristic of the microbial cell. Survival mechanisms in response to osmotic downshift or upshift allow passive water loss or uptake. In response to altered osmolarity, yeasts cells develop mechanisms to adjust to high external osmolarity and to maintain or to recover an inside-directed driving force for water adaptation. These mechanisms are based on sensing the osmotic changes [50] and accumulating chemically inert osmolytes, for example glycerol. The high osmolarity glycerol (HOG) signalling system plays a central role in the osmotic adaptation of yeasts. *Saccharomyces cerevisiae* monitors osmotic changes through the sensor histidine kinase (*Slu1*) localised at the plasma membrane. Under optimal environmental conditions, *Slu1* is active and inhibits signalling. Upon loss of turgor pressure, *Slu1* is inactivated, and this results in the activation of the nitrogen-activated protein (NAO) kinase cascade and phosphorylation of the NAP kinase (Hog1). Active Hog1 accumulates in the nucleus, where it affects gene expression. Two HOG target genes encode for enzymes involved in the synthesis of glycerol. Because of the presence of three hydroxylic functions that attract water clouds, glycerol serves as an osmolyte to increase the intra-cellular osmotic pressure [50]. Since glycerol is more reduced than the substrate glucose, its s ynthesis also affects the redox metabolism. Therefore, the redox metabolism needs to be adjusted accordingly. The pathway for the synthesis of glycerol is shown in Fig. 6.3. The NAD-dependent glycerol-3-phosphate dehydrogenase (Gpd) and the glycerol-3-phosphatase (Gpp) catalyse the two step reactions. *Saccharomyces cerevisiae* also possesses genes that might encode the enzyme glycerol dehydrogenase (*GCY1* and *YPR1*) and the dihydroxyacetone kinase (*DAK1* and *DAK2*). These two enzymes are involved in the pathway for glycerol degradation [50]. The pathway for glycerol via Gpd and Gpp converts NADH to NAD, while the conversion of glycerol to dihydroxyacetone phosphate via Gcy1p and Dak1p reduces NADP to NADPH. Hence, the glycerol-dihydroxyacetone phosphate cycle mainly acts as transhydrogenase for the interconversion of NADH to NADPH. Since stress conditions require high levels of NADPH to manage reactive oxygen species, this pathway assumes an important role. The capacity for other NADPH-generating reactions, such as those of the pentose phosphate pathway, is also increased under stress conditions. When salt-stressed yeast is inoculated in dough with high sugar content, the fermentation time was significantly reduced and bread-specific volume increased due to glycerol accumulation and a less dense gluten network. Moreover, two-step industrial fermentation, including a pre-adaptation to osmotic stress, in order to enhance flavour, has been proposed [51].

6.2.2.4 Membrane Lipids as Modulators of Stress Tolerance in Yeasts

Biological membranes are the barrier that separates cells from their environment and are the primary target for damage during environmental stress. Sudden changes of environmental conditions cause alteration in the organisation and structure of membrane lipids, and alter the function of many cellular activities. Homeoviscous adaptation to low temperature maintains the molecular order of membrane lipids as well as the activity of membrane-associated enzymes and transporters. To date, most of the research in this field has focused on the connection between the physical state of the membrane and cold and ethanol tolerance. In organisms producing ethanol, including yeasts, intra-cellular ethanol freely diffuses into an external medium. At high concentration, the ethanol present in the external medium acts as a chemical stress. A long-standing conundrum is the mechanism by which cells growing in the presence of a high concentration of ethanol modify their membrane composition. The plasma membrane lipid is the main site of impediment and interaction with ethanol. The altered lipid composition, following the exposure to ethanol, combats the deleterious effect. The ethanol-dependent modification of phospholipids' fatty acid composition was also shown in *S. cerevisiae* [52]. The addition of ethanol (0.5–1.5 M) to *S. cerevisiae* leads to a progressive decrease of the proportion of saturated fatty acids (SFAs) (mainly 16:0) and to a corresponding increase of the mono-unsaturated fatty acid residues, especially 18:1. This increased unsaturation favours an increase of the fluidity of the membrane. On the basis of similar observations in *Escherichia coli*, it was suggested that increased unsaturation was not due to the additional synthesis of unsaturated fatty acids (UFAs), but was the result of the decrease of levels of SFAs. An exception to this hypothesis was the increase of 18:0 in the presence of ethanol [52–54]. It is assumed that the presence of ethanol

Table 6.2 Total cell fatty acid composition in 48 h *S. cerevisiae* cells in relation to different incubation temperatures (15, 20, 30, and 40°C) [55]

	Strain							
	S. cerevisiae BG7FI				*S. cerevisiae* 635			
	Temperature (°C)				Temperature (°C)			
Fatty acid[a]	15	20	30	40	15	20	30	40
C6:0	0.00	0.38	0.02	0.00	0.00	0.00	0.00	0.00
C8:0	0.00	0.46	0.21	0.06	0.00	0.16	0.00	0.33
C10:0	1.44	6.47	2.05	0.49	1.06	2.55	6.06	0.76
C10:1	0.00	0.54	0.00	0.00	0.00	0.27	0.00	0.00
C12:0	3.72	7.86	8.11	3.72	3.43	8.59	18.21	3.09
C12:1	0.11	0.61	0.93	0.16	0.24	0.57	0.76	0.37
C14:0	4.02	2.08	1.52	1.63	5.28	4.45	3.56	1.90
C14:1	0.81	0.76	0.62	0.23	1.78	1.57	1.25	0.20
C16:0	33.19	37.05	41.28	38.34	25.84	32.57	33.77	33.73
C16:1	33.62	26.13	25.06	30.46	37.26	28.85	19.47	30.67
C18:0	8.13	8.73	11.19	12.83	6.00	7.17	8.55	11.31
C18:1n9	12.41	8.29	7.82	9.67	17.17	12.50	7.03	15.34
C18:1n11	0.44	0.24	0.51	2.01	0.48	0.35	0.25	1.64
C18:2	2.12	0.39	0.68	0.42	1.47	0.41	1.09	0.64
Unsaturation degree[b]	0.51	0.37	0.36	0.43	0.59	0.45	0.31	0.49
C16:1/C16:0	1.01	0.70	0.61	0.79	1.44	0.88	0.58	0.91
C18:1/C18:0	1.58	0.98	0.74	0.91	2.94	1.79	0.85	1.50
Chain length[c]	16.16	15.46	15.86	16.27	16.15	15.74	15.12	16.33

[a]Expressed as a percentage of total fatty acids
[b]Δ/mol = [1(mol%monoenes)/100 + 2(mol%dienes)/100 + 3(mol%trienes)]/100
[c]Σ(PC)/100, where P is percentage of a fatty acid and C is its carbon atom number

on both sides of the plasma membrane causes the separation of the two monolayers and the concurrent increased synthesis of C18 fatty acid residues, which helps in the maintenance of membrane integrity [52].

The decrease of the degree of fatty acid unsaturation along with the temperature increase is widely documented. The incorporation of exogenous oleic acid has been reported to increase the tolerance of yeasts to the lethal effect of temperatures between 41°C and 47°C [55]. Table 6.2 shows the modulation of the membrane fatty acid composition of two strains of *S. cerevisiae* in a temperature range from 15 to 40°C [55]. For both strains, the relationship between unsaturation level and temperature showed two maximum values at 15 and 40°C. The latter corresponds to the maximum fermentation temperature for these strains. These results suggest that the role of the unsaturation level has to be considered not only in terms of the contribution to membrane fluidity when temperature decreases but also as a mechanism to protect the cell membrane from damage generated by oxidative and thermal stresses [55, 56]. It was postulated that an oxygen-consuming desaturase prevents the accumulation of oxygen and reactive oxygen species such as H_2O_2 in yeasts. During sourdough fermentation, yeasts are exposed to H_2O_2 [57]. It has been observed that the co-inoculation of *S. cerevisiae* and *L. sanfranciscensis* into dough gives rise to the release of medium-chain fatty acids and of 2(5H)-furanones presumably originating from peroxidation of membrane-associated UFAs, followed by a sequence of β-oxidation reactions [58, 59]. The biosynthesis or integration of UFAs in the yeast membrane is reported as a mechanism to detoxify H2O2 and protect cells from oxidative stress [59] and this has already been observed in *L. helveticus* [60]. ROS with consequent peroxidation of membrane fatty acids has been observed in *S. cerevisiae* after a cycle of freezing/thawing [61].

Sterols are membrane-associated lipids that play an important role in the tolerance to physico-chemical stresses including ethanol exposure. The ability of sterols to increase stress tolerance is well documented [62]. Oxygen is required for the cyclization of squalene to sterols. Thus, the presence of oxygen during the early phase of yeast fermentation is considered a limiting factor for growth and fermenting due to the membrane-stabilising effect of sterols. Also the synthesis of UFAs is prevented by anaerobic conditions. When sterol and UFA biosynthesis is prevented by anaerobic conditions, yeasts integrate sterols and UFAs from the fermentation media in their membranes. Cereals contain a large variety of nutrients including phytosterols, which serve as a source of membrane sterols for yeasts also in anaerobic conditions.

6.3 Minor Yeast Metabolites in Sourdough

In addition to ethanol and ClO_2, yeasts generate a large spectrum of metabolites during sourdough fermentation. Yeasts are responsible for characterising the aroma profile, or so-called "fermented taste" in the production of bread and alcoholic beverages. This profile consists of a complex mixture of flavour compounds but the characteristic impact on flavour is determined mainly by fusel alcohols and their derivatives. Commonly known fusel alcohols are 3-methyl-1-butanol, 2-methyl-1-butanol and 2-methyl-1-propanol. When ammonia is used as the nitrogen source, fusel alcohols can be synthetized via the isoleucine, valine and leucine (ILV) pathway [63]. However, their concentration drastically increases when branched chain amino acids (BCAAs) are present in the media. Yeasts convert free amino acids mainly by the Ehrlich pathway [64] (Table 6.3). The Ehrlich pathway assumes the conversion of the BCAAs to fusel alcohols by three enzymatic steps. The first step is a transaminase, in which the amino group of BCAAs is transferred to 2-oxoglutarate, resulting in branched chain oxoacids and glutamate. The second step is the decarboxylation step that converts the branched chain oxoacid to the branched chain aldehydes. The last step is the reduction step in which branched chain aldehydes are reduced to branched chain alcohols or so-called fusel alcohols ([65] Fig. 6.5). For instance, *S. cerevisiae* and other sourdough yeasts convert leucine and phenylalanine to 3-methylbutanol and 2-phenylethanol, respectively [64]. The Strecker reaction during baking also generates α-dicarbonyl compounds such as methylglyoxal (2-oxopropanal), which leads to the corresponding aldehydes and acids [66].

The presence of yeasts (*Saccharomyces* and *Hansenula* genera) during sourdough fermentation favoured an increased synthesis of alcohols, esters and some carbonyl compounds compared to sourdough fermentation without the addition of yeasts [67].

Table 6.3 Higher alcohols and their precursor amino acids

Amino acids	Ketoacids	Aldehydes	Organic acids	Alcohols	Esters
Ile	α-Keto-3-methyl-pentanoic acid	2-Methylbutanal	2-Methyl butyric acid	2-Methylbutanol	
Leu	α-Ketoisocaproic acid	3-Methylbutanal; 2-Methylpropanal	3-Methylbutyric acid 2-Methyl propanoic acid	3-Methylbutanol 2-Methylpropanol	Ethyl-3-methylbutanoate
Val	α-Ketoisovaleric acid	2-Methylpropanal	2-Methyl propanoic acid	2-Methylpropanol	Ethyl isobutanoate
Phe	Phenyl pyruvate	Benzaldehyde	Benzoic acid	Phenylmethanol	Ethyl benzoate
		Phenylacetaldehyde	Phenylacetic acid	Phenylethanol	Phenylethyl-acetate
Trp	Indole-3-pyruvate	Indole-3-acetaldehyde	Indole-3-acetic acid		
Met	α-Keto methylthio butyrate	Methional	Methylthiobutyric acid	Methionol	Ethyl-3-methylthio propionate
		Methylthio-acetaldehyde		Methanethiol	Methylthioacetate

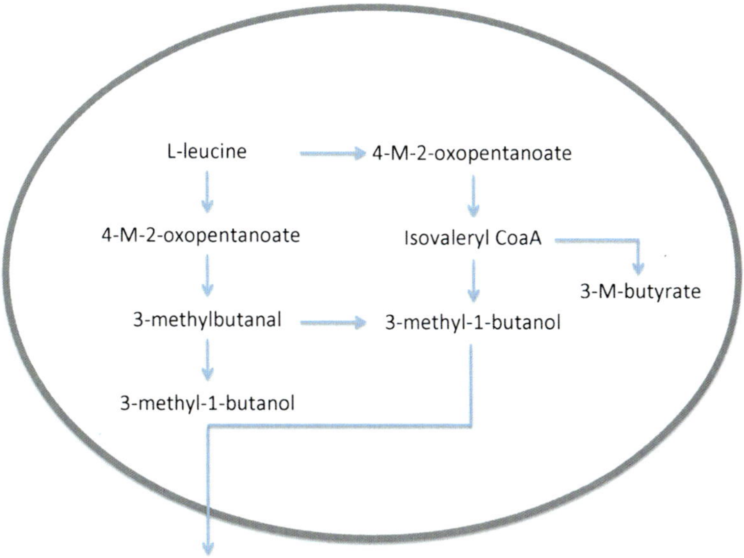

Fig. 6.5 Schematic representation of metabolic routes of L-leucine in yeasts, leading to 4-methyl-2-oxopentanoate, 3-methyl-1-butanol and 3-methylbutyrate (Adapted from [65])

Especially, the levels of ethanol, methylpropanol, 2- and 3-methylbutanol as well as of ethyl acetate and diacetyl were considerably increased in sourdoughs with the addition of yeasts. Diacetyl is synthesised either by lactic acid bacteria or yeasts [68]. Although it is difficult to distinguish between the contribution of the two groups of microorganisms, processing parameters such as high water content and temperature may favour the growth of yeasts [69–71] and, consequently, their contribution to the synthesis of volatile compounds during fermentation. There was a clear difference between the volatile compound profiles of wheat dough fermented with *Dekkera bruxellensis* compared to *S. cerevisiae*. Contrarily to *S. cerevisiae*, *Dekkera* sp. did not generate the same levels of 3-methylbutanol but formed high concentrations of esters such as the odour-intense ethyl 2-methylbutanoate [72].

Table 6.4 shows the metabolites synthesised by *L. sanfranciscensis* and *S. cerevisiae* in doughs with different dough yields [51]. When the dough was inoculated with *L. sanfranciscensis*, isobutanol, acetoin, 1-hexanol, ethyl octanoate and butyric acid were found. When the dough was inoculated with *S. cerevisiae* at a dough yield value of 146, the highest synthesis of volatile compounds was found, which included alcohols and short- and medium-chain fatty acids.

6.4 Quorum Sensing on Sourdough Yeasts

Several reports have shown quorum sensing-like phenomena in fungal species. The morphological transition, from the filamentous and mycelial form to the yeast form, or vice versa, was always found. The regulation of the switch between the filamentous

Table 6.4 Selected metabolites produced by *L. sanfranciscensis* and *S. cerevisiae* inoculated in dough with various dough yields (data expressed as peak chromatographic area) [51]

	L. sanfranciscensis		*S. cerevisiae*	
	DY 220	DY 146	DY 220	DY 146
Compound				
Ethanol	7.72	7.80	8.10	8.34
Isobutanol	6.19	7.46	7.15	7.31
Isoamyl alcohol	7.60	7.47	7.77	7.97
Acetoin	6.42	6.69	5.63	6.19
2,3-Butanediol	n.d.	n.d.	5.55	5.35
Butyric acid	6.42	6.48	6.59	6.57
Ethyl decanoate	n.d.	n.d.	6.12	5.44
Ethyl-9-decenoate	5.78	6.29	n.d.	n.d.
Hexanoic acid	7.64	7.34	6.79	6.95
Phenylethanol	6.23	6.28	7.87	7.93
Ethyl hexadecanoate	n.d.	n.d.	6.17	6.41

Standard deviation values of the three repetitions in this experiment were less than 10%
n.d. Below detection limit

and yeast forms of the parasitic fungus *Histoplasma capsulatum* provided the first example of an apparent quorum-sensing mechanism in eukaryotes [73].

Saccharomyces cerevisiae may show the morphological transition from the yeast to the filamentous forms in response to environmental cues. Growth under nitrogen-limiting conditions is an example, which activates several signal transduction pathways. One of these routes is the Ras-cAMP-dependent protein kinase (PKA) pathway. Chen and Fink [74] provided some intriguing molecular clues to PKA signalling. The conditioned medium of *S. cerevisiae* stationary-phase cultures markedly induces the filamentous growth. Induction was partially related to the fivefold stimulation of the transcription of *FLO11*, a gene essential for filamentous growth. The active molecules were purified and were found to be phenylethanol and tryptophol, two aromatic alcohols derived from phenylalanine and trytophan, respectively, which are always present in yeast-fermented foods. The addition of both alcohols causes the more vigorous filamentous growth and the induction of *FLO11*. On the basis of the bacterial paradigm, these compounds have the features of quorum-sensing molecules. The synthesis of these molecules is the highest during the stationary phase of growth, the addition of tryptophol induces the expression of genes (*ARO9* and *ARO10*), which are required to convert tryptophan to tryptophol, and mutants unable to synthesise these alcohols show a markedly decreased filamentous growth and *FLO11* expression.

Preliminary results on inter-species communication during sourdough fermentation have been reported [51]. As shown in Fig. 6.6, the exposure of *C. milleri* to the cell-free wheat flour hydrolyzed (WFH) medium, previously inoculated with *L. sanfranciscensis*, induced morphological changes and early autolysis of the yeast.

Figure 6.7 shows another possible effect of the inter-species signalling mechanism, which is mediated by lactic acid bacteria metabolites. After 2 h of exposure to

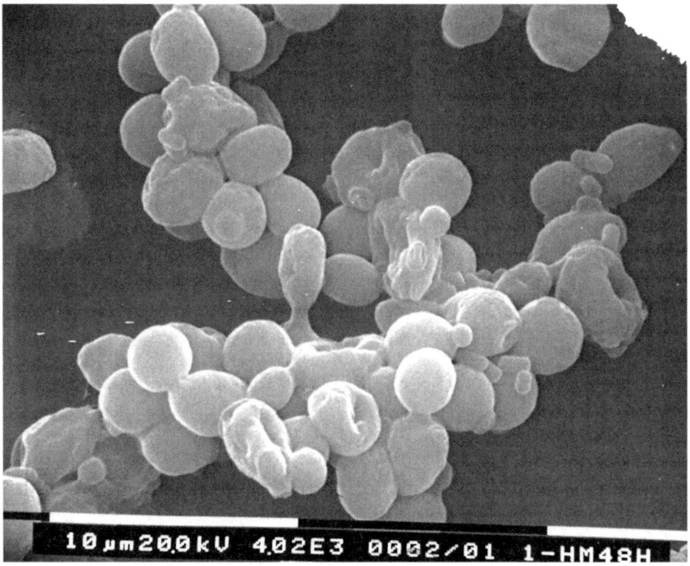

Fig. 6.6 Cells of *S. cerevisiae* exposed for 4 h to cell-free liquid Wheat Flour Hydrolyzed (WFH) previously inoculated with *L. sanfranciscensis*

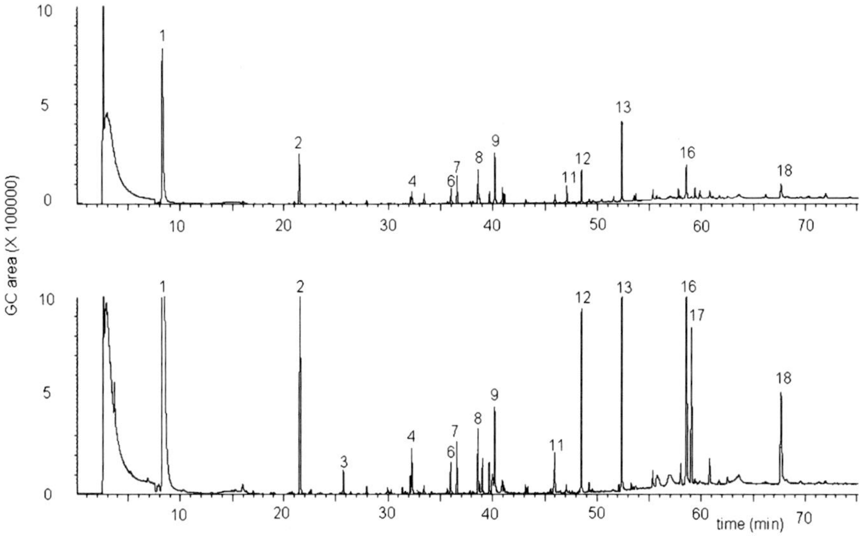

Fig. 6.7 GC-SPME profile of volatile compounds produced by *C. milleri* when exposed for 4 h to cell-free CM in which *L. sanfranciscensis* (LSCE1) had been incubated for 4 h in WFH with (**a**) and without starch (**b**). *1* Ethanol, *2* isoamyl alcohol, *3* acetoin, *4* ethyl octanoate, *5* acetic acid, *6* 1-octanol, *7* isobutyric acid, *8* butyric acid, *9* isovaleric acid, *10* ethyl-9-decenoate, *11* hexanoic acid; *12* phenylethanol, *13* octanoic acid, *14* c-octalactone, *15* c decalactone, *16* decanoic acid, *17* ethyl-9-hexadecenoate, and *18* dodecanoic acid [51]

the cell-free wheat flour hydrosylate fermented *L. sanfranciscensis*, *C. milleri* released metabolites belonging to the family of lactones. Glutamic acid was suggested as the presumptive precursor of certain γ-lactones in yeasts [75]. Δ-Lactones were obtained from 11-hydroxy fatty acids during fermentation with *S. cerevisiae*, *Sporobolomyces*, *Geotrichuum* and *Candida* spp.

6.5 Baker's Yeast in the Bread-Making Industry

The production of baker's yeast biomass represents a highly competitive multi-billion dollar global industry. The large variety of bread-making processes and recipes used around the world place considerable demands on baker's yeasts. These demands translate into technological and economic challenges for baker's yeast industries. During production and dough fermentation, cells of baker's yeasts are exposed to environmental stresses such as freeze-thaw, dehydration and oxidation. Damage to cell macromolecules (e.g. proteins, nucleic acids and membranes) is avoided through microbial adaptation and tolerance to multiple stress conditions [76].

6.5.1 Baker's Yeast Production

The first stage in the production of baker's yeast consists of growing the yeast from a pure culture in a series of fermentation vessels. Baker's yeast is recovered from the last step of fermentation through centrifugation to concentrate the biomass. Subsequently, filtration further concentrates the yeast cells. The collected pellet is blended in mixers with small amounts of water, emulsifiers, and oil. Finally, the mixed press cake is extruded, cut and either wrapped for shipment or dried to form dry yeast.

6.5.1.1 Raw Materials

The species *S. cerevisiae* is used for producing compressed yeast. Different strains of *S. cerevisiae* are required to produce each of the two dry yeast products, Active Dry Yeast (ADY) and Instant Dry Yeast (IDY). Instant dry yeast is produced from a faster-reacting yeast strain than that used for ADY. Cane molasses and beet molasses are the principal substrates for yeast growth. Molasses contains 45–55% (wt/wt) of fermentable carbohydrates (sucrose, glucose and fructose). The amount and type of cane and beet molasses used depend on the types and costs, and on the presence of inhibitors and toxins. Usually, a blend consisting of both cane and beet molasses is used. Once the molasses mixture is blended, the pH is adjusted to 4.5–5.0 to prevent bacterial growth. The substrate is clarified to remove sludge and is then sterilised with high-pressure steam. After sterilisation, it is diluted with water and held in tanks until use. Additional nutrients, including nitrogen, potassium, phosphate, magnesium and calcium, with traces of iron, zinc, copper, manganese and molybdenum, are added. Usually, nitrogen is supplied by adding ammonium salts, aqueous ammonia or anhydrous ammonia. Phosphates and magnesium are added in the form of phosphoric acid or phosphate salts and magnesium salts. Vitamins such as biotin, inositol, pantothenic acid and thiamine are also required. The latter is added to the feedstock, while the others are already present in the molasses malt.

6.5.1.2 Fermentation

Yeast cells are grown in a series of fermentations to decrease the synthesis of ethanol and CO_2. The fermentation is operated under aerobic conditions. The initial stage of propagation takes place in the laboratory. A portion of the pure yeast culture is mixed with molasses malt in a sterilised flask, and the yeast is allowed to grow for 2–4 days. The entire content of the flask is used to inoculate the first fermentation vessel. Batch fermentations are carried out, where the yeast is allowed to grow for 13–24 h. Usually, one or two vessels are used for this stage of the process. The batch fermentations are essentially a propagation of the flask fermentation, except for the use of sterile aeration and aseptic transfer for the subsequent stage. Following pure culture fermentations, the yeast mixture is transferred into an intermediate vessel under batch or fed-batch conditions. The following fermentation is a stock fermentation stage (Fig. 6.8). The content from the intermediate vessel is pumped into the stock vessel, which is equipped for increasing feeding under aeration. This stage is termed stock fermentation, since after the fermentation is completed, the yeast biomass is separated from the bulk of the vessel through centrifugation, which produces a stock of yeast for the next stage. The next stage, pitch fermentation, also produces a stock of yeast. Aeration is vigorous, and molasses and other nutrients are increasingly fed into the vessel. The liquor from this vessel is usually shared into several parts for pitching the final trade fermentation. Alternatively, the yeast is separated by centrifugation and stored for several days before use. The final trade fermentation has the highest degree of aeration. A large air supply is required and the vessels are often started in a staggered fashion to reduce the size of the air compressors. The initial fermentation lasts 11–15 h. After which the molasses have been fed into the vessel, the liquid is aerated for 0.5–1.5 h to allow a further maturation of the yeast. This step has a stabilising effect for the subsequent refrigerated storage. The content of the yeast biomass increases stage by stage: ca. 120 kg in the intermediate vessel, 420 kg in the stock vessel, 2,500 kg in the pitch vessel, and 15,000–100,000 kg in the trade vessel. Once the optimum quantity of yeast is grown, the biomass is recovered from the final trade vessel by centrifugation. The centrifuged yeast biomass is further concentrated by a filter press or rotary vacuum filter. The filter press forms a filter cake containing 27–32% of solids. The rotary vacuum filter forms a cake with ca. 33% of solids. The filter cake is then blended in mixers with small amounts of water, emulsifiers, and cutting oils to form the end-product. The final packaging steps vary depending on the type of yeast product.

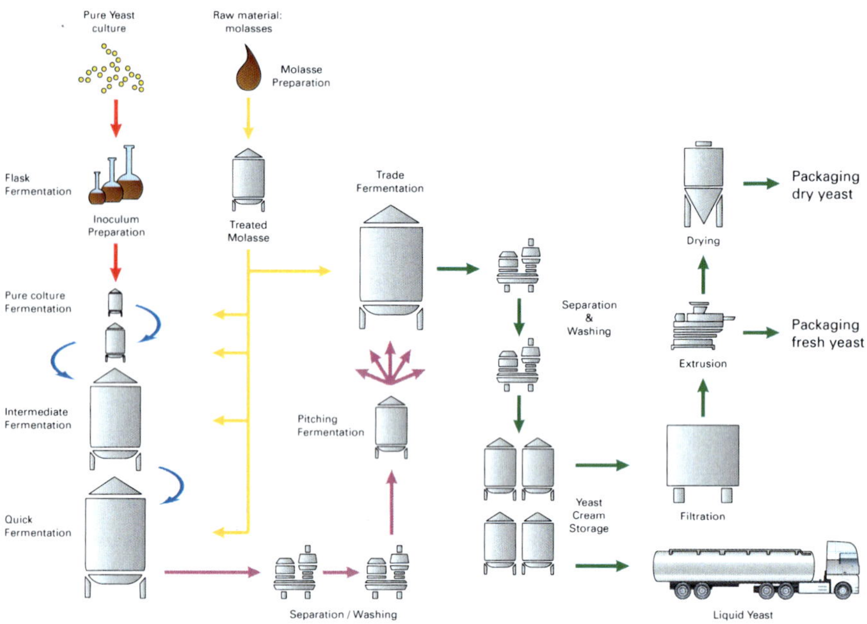

Fig. 6.8 Process layout of the different commercial baker's yeast formulas

For the compressed yeast formula, emulsifiers are added giving the yeast a white, creamy appearance and inhibiting water spotting of the yeast cake. A small amount of oil, usually soybean or cottonseed oil, is added to help the extrusion of the yeast through nozzles to form continuous ribbons of yeast cake. The ribbons are cut, and the yeast cake is wrapped and cooled to less than 8°C, being ready for the market under refrigerated conditions. For dry yeast preparation, after filtration the biomass is sent to an extruder, where emulsifiers and oils (different from those used for compressed yeast) are added to texturise it and to help extrusion. After extrusion in thin ribbons, the yeast is cut and dried under batch or continuous drying systems. Following drying, the yeast is vacuum-packed or packed under nitrogen gas before heat sealing. The shelf-life at room temperature is 1–2 years.

6.5.2 General Characteristics of Fresh and Dry Baker's Yeast

Baker's yeast consists of living cells of *S. cerevisiae*. To describe the characteristics of baker's yeast, two types of formula are considered: fresh and dry yeast.

6.5.2.1 Fresh Baker's Yeast

Fresh baker's yeast is commercialised as block or compressed yeast, granulated yeast and liquid (cream) yeast. Compressed yeast is available in blocks with variable consistency. The consistency varies from high plasticity (kneadable) to friable (crumbly texture). Granulated yeast is in the form of small granules. Liquid yeast is a suspension of yeast cells in water with a cream-like viscosity. The dry matter content of fresh baker's yeast varies depending on the formula, the fermentation performance, and the consistency/friability. Liquid and compressed yeast with high plasticity and friability have a dry matter content of 15–21, 27–31 and 30–35%, respectively. The dry matter of granulated yeast is 31–37%. The nitrogen content of the dry matter has a value of ca. 8.0%. The ash content of the dry matter is ca. 6%, and the value of pH is usually ca. 5.0. Fresh baker's yeast is suitable to be used directly in dough. As it is a perishable product, it should be stored under refrigerated conditions.

6.5.2.2 Dry Baker's Yeast

Dry baker's yeast is sold as active dry yeast (ADY) and instant dry yeast (IDY). ADY is characterised by spherical particles with a diameter of 0.2–3 mm. The yeast is reactivated under warm water at ca. 38°C (not exceeding 45°C) and then used as fresh yeast. IDY consists of porous cylindrical particles with a diameter of 0.5 mm and length up to a few millimetres. Rehydration in water is not necessary and it may be added directly to the flour. Depending on the activity and recipe, IDY is added to the flour at a level that corresponds to one-third of that of fresh yeast. Direct contact of the yeast with salt, fat or sugar has to be avoided to prevent osmotic stress and dispersion during rehydration of the flour. Careful rehydration of the flour with water at room temperature is needed. The dry matter ranges from 92–96 to 93–97% for ADI and IDY, respectively. The density is 75–95 and 55–80% for ADI and IDY, respectively. The nitrogen content of the dry matter has a value of ca. 8.0% and the ash content of the dry matter is ca. 6%,with a pH of usually ca. 6.0.

References

1. Gobbetti M (1998) The sourdough microflora: interaction of lactic acid bacteria and yeasts. Trends Food Sci Technol 9:267–74
2. Mäntynen VH, Korhola M, Gudmundsson H, Turakainen H, Alfredsson GA, Salovaara H, Lindström K (1999) A polyphasic study on the taxonomic position of industrial sour dough yeasts. Syst Appl Microbiol 22:87–96
3. Pulvirenti A, Solieri L, Gullo M, De Vero L, Giudici P (2004) Occurrence and dominance of yeast species in sourdough. Lett Appl Microbiol 38:113–117
4. Rocha JM, Malcata FX (1999) On the microbiological profile of traditional Portuguese sourdough. J Food Protect 62:1416–1429

5. Gullo M, Romano AD, Pulvirenti A, Giudici P (2002) *Candida humilis*–dominant species I sourdoughs for the production of durum wheat bran flour bread. Int J Food Microbiol 80:55–59
6. Meroth CB, Hammes WP, Hertel C (2003) Identification and population dynamics of yeasts in sourdough fermentation processes by PCR-denaturing gradient gel electrophoresis. Appl Environ Microbiol 69:7453–7461
7. Gänzle MG, Ehmann MA, Hammes WP (1998) Modelling of growth of *Lactobacillus sanfranciscensis* and *Candida milleri* in response to process parameters of the sourdough fermentation. Appl Environ Microbiol 64:2616–2623
8. Succi M, Reale A, Andrighetto C, Lombardi A, Sorrentino E, Coppola R (2003) Presence of yeasts in southern Italy sourdoughs from *Triticum aestivum* flour. FEMS Microbiol Lett 225:143–148
9. Valmorri S, Tofalo R, Settanni L, Corsetti A, Suzzi G (2010) Yeast microbiota associated with spontaneous sourdough fermentations in the production of traditional wheat sourdough breads of the Abruzzo region (Italy). Antonie Van Leeuwenhoek 97:119–129
10. Walker GM (1998) Yeast physiology and biotechnology. Wiley, Hoboken
11. Gianotti A, Vannini L, Gobbetti M, Corsetti A, Gardini F, Guerzoni ME (1997) Modelling of the activity of selected starters during sourdough fermentation. Food Microbiol 14:327–337
12. Wehrle K, Arendt EK (1998) Rheological changes in wheat sourdough during controlled and spontaneous fermentation. Cereal Chem 75:882–886
13. Gasch AP, Spellman PT, Kao CM, Carmel-Harel O, Eisen MB, Storz G, Botstein D, Brown PO (2000) Genomic expression programs in the response of yeast cells to environmental changes. Mol Biol Cell 11:4241–4257
14. Gasch AP (2002) The environmental stress response: a common yeast response to environmental stresses. In: Hohmann S, Mager P (eds) Yeast stress responses, vol 1. Springer, Heidelberg, pp 11–70
15. Saldanha AJ, Brauer MJ, Botstein D (2004) Nutritional homeostasis in batch and steady-state culture of yeast. Mol Biol Cell 15:4089–4104
16. Marzluf GA (1997) Genetic regulation of nitrogen metabolism in the fungi. Microbiol Mol Biol Rev 61:17–32
17. Levy S, Ihmels J, Carmi M, Weinberger A, Friedlander G, Barkai N (2007) Strategy of transcription regulation in the budding yeast. PLoS One 2:e250
18. Paramithiotis S, Gioulatos S, Tsakalidou E, Kalantzopoulos G (2006) Interactions between *Saccharomyces cerevisiae* and lactic acid bacteria in sourdough. Process Biochem 41:2429–2433
19. Vernocchi P, Valmorri S, Gatto V, Torriani S, Gianotti A, Suzzi G, Guerzoni ME, Gardini F (2004) A survey on yeast microbiota associated with an Italian traditional sweet-leavened baked good fermentation. Food Res Int 37:469–476
20. Thiele C, Gänzle MG, Vogel RF (2002) Contribution of sourdough lactobacilli, yeast, and cereal enzymes to the generation of amino acids in dough relevant for bread flavour. Cereal Chem 79:45–51
21. Brandt MJ, Hammes WP, Gänzle MG (2004) Effects of process parameters on growth and metabolism of *Lactobacillus sanfranciscensis* and *Candida humilis* during rye sourdough fermentation. Eur Food Res Technol 218:333–338
22. Lien-Te Y, Mei-Li W, Albert Linton C, Tzou-Chi HL (2009) A novel steamed bread making process using salt-stressed baker's yeast. Int J Food SciTech 44:2637–2643
23. Verduyn C, Zomerdijk TPL, Van Dijken JP, Scheffers WA (1984) Continuous measurement of ethanol production by aerobic yeast suspension with an enzyme electrode. Appl Microbiol Biotechnol 19:181–185
24. Kappeli O (1986) Regulation of carbon metabolism in *Saccharomyces cerevisiae* and related yeasts. Adv Microb Physiol 28:181–209
25. Ozcan S, Johnston M (1999) Function and regulation of yeast hexose transporters. Microbiol Mol Biol Rev 63:554–569

26. Reifenberger E, Boles E, Ciriacy M (1997) Kinetic characterization of individual hexose transporters of *Saccharomyces cerevisiae* and their relation to the triggering mechanisms of glucose repression. Eur J Biochem 245:324–333
27. Barnett JA, Entian KD (2005) A history of research on yeasts 9: regulation of sugar metabolism. Yeast 22:835–894
28. Rolland F, Winderickx J, Thevelein JM (2002) Glucose-sensing and–signalling mechanisms in yeast. FEMS Yeast Res 2:183–201
29. Zaman S, Lippman SI, Zhao X, Broach JR (2008) How *Saccharomyces* responds to nutrients. Annu Rev Genet 42:27–81
30. Santangelo GM (2006) Glucose signaling in *Saccharomyces cerevisiae*. Microbiol Mol Biol Rev 70:253–282
31. Turcotte B, Liang XB, Robert F, Soontorngun N (2009) Transcriptional regulation of nonfermentable carbon utilization in budding yeast. FEMS Yeast Res 10:2–13
32. Fendt SM, Sauer U (2010) Transcriptional regulation of respiration in yeast metabolizing differently repressive carbon substrates. BMC Syst Biol 4:12
33. Goddard MR (2008) Quantifying the complexities of *Saccharomyces cerevisiae*'s ecosystem engineering via fermentation. Ecology 89:2077–2082
34. Rokhlenko O, Wexler Y, Yakhini Z (2006) Similarities and differences of gene expression in yeast stress conditions. Bioinformatics 23:184–190
35. Ball CA, Dolinski K, Dwight SS, Harris MA, Issel-Tarver L, Kasarskis A, Scafe CR, Sherlock G, Binkley G, Jin H, Kaloper M, Orr SD, Schroeder M, Weng S, Zhu Y, Botstein D, Cherry JM (2000) Integrating functional genomic information into the *Saccharomyces* genome database. Nucleic Acids Res 28:77–80
36. Bose S, Dutko JA, Zitomer RS (2005) Genetic factors that regulate the attenuation of the general stress response of yeast. Genetics 169:1215–1226
37. Barber B, Ortola C, Barber S, Fernandez F (1992) Storage of packaged white bread. III. Effects of sourdough and addition of acids on bread characteristics. Z Lebensm Unters Forsch 194:442–449
38. Serrazanetti DI, Guerzoni ME, Corsetti A, Vogel RF (2009) Metabolic impact and potential exploitation of the stress reactions in lactobacilli. Food Microbiol 26:700–711
39. Hüfner E, Britton RA, Roos S, Jonsson H, Hertel C (2008) Global transcriptional response of *Lactobacillus reuteri* to the sourdough environment. Syst Appl Microbiol 31:323–338
40. Häggman M, Salovaara H (2008) Effect of fermentation rate on endogenous leavening of *Candida milleri* in sour rye dough. Food Res Int 41:266–273
41. Tai SL, Daran-Lapujade P, Walsh MC, Pronk JT, Daran JM (2007) Acclimation of *Saccharomyces cerevisiae* to low temperature: a chemostat-based transcriptome analysis. Mol Biol Cell 18:5100–5112
42. Sahara T, Goda T, Ohgiya S (2002) Comprehensive expression analysis of time-dependent genetic responses in yeast cells to low temperature. J Biol Chem 277:50015–50021
43. Schade B, Jansen G, Whiteway M, Entian KD, Thomas DY (2004) Cold adaptation in budding yeast. Mol Biol Cell 15:5492–5502
44. Homma T, Iwahashi H, Komatsu Y (2003) Yeast gene expression during growth at low temperature. Cryobiology 46:230–237
45. Murata Y, Homma T, Kitagawa E, Momose Y, Sato MS, Odani M, Shimizu H, Hasegawa-Mizusawa M, Matsumoto R, Mizukami S, Fujita K, Parveen M, Komatsu Y, Iwahashi H (2006) Genome-wide expression analysis of yeast response during exposure to 4 degrees C. Extremophiles 10:117–128
46. Castrillo JI, Zeef LA, Hoyle DC, Zhang N, Hayes A, Gardner DC, Cornell MJ, Petty J, Hakes L, Wardleworth L, Rash B, Brown M, Dunn WB, Broadhurst D, O'Donoghue K, Hester SS, Dunkley TP, Hart SR, Swainston N, Li P, Gaskell SJ, Paton NW, Lilley KS, Kell DB, Oliver SG (2007) Growth control of the eukaryote cell: a systems biology study in yeast. J Biol 6:4.
47. Kandror O, Bretschneider N, Kreydin E, Cavalieri D, Goldberg AL (2004) Yeast adapt to near-freezing temperatures by STRE/Msn2,4-dependent induction of trehalose synthesis and certain molecular chaperones. Mol Cell 13:771–781

48. De Melo HF, Bonini BM, Thevelein J, Simões DA, Morais MA Jr (2010) Physiological and molecular analysis of the stress response of *Saccharomyces cerevisiae* imposed by strong inorganic acid with implication to industrial fermentations. J Appl Microbiol 109:116–127
49. Peres MFS, Tininis CRCS, Souza CS, Walker GM, Lalule C (2005) Physiological responses of pressed baker's yeast cells pre-treated with citric, malic and succinic acids. World J Microb Biot 21:537–543
50. Hohmann S (2002) Osmotic stress signaling and osmoadaptation in yeasts. Microbiol Mol Biol Rev 66:300–372
51. Vernocchi P, Ndagijimana M, Serrazanetti DI, Gianotti A, Vallicelli M, Guerzoni ME (2008) Influence of starch addition and dough microstructure on fermentation aroma production by yeasts and lactobacilli. Food Chem 108:1217–1225
52. Beavan MJ, Charpentier C, Rose AH (1982) Production and tolerance of ethanol in relation to phospholipid fatty-acyl composition in *Saccharomyces cerevisiae* NCYC 431. J Gen Microbiol 128:1447–1455
53. Ingram LO (1976) Adaptation of membrane lipids to alcohols. J Bacteriol 125:670–678
54. Buttke TM, Ingram LO (1978) Mechanism of ethanol-induced changes in lipid composition of *Escherichia coli*: inhibition of saturated fatty acid synthesis in vivo. Biochemistry 17:637–644
55. Sinigaglia M, Gardini F, Guerzoni ME (1993) Relationship between thermal behaviour, fermentation performance and fatty acid composition in two strains of *Saccharomyces cerevisiae*. Appl Microbiol Biotechnol 39:593–598
56. Guerzoni ME, Ferruzzi M, Sinigaglia M, Criscuoli GC (1997) Increased cellular fatty acid desaturation as a possible key factor in thermotolerance in *Saccharomyces cerevisiae*. Can J Microbiol 43:569–576
57. Vermeulen N, Gänzle MG, Vogel RF (2007) Glutamine deamidation by cereal-associated lactic acid bacteria. J Appl Microbiol 103:1197–1205
58. Ndagijimana M, Vallicelli M, Cocconcelli PS, Cappa F, Patrignani F, Lanciotti R, Guerzoni ME (2006) Two 2[5 H]-furanones as possible signaling molecules in *Lactobacillus helveticus*. Appl Environ Microbiol 72:6053–6061
59. Guerzoni ME, Vernocchi P, Ndagijimana M, Gianotti A, Lanciotti R (2007) Generation of aroma compounds in sourdough: effects of stress exposure and lactobacilli–yeasts interactions. Food Microbiol 24:139–148
60. Guerzoni ME, Lanciotti R, Cocconcelli PS (2001) Alteration in cellular fatty acid composition as a response to salt, acid, oxidative and thermal stresses in *Lactobacillus helveticus*. Microbiology 147:2255–2264
61. Rodriguez-Vargas S, Sanchez-Garcia A, Martinez-Rivas JM, Prieto JA, Randez-Gil F (2007) Fluidization of membrane lipids enhances tolerance of freezing and salt stress by *Saccharomyces cerevisiae*. Appl Environ Microbiol 73:110–116
62. Thomas SD, Hossack JA, Rose AH (1978) Plasma-membrane lipid composition and ethanol tolerance in *Saccharomyces cerevisiae*. Arch Microbiol 117:239–245
63. Derrick S, Large PJ (1993) Activities of the enzymes of the Ehrlich pathway and formation of branched-chain alcohols in *Saccharomyces cerevisiae* and *Candida utilis* grown in continuous culture on valine or ammonium as sole nitrogen source. J Gen Microbiol 139:2783–2792
64. Schieberle P (1996) Intense aroma compounds—useful tools to monitor the influence of processing and storage on bread aroma. Adv Food Sci 18:237–244
65. Schoondermark-Stolk SA, Jansen M, Veurink JH, Verkleij AJ, Verrips CT, Euverink GJ, Boonstra J, Dijkhuizen L (2006) Rapid identification of target genes for 3-methyl-1-butanol production in Saccharomyces cerevisiae. Appl Microbiol Biotechnol 70:237–246
66. Hofmann T, Schieberle P (2000) Formation of aroma-active Strecker-aldehydes by a direct oxidative degradation of Amadori compounds. J Agric Food Chem 48:4301–4305
67. Damiani P, Gobbetti M, Cossignani L, Simonetti MS, Rossi J (1996) The sourdough microflora. Characterization of hetero- and homofermentative lactic acid bacteria, yeasts and their interactions on the basis of the volatile compounds produced. Lebensm Wiss Technol 29:63–70

68. Torner MJ, Martinez-Anaya MA, Antuna B, de Barber CB (1992) Headspace flavor compounds produced by yeasts and lactobacilli during fermentation of preferments and bread doughs. Int J Food Microbiol 15:145–152
69. Gobbetti M, Simonetti MS, Corsetti A, Santinelli F, Rossi J, Damiani P (1995) Volatile compound and organic acid production by mixed wheat sour dough starters: influence of fermentation parameters and dynamics during baking. Food Microbiol 12:497–507
70. Hansen J, Kielland-Brandt MC (1996) Modification of biochemical pathways in industrial yeasts. J Biotechnol 49:1–12
71. Lund B, Hansen A, Lewis MJ (1989) The influence of dough yield on acidification and production of volatiles in sour doughs. Food Sci Technol 22:150–153
72. Hansen A, Schieberle P (2005) Generation of aroma compounds during sourdough fermentation: applied and fundamental aspects. Trends Food Sci Tech 16:85–94
73. Kügler S, Sebghati TC, Eissenberg LG, Goldmanm WE (2000) Phenotypic variation and intracellular parasitism by *Histoplasma capsulatum*. Proc Natl Acad Sci 97:8794–8798
74. Chen H, Fink GR (2006) Feedback control of morphogenesis in fungi by aromatic alcohols. Genes Dev 120:1150–1161
75. Wurz REM, Kepner RE, Webb AD (1988) The biosynthesis of certain gamma-lactones from glutamic acid by film yeast activity on the surface of flor sherry. Am J Enol Vitic 39:234–238
76. Shima J, Takagi H (2009) Stress-tolerance of baker's-yeast (*Saccharomyces cerevisiae*) cells: stress-protective molecules and genes involved in stress tolerance. Biotechnol Appl Biochem 53:155–164

Chapter 7
Physiology and Biochemistry of Lactic Acid Bacteria

Michael Gänzle and Marco Gobbetti

7.1 Introduction

In the past decades, studies on the physiology and biochemistry of sourdough lactic acid bacteria provided insight into the microbial ecology of sourdough as well as the effect of the metabolic activity of lactic acid bacteria on flavor, texture, shelf-life, and nutritional properties of leavened baked goods. Lactic acid bacteria are the dominant microorganisms of sourdough. Their metabolic versatility favors adaptation to the various processing conditions and the metabolic interactions with autochthonous yeasts determine mechanisms of proto-cooperation during sourdough fermentation [1–3]. *Lactobacillus* species are most frequently found in sourdough fermentations although species belonging to the genera *Pediococcus*, *Enterococcus*, *Lactococcus*, *Weissella* and *Leuconostoc* were also identified ([4–6], see Chap. 5). A large number of *Lactobacillus* species were first identified from sourdoughs or fermentation processes of cereals [5]. This chapter gives an overview of the general growth and stress parameters, carbohydrate and amino acid metabolism, synthesis of exopolysaccharides and antimicrobial compounds, and the conversion of phenolic compounds and lipids of lactic acid bacteria during sourdough fermentation.

M. Gänzle (✉)
Department of Agricultural, Food and Nutritional Science, University of Alberta
Edmonton, Canada
e-mail: mgaenzle@ualberta.ca

M. Gobbetti
Department of Soil, Plant and Food Science, University of Bari Aldo Moro, Bari, Italy

M. Gobbetti and M. Gänzle (eds.), *Handbook on Sourdough Biotechnology*,
DOI 10.1007/978-1-4614-5425-0_7,

7.2 General Growth and Stress Parameters

The large diversity of lactic acid bacteria associated with sourdough fermentation (Chap. 5), is matched by a comparable diversity of general growth and stress parameters. Different processes of sourdough fermentation select for organisms with different growth parameters. Generally, type I sourdoughs, which are characterized by frequent back-slopping to achieve leavening without addition of baker's yeast, select for organisms growing rapidly in cereal substrates. Under these conditions, *Lactobacillus sanfranciscensis* is often found as the predominant organism. Type II sourdoughs, which are characterized by long fermentation times and high fermentation temperatures, select for acid-tolerant organisms and *L. pontis*, *L. fermentum*, *L. reuteri* and related organisms are frequently found [6–8].

This effect of fermentation parameters on sourdough microbiota is reflected by the response of the growth of sourdough lactic acid bacteria to pH, temperature, and NaCl concentration. Mathematical models for the growth of sourdough lactic acid bacteria were developed for *L. sanfranciscensis* [9], *L. pontis* [10], *L. amylovorus* [10, 11] and *L. plantarum* [12]. The optimum temperature for growth is species specific; mesophilic organisms grow optimally between 30 and 35°C, while thermophilic organisms do not grow at ambient temperature (25°C) and grow optimally between 40 and 45°C [9–11, 13]. Traditional sourdough fermentations in Europe are typically carried out in the temperature range of 25–35°C (Chap. 4) and are thus dominated by mesophilic lactic acid bacteria. Many industrial processes and cereal fermentations in tropical climates are conducted at higher temperatures and accordingly select for thermophilic lactic acid bacteria [7, 8].

The optimum pH of sourdough lactic acid bacteria is typically between 5.0 and 6.0 [9–11], matching the pH of sourdough after inoculation with 5–20% of a previous batch of sourdough. Remarkably, the pH of wheat flour, about 6.2–6.5, is close to the maximum pH permitting growth of *L. sanfranciscensis,* pH 6.7 [9]. The minimum pH of growth of *L. sanfranciscensis* and *L. pontis* was determined as pH 3.9 and pH 3.5, respectively. Growth and lactic acid production by *L. plantarum* continues until a pH of 3.1 is reached [12]. The higher pH tolerance of *L. pontis* corresponds to its competitiveness in type II sourdoughs, which select for acid-tolerant lactic acid bacteria. The acid tolerance of *L. pontis* and related organisms adapted to type II sourdoughs is dependent on conversion of arginine and glutamate, which consume intracellular protons, and on the formation of exopolysaccharides. These metabolic pathways are discussed in more detail below.

Lactic acid bacteria tolerate concentrations of undissociated acetic and lactic acids that exceed by far those concentrations that are typically encountered in sourdough [9–11]. Growth of *L. sanfranciscensis* and *L. pontis* is observed at lactate concentrations of up to 300 and 500 mmol/L, respectively [9, 10]; both organisms also tolerate high concentrations of acetic acid. This high tolerance for organic acids contrasts with the response of yeasts, which are inhibited by undissociated organic acids but not by low pH. Moreover, because the pH but not the organic acid concentration limits the growth of lactic acid bacteria in sourdough, the selection of cereal substrates with a high buffering capacity, for example whole wheat flour or bran, allows the production of sourdough or sourdough products with a high concentration of organic acids and a corresponding high total titrable acidity.

The response of sourdough lactic acid bacteria to high salt concentrations is also species specific. Generally, obligate heterofermentative lactobacilli are more sensitive to NaCl in comparison to other lactobacilli. For example, the heterofermentative *L. sanfranciscensis* and *L. pontis* are inhibited by 4% NaCl whereas *L. plantarum* and *L. amylovorus* tolerate up to 6% NaCl [9, 10, 12].

7.3 Metabolism of Carbohydrates

Lactic acid bacteria belong to three metabolic categories: (1) obligately homofermentative organisms, which ferment hexoses through the EMP (Embden-Meyerhof-Parnas) pathway to lactate as the major end product of carbohydrate metabolism (Fig. 7.1). Pentoses are not fermented. Examples of obligately homofermentative lactobacilli occurring in sourdough include *L. delbrueckii*, *L. acidophilus*, *L. farciminis*, *L. amylovorus*, and *L. mindensis*. (2) Obligately heterofermentative organisms, which ferment hexoses and pentoses through the 6-PG/PK (6-phosphogluconate/phosphoketolase) pathway and synthesize equimolecular amounts of lactate and ethanol or acetate. CO_2 is additionally produced from hexoses (Fig. 7.2). Key examples of obligately heterofermentative lactobacilli in sourdough are *L. sanfranciscensis*, *L. rossiae*, *L. brevis*, *L. pontis*, and *L. fermentum*. (3) Facultatively heterofermentative organisms, which ferment hexoses through the EMP pathway, and pentoses and gluconate through the 6-PG/PK pathway. Examples of facultatively homofermentative lactobacilli occurring in sourdough include *L. plantarum*, *L. alimentarius*, *L. paralimentarius*, and *L. curvatus* [13].

Homofermentative metabolism of hexoses under anaerobic conditions yields 2 mol ATP per mol hexose. In contrast, heterofermentative metabolism of hexoses under anaerobic conditions yields only 1 mol ATP per mol hexose unless co-substrates are present (Fig. 7.1). Contrarily to other fermented foods where obligately homofermentative species have the major role, obligately heterofermentative lactic acid bacteria are dominant in sourdough fermentations [6]. Several factors determine the dominance of heterofermentative strains: (1) the metabolism of maltose via maltose phosphorylase activity, simultaneous fermentation of hexoses and pentoses through the 6-PG/PK pathway, and the use of fructose and other substrates as an external acceptor of electrons; (2) the optimal pH and temperature which often coincides with the values of sourdough fermentation; (3) the capacity of showing alternative phenotype responses and to markedly adapt under various environmental stresses; and (4) the synthesis of a large spectrum of antimicrobial compounds [6]. The variable sources of fermentable carbohydrates in sourdough determine phenotypic responses that include the use of external acceptors of electrons, the preferential and/or simultaneous use of nonconventional energy sources, and the interaction with exogenous or endogenous enzymes from the flour.

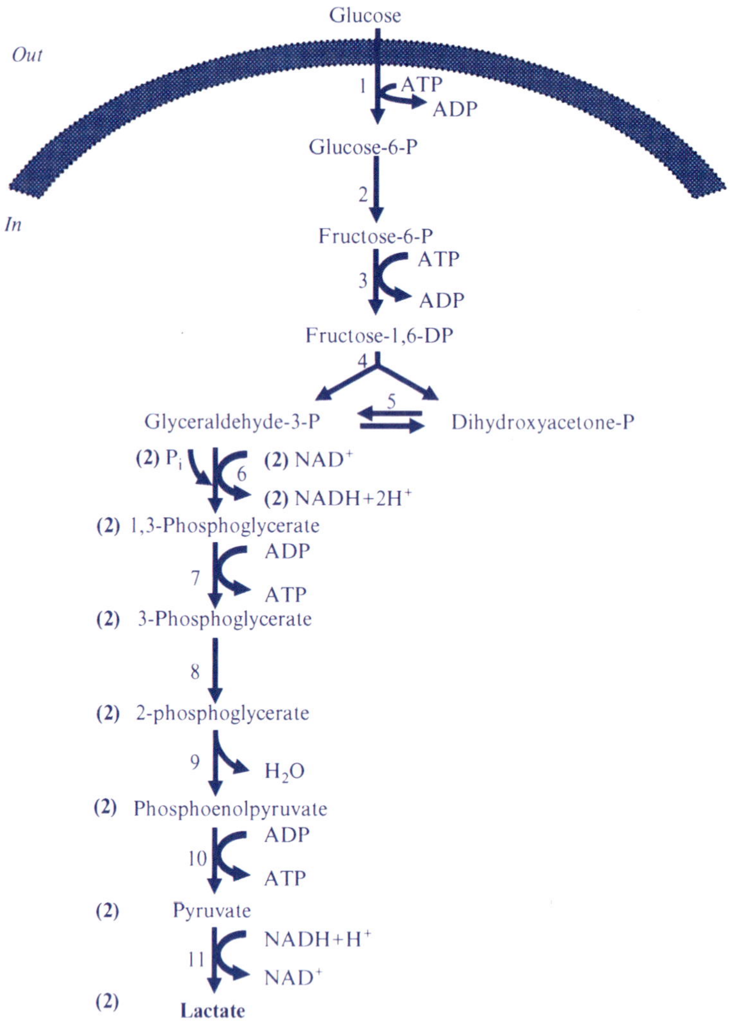

Fig. 7.1 Embden-Meyerhof-Parnas (EMP) pathway; homolactic fermentation. The final products of glucose metabolism are in *bold*. (*2*) indicates the formation of two moles of each compound. *1* Glucokinase, *2* glucose-6-phosphate isomerase, *3* phosphofructokinase, *4* fructose 1,6-bisphosphate aldolase, *5* triosephosphate isomerase, *6* glyceraldehyde 3-phosphate dehydrogenase, *7* 3-phosphoglycerate kinase, *8* phosphoglycerate mutase, *9* enolase, *10* pyruvate kinase, *11* lactate dehydrogenase

7.3.1 Use of External Acceptors of Electrons

Because the energy yield of heterofermentative metabolism of hexoses is low, the competitiveness of obligate heterofermentative lactobacilli depends on the use of external electron acceptors. The practical relevance of the use of external acceptors of electrons is represented by the substantial modification of the

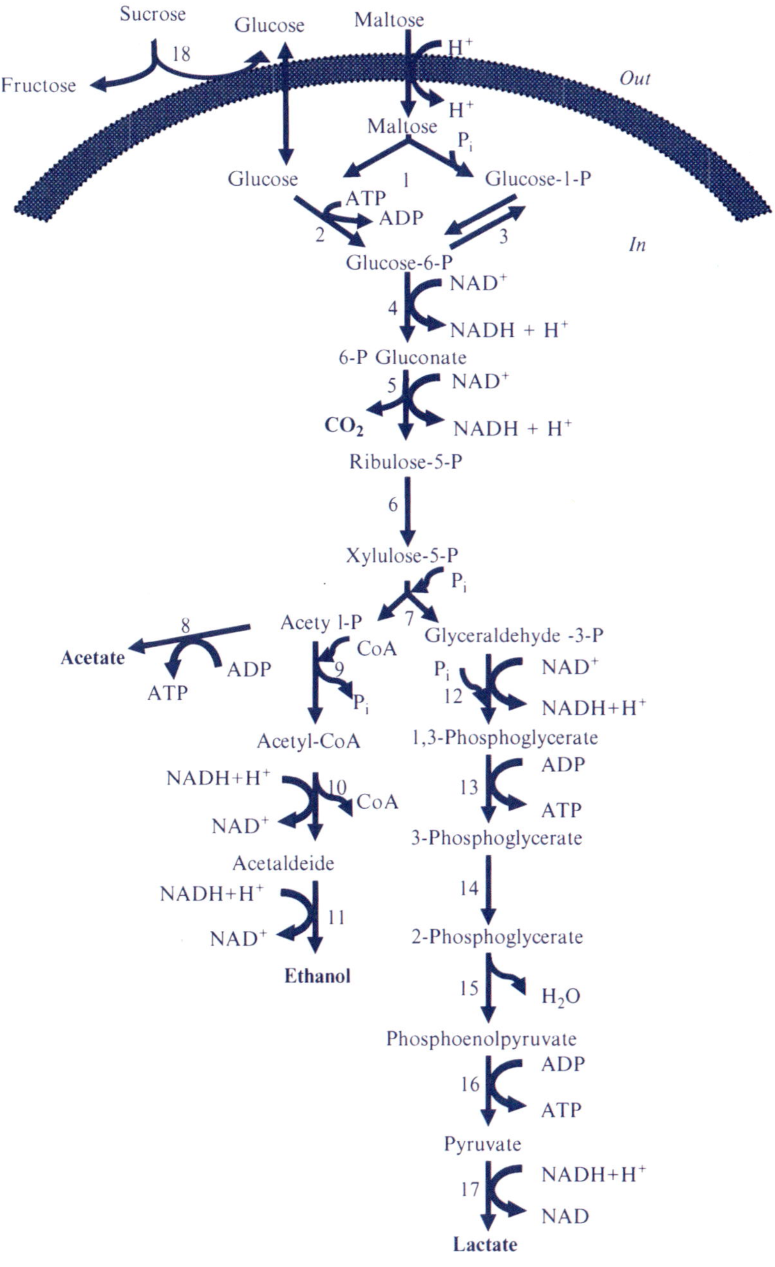

Fig. 7.2 6-Phosphogluconate/phosphoketolase pathway (6-PG/PK); heterolactic fermentation. The final products of glucose metabolism are in *bold* (Adapted from [3]). *1* Maltose phosphorylase, *2* hexokinase, *3* phosphoglucomutase, *4* glucose-6-phosphate dehydrogenase, *5* 6-phosphogluconate decarboxylase, *6* epimerase, *7* phosphoketolase, *8* acetate kinase, *9* phosphotransacetylase, *10* aldehyde dehydrogenase, *11* alcohol dehydrogenase, *12* glyceraldehyde 3-phosphate dehydrogenase, *13* 3-phosphoglycerate kinase, *14* phosphoglycerate mutase, *15* enolase, *16* pyruvate kinase, *17* lactate dehydrogenase, *18* levansucrase

Fig. 7.3 Examples of some reactions that allow NADH + H^+ co-factor reoxidation (Adapted from [3]). *1* Mannitol dehydrogenase, *2* alcohol dehydrogenase, *3* glutathione dehydrogenase, *4* NADH oxidase. *GSSG*, oxidized glutathione; *GSH* reduced glutathione; *R=O*, aldehyde (e.g., hexanal); *R-OH*, corresponding alcohol (e.g., hexanol)

fermentation quotient (see Chap. 4) which positively influences the sensory and shelf-life characteristics of sourdough baked goods [2]. Additional ATP is synthesized when acetyl-phosphate is employed for acetate synthesis through acetate kinase (Fig. 7.2). The synthesis of acetate and ATP as alternative metabolites from acetyl-phosphate requires the availability of co-substrates to oxidize NADH that were generated in the upper branch of the 6-PG/PK pathway. When external acceptors of electrons are available, the recycling of NADH is achieved without the need to synthesize ethanol from acetyl-phosphate. Most heterofermentative lactic acid bacteria are capable of fructose reduction to mannitol to achieve co-factor regeneration (Fig. 7.3). Fructose is quantitatively converted to mannitol by most heterofermentative lactic acid bacteria under acidic conditions [3, 14]. When maltose-negative and maltose-positive sourdough lactic acid bacteria are associated, fructose conversion may have a further role [2]. Most strains of *Weissella* spp. differ from other heterofermentative lactic acid bacteria because they do not convert fructose to mannitol with concomitant acetate formation [15]. The activity of the mannitol dehydrogenase of *L. sanfranciscensis* LTH2581 is optimal at 35°C and pH 5.8–8.0 [16]. Once synthesized, mannitol could be further used as an energy source by strains of *L. plantarum*. This occurs under anaerobiosis and in the presence of ketoacids (e.g., pyruvate) as electron acceptors [17].

Oxygen is also used as an external electron acceptor and is reduced to H_2O with H_2O_2 as an intermediate ([2, 18]; Fig. 7.3). Aerobiosis also induced the expression of a 12.5-kDa superoxide dismutase (SOD), probably Mn^{2+} dependent [18]. Overall, lactic acid bacteria possess various enzymes that are involved in the detoxification of oxygen radicals. NADH-peroxidases and the system involved in the transport of L-cysteine are specifically used by sourdough lactobacilli to detoxify H_2O_2 [19]. The latency phase of growth and cell yield of *L. sanfranciscensis* CB1 are positively influenced by traces of oxygen and Mn^{2+} [18]. When aldehydes are available in the environment, *L. sanfranciscensis* showed R-specific activity by NADH-dependent alcohol dehydrogenase [20, 21] (Fig. 7.3). For instance, the reduction of hexanal to hexanol activates the acetate kinase pathway and the synthesis of acetic acid. Also the reduction of oxidized glutathione (GSSG) into reduced glutathione (GSH), via glutathione dehydrogenase, protects against oxidative stress, and allows the synthesis of acetic acid [22] (Fig. 7.3).

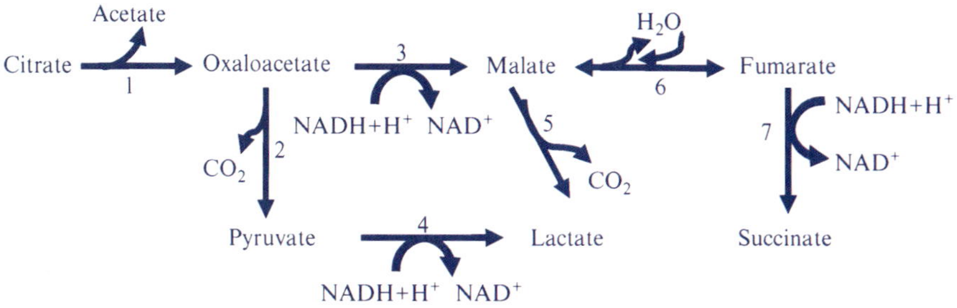

Fig. 7.4 Citric acid metabolism (Adapted from [3]). *1* Citrate lyase, *2* oxaloacetate decarboxylase, *3* malate dehydrogenase, *4* lactate dehydrogenase, *5* malolactic enzyme, *6* fumarase, *7* succinate dehydrogenase

7.3.2 Metabolism of Organic Acids

Sourdough lactic acid bacteria exhibit strain-dependent metabolism of citrate, malate, and fumarate; a few species are also capable of anaerobic lactate metabolism. Citrate, malate, and lactate conversion consumes intracellular protons and thus increases the acid tolerance of lactic acid bacteria. The initial steps of citrate metabolism by lactic acid bacteria are transport by citrate permease and the reaction catalyzed via citrate lyase (Fig. 7.4). Two alternative destinations are possible for oxaloacetate: the first allows the synthesis of succinic acid; the second proceeds via decarboxylation into pyruvate [23]. *Lactococcus lactis* converts part of the pyruvate into α-acetolactate. This reaction takes place when external acceptors of electrons (e.g., citrate) are available, resulting in a surplus of pyruvate with respect to the amount needed to regenerate NADH through lactate dehydrogenase. α-Acetolactate is reduced to acetoin or nonenzymatically converted to diacetyl, an important flavor compound of the bread crumb [24]. Diacetyl formation is not observed in obligate heterofermentative lactobacilli and conversion of citrate to succinate is the most common route for other sourdough lactic acid bacteria. Nevertheless, the synthesis of diacetyl was found in sourdoughs fermented with *L. plantarum*, *L. farciminis*, *L. alimentarius* and *L. acidophilus* [25]. *Lactobacillus sanfranciscensis* uses the route of the pyruvate to convert citrate into lactate and acetate, also including the co-fermentation with maltose [26, 27]. The regeneration of the co-factor during the reaction catalyzed by the lactate dehydrogenase allows the additional synthesis of acetate. Lactate formation from citrate does not cause a decrease of the value of pH, and, therefore, the citrate metabolism of *L. sanfranciscensis* during sourdough fermentation is not limited by the low pH [3]. The use of citrate during sourdough fermentation thus favors an increase of lactate and acetate concentrations. Malate and fumarate are also converted to lactate by *L. sanfranciscensis* [3, 26] (Fig. 7.4). In sourdough, the synthesis of pyruvate and lactate may also result from the catabolism of nonconventional substrates (e.g., amino acids). For instance, serine could be deaminated into ammonium and pyruvate. This latter may be, in turn, reduced to lactate. Pyruvate may be synthesized directly (e.g., alanine) or indirectly (e.g., aspartic acid) through the reaction of transamination [17].

Lactate conversion has been described for *L. parabuchneri*, an isolate from ting, a lactic-fermented sorghum sourdough [28]. Lactate conversion to propanediol proceeds through NADH-dependent reduction of lactate to lactaldehyde and 1,2 propanediol. The reduction of two NADH to NAD^+ allows the concomitant formation of acetate from lactate [29]. Lactate conversion to 1,2-propanediol occurs mainly in the stationary phase of growth and the consumption of lactate improves the stationary phase survival of *L. parabuchneri* [30].

The practical relevance of the use of nonconventional energy sources is more complex and less diffuse within the population of sourdough lactic acid bacteria. In some cases, the addition (e.g., citrate) is needed for conditioning of the microbial metabolism. In other cases (e.g., deamination or transamination of amino acids), it is possible to note a large phenotype variability depending on the biotypes.

7.3.3 Preferential and/or Simultaneous Use of Energy Sources

Lactic acid bacteria use various sources of energy according to hierarchical features that are determined through mechanisms of global control, mainly regulated at the transcription level [31]. When bacteria are subjected to a mixture of energy sources, they preferentially use the substrate that ensures the highest cell yield. When growing on glucose-containing mixtures, *L. amylovorus* DCE 471 always consumed glucose most rapidly, which seemed to steer growth during the early phase. Maltose consumption started only when low levels of glucose were reached [32]. The presence of glucose repressed the fermentation of fructose, maltose, and sucrose of *L. paralimentarius* and *Weissella cibaria* [33]. However, maltose, sucrose and fructose metabolism is not repressed by glucose in the sourdough-adapted species *L. reuteri* and *L. sanfranciscensis*. Levansucrase, the only enzyme responsible for sucrose metabolism in *L. sanfranciscensis* and a major contributor to sucrose metabolism in *L. reuteri* is constitutively expressed or regulated independent of the carbohydrate supply [35, 39]. In *L. reuteri*, sucrose phosphorylase is induced by sucrose or raffinose but not repressed by glucose [40]. The efficient and preferential metabolism of maltose and sucrose by several obligate heterofermentative lactobacilli from sourdough likely reflects adaptation to cereal substrates where sucrose and raffinose are the major carbon sources in the resting grain while maltose is the major carbon source that is liberated by cereal enzymes during fermentation.

Metabolism via maltose phosphorylase or sucrose phosphorylase activity allows a higher energy yield because the chemical energy of the glycosidic bond is employed for ATP synthesis [34, 35]. Both enzymes are frequently found in heterofermentative lactic acid bacteria from sourdough. Expression of maltose phosphorylase in *L. sanfranciscensis* and *L. reuteri* is constitutive and not repressed by glucose, in contrast, hexokinase activity is observed only if glucose is available. During growth on maltose, *L. sanfranciscensis* phosphorylyses maltose and accumulates glucose in the environment, according to the molar ratio of ca. 1:1 (Fig. 7.2) [34, 36]. Glucose accumulated by *L. sanfranciscensis* is available for maltose-negative yeasts and lactic acid bacteria. Nevertheless, the release of glucose from the cell takes place only when the activity of the enzyme hexokinase is not induced [37]. The accumulation of glucose was not found for some strains of *L. sanfranciscensis* that were cultivated in the presence of maltose and fructose. Therefore, it was hypothesized that once liberated the glucose is used through the activity of the enzyme hexokinase, which is in turn induced by the presence of fructose [38].

Some strains of *L. sanfranciscensis* have the capacity to express β-glucosidase activity, but this activity is repressed by glucose [41]. Lactobacilli from sourdough also show simultaneous rather than consecutive fermentation of pentoses and hexoses. Compared to growth on maltose as the only energy source, *L. alimentarius* 15 F, *L. brevis* 10A, *L. fermentum* 1 F e *L. plantarum* 20B showed the highest growth and acidification rates, and the highest cell yield when cultivated in the presence of xylose, ribose, and arabinose [42]. Other sourdough lactic acid bacteria showed the highest performance in terms of growth and synthesis of acetic acid when cultivated in the presence of a mixture of pentose carbohydrates [43]. The presence of pentoses induces the synthesis of the phosphoketolase enzyme in facultatively heterofermentative lactic acid bacteria (e.g., *L. plantarum*) and allows the second half of the 6-PG/PK pathway to proceed (Fig. 7.2). The selection of lactobacilli with the capacity to ferment pentose carbohydrates is thus a suitable alternative to sucrose addition to increase acetate formation.

7.4 Proteolysis and Catabolism of Free Amino Acids

7.4.1 Proteolysis

Lactic acid bacteria are characterized by multiple amino acid auxotrophies. Lactic acid bacteria depend on substrate-derived proteases or on the activity of their proteolytic system to satisfy the nitrogen metabolism [44]. The proteolytic systems of lactic acid bacteria includes serine cell-envelope-associated proteinase (CEP – PrtP) which is associated to the cell wall, oligopeptide and amino acid transporters, and a large number of intracellular peptidases [45] (Fig. 7.5). Contrary to lactic acid bacteria in dairy fermentations, *L. sanfranciscensis* ATCC 27651^{T} and most other sourdough lactobacilli do not possess a cell-envelope-associated proteinase and depend on cereal-associated proteases [46, 47]. The comparison between sourdoughs and chemically acidified doughs showed that the degradation of the native proteins is mainly due to the activity of cereal endogenous proteinases [46, 48, 49]. The acidification by sourdough lactic acid bacteria favors the activation of aspartate-proteinases from cereals, which have an optimal pH of activity that ranged between 3.0 and 4.5 [50]. Nevertheless, selected strains of sourdough lactobacilli showed the capacity to hydrolyze albumins, globulins, and gliadins during fermentation [51–53]. Oligopeptides (4–40 amino acids) are transported inside the bacterial cell and hydrolyzed through a complex system of peptidases. An overview of the intracellular peptidases of *L. sanfranciscensis* is shown in Fig. 7.6. Several intracellular peptidases of *L. sanfranciscensis* CB1 were biochemically characterized: a 65-kDa metal-dipeptidase (PepV), with

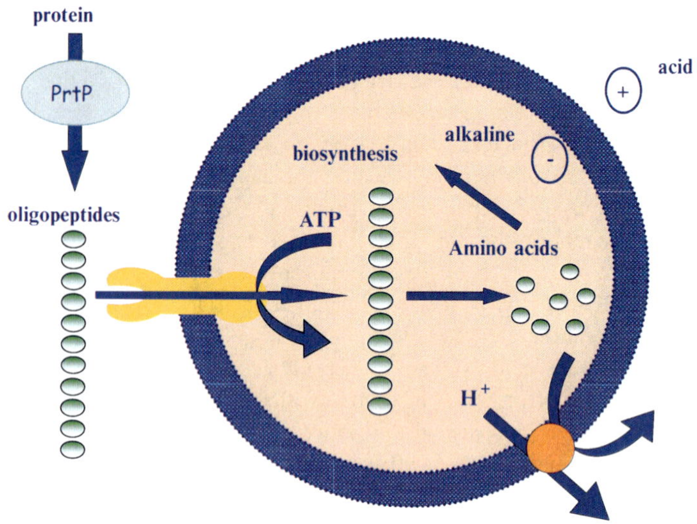

Fig. 7.5 Proteolytic system of lactic acid bacteria (Adapted from [44]). *PrtP*, Cell-envelope-associated proteinase (CEP)

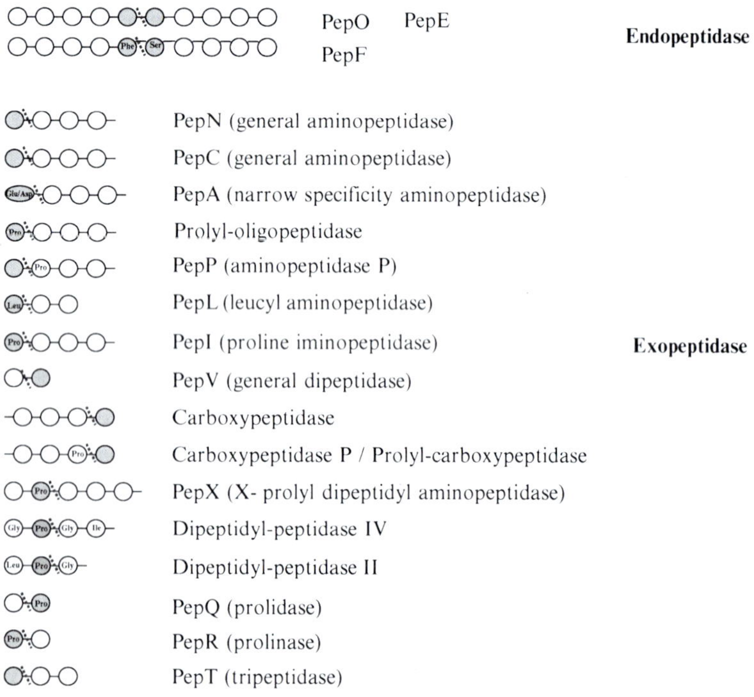

Fig. 7.6 Peptidase system of lactic acid bacteria (e.g. *Lactobacillus sanfranciscensis*) and substrate specificity (Adapted from [50])

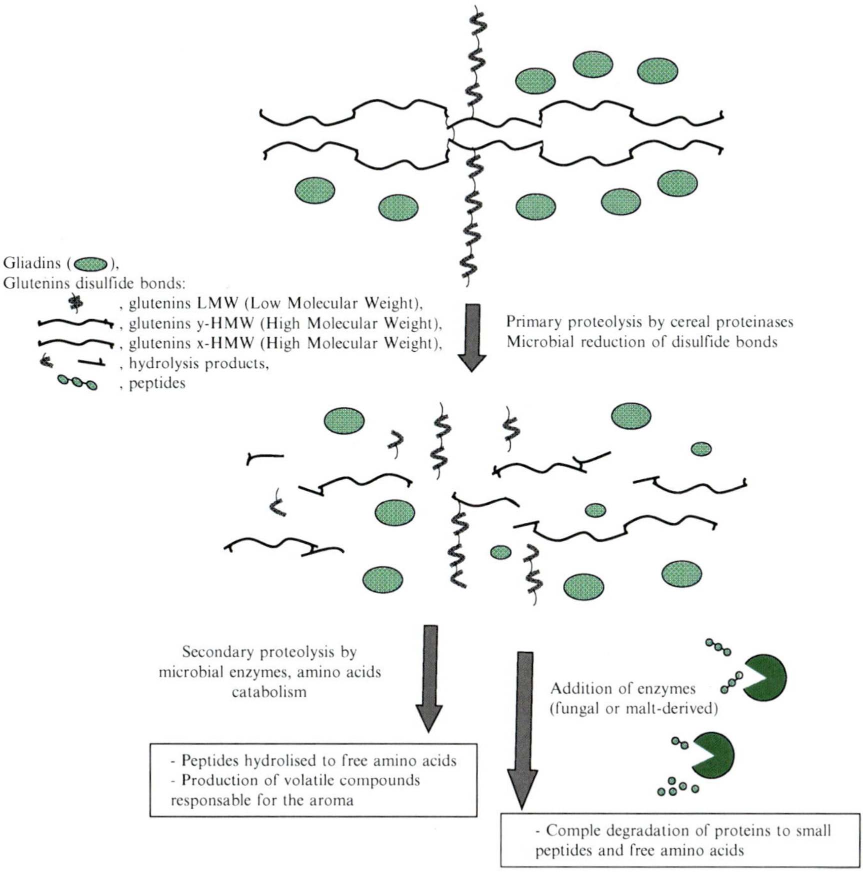

Fig. 7.7 Proteolytic scheme that occurs during sourdough fermentation. Figure shows substrates, enzymes and the activities of primary and secondary proteolysis (Adapted from [50])

elevated specificity towards dipeptides containing hydrophobic amino acids; a 75-kDa general aminopeptidase (PepN); and an X-prolyl-dipeptidyl aminopeptidase (PepX), which exclusively hydrolyzes substrates with the N-terminal sequence X-Pro [51, 54]. The capacity to hydrolyze oligopeptides adjacent to proline residues was shown in selected sourdough lactobacilli. However, the proteolytic system of lactic acid bacteria does not cleave peptides with the sequence motif XPP [55]. Overall, sourdough fermentation increases amino acid concentrations compared to doughs that are chemically acidified [48, 52]. Proteolysis during sourdough fermentation proceeds in two stages (Fig. 7.7): (1) primary proteolysis that releases oligopeptides, and is mainly operated by cereal endogenous proteinases that are activated at low pH and by accumulation of thiols; and (2) secondary proteolysis that releases free amino acids and small-sized peptides, and is exclusively operated through peptidase activity of lactic acid bacteria.

The degradation of the cereal proteins has a fundamental importance for the rheology and sensory features of leavened baked goods. High molecular weight glutenins (HMWG) and gliadins, following the decreasing order of solubility of fractions γ, α, and ω, are hydrolyzed into alcohol-soluble oligopeptides [47]. The partial hydrolysis of glutenins during sourdough fermentation results in the depolymerization and solubilization of the glutenin macropolymer (GMP), thus affecting the viscoelastic properties of the dough. The degree of polymerization of GMP is also influenced by reducing agents. The reduced glutathione (GSH) is the most important reducing agent in wheat dough. GSH may be subjected to an exchange reaction with the thiol groups of the gluten proteins, thus reducing the intermolecular disulfide bond and favoring the decrease of the molecular mass of the GMP [50]. Sourdough heterofermentative lactic acid bacteria possess glutathione reductase activity (see Sect. 3.1), which reduces the extracellular oxidized GSSG to GSH. The continuous recycling of GSSG into GSH keeps the level of SH groups in the dough elevated and increases the number of SH groups in the gluten proteins [56].

The proteolytic system of sourdough lactic acid bacteria contributes to the accumulation of bioactive peptides in sourdough. Fermentation of rye malt sourdoughs with *L. reuteri* resulted in the accumulation of angiotensin-converting-enzyme inhibitory peptides [55]; the formation of peptides with antioxidant activity was also demonstrated [57]. Sourdough fermentation in combination with fungal enzymes was also demonstrated to decrease the concentration of gluten to levels of less 10 ppm, which are tolerated by celiac patients [52, 58]. This hydrolyzed wheat flour was used for the manufacture of baked goods and administered to celiac patients for 60 days. As shown by hematology, immunology and histological analyses, the consumption of 10 g of equivalent gluten per day was absolutely safe for celiac patients [59, 60]. Nine peptidases were partially purified from the pooled cytoplasm extract of the above-selected sourdough lactobacilli and used to hydrolyze the 33-mer epitope, the most common immunogenic peptide generated during digestion of *Triticum* species. At least three peptidases, general aminopeptidase type N (PepN), X-prolyl dipeptidyl aminopeptidase (PepX), and endopeptidase (PepO) were necessary to detoxify the 33-mer without generation of related immunogenic peptides. After 14 h of incubation, the combination of at least six different peptidases totally hydrolyzed the 33-mer into free amino [61]. Peptidase activities of sourdough lactobacilli show considerable diversity at the strain level [55, 62] and the expression of genes coding for peptidases and peptide transport is under control of the concentration of peptides that are present during sourdough fermentation [46].

7.4.2 Amino Acid Metabolism

Peptides and free amino acids represent the substrates for microbial conversion, or are transformed during baking to volatile flavor compounds (Table 7.1). Catabolism of free amino acids by sourdough lactic acid bacteria not only has sensory implications but also increases acid resistance and the microbial energy yield under

Table 7.1 Examples of amino acid precursors and derived carbonylic compounds, which are generated during sourdough fermentation and/or during baking

Amino acid precursor	Derived carbonylic compound
Leucine	3-Methylbutanol
Isoleucine	2-Methylbutanol
Valine	2-Methylpropional
Alanine	Acetaldehyde
Methionine	Methional
Phenylalanine	Phenylacetaldehyde
Threonine	2-Hydroxypropional

starvation conditions [63, 64]. Major pathways for amino acid conversion by lactic acid bacteria include decarboxylation reactions, transamination, and metabolism by lyases (for review, see [65]). These pathways lead to the synthesis of ketoacids, aldehydes, acids and alcohols, which are important flavor compounds of baked goods (Table 7.1) [66]. Major pathways that are relevant in sourdough fermentation are outlined in more detail below.

The catabolism of arginine (Arginine Deiminase, ADI) was studied in depth on sourdough lactic acid bacteria (Fig. 7.8). Three enzymes are involved, arginine deaminase (ADI), ornithine carbamoyl transferase (OTC), and carbamate kinase (CK). A fourth protein, located at the cell membrane which acts as transporter, allows the antiporter exchange between arginine and ornithine [67]. The activity of ADI, OTC, and CK enzymes is a species-specific property of several obligately heterofermentative species, including *L. rossiae*, *L. reuteri*, *L. brevis*, *L. hilgardii*, *L. fermentum*, *L. pontis*, and *L. fructivorans* [68]. Generally, arginine is quantitatively converted to ornithine by ADI-positive lactic acid bacteria during sourdough fermentation [48]. The activity of ADI, OTC, and CK of *L. rossiae* CB1 is well adapted to the acidity (pH 3.5–4.5) and temperature (30–37°C) conditions of sourdough fermentation [69]. The ADI pathway in *L. fermentum* IMDO 130101 is upregulated in response to temperature and salt stress conditions [70]. During sourdough fermentation, the expression of the ADI pathway favors: (1) microbial growth and survival, which determines a constant composition of the microbial population in the sourdough; (2) the enhanced tolerance of lactic acid bacteria to acidity by contributing to homeostasis of the intracellular pH; and (3) an increased synthesis of ornithine which is converted, during baking, into 2-acetyl-1-pyrroline, the compound responsible for the typical flavor of the bread crust.

Glutamine is the most abundant amino acid of wheat proteins; *L. sanfranciscensis* and other sourdough lactobacilli convert glutamine into glutamate (Fig. 7.9). Glutamine conversion to glutamate improves the adaptation of lactobacilli to sourdough acidity due to consumption of protons and liberation of ammonia which increases the extracellular pH. The synthesis of glutamate also has a positive effect on the sensory properties of leavened baked goods [71]. Glutamate is alternatively decarboxylated to γ-aminobutyrate (GABA), or converted to α-ketoglutarate (αKG) by glutamate dehydrogenase (EC 1.4.1.3). Heterofermentative lactobacilli preferably

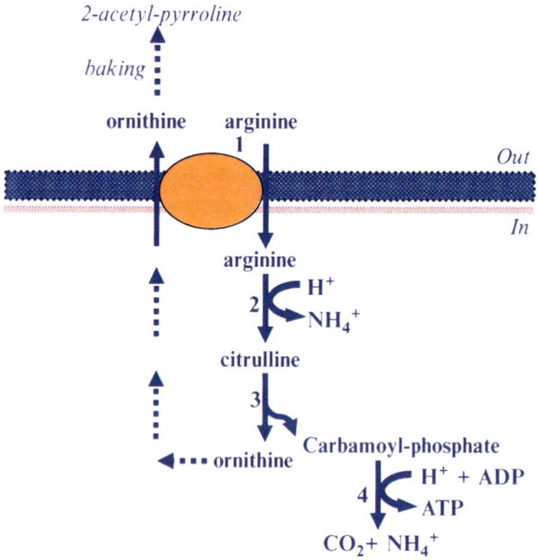

Fig. 7.8 Arginine deiminase catabolism (ADI) in sourdough lactic acid bacteria (Adapted from [3]). *1* Arginine-ornithine antiporter, *2* arginine deaminase, *3* ornithine transcarbamylase, *4* carbamate kinase

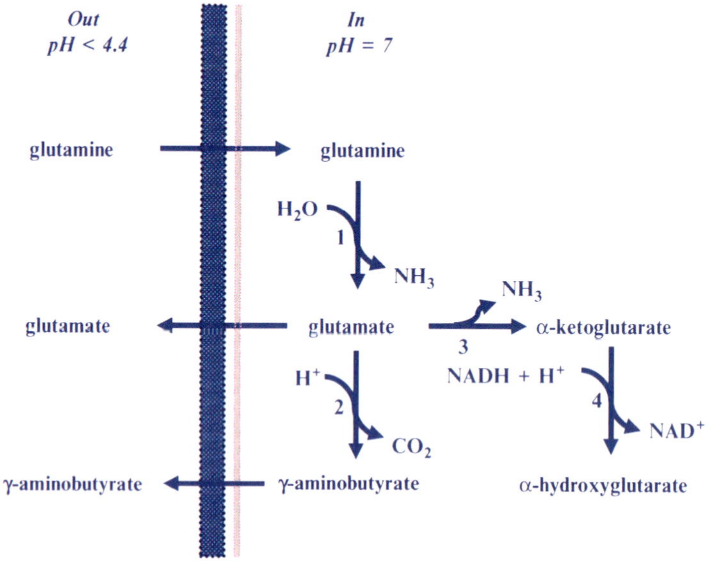

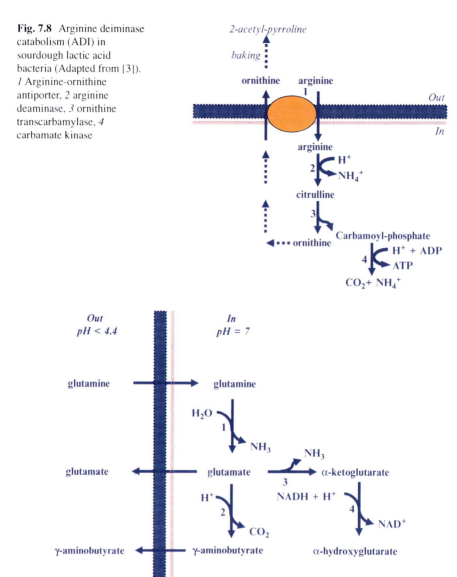

Fig. 7.9 Glutamine and glutamate metabolism of sourdough lactic acid bacteria. *1* Glutaminase, *2* glutamate decarboxylase, *3* glutamate dehydrogenase, *4* alcohol dehydrogenase

employ αKG as an electron acceptor to regenerate NADH [20]. αKG also acts as the preferred amino acceptor in transamination reactions with leucine, phenylalanine, and other amino acids [72]. After peptide transport, the transamination reactions are the second limiting factor for the conversion of amino acids by lactic acid bacteria [73]. Consequently, the addition of α-ketoglutarate into the food matrix or the selection of glutamate dehydrogenase positive strains considerably increases the

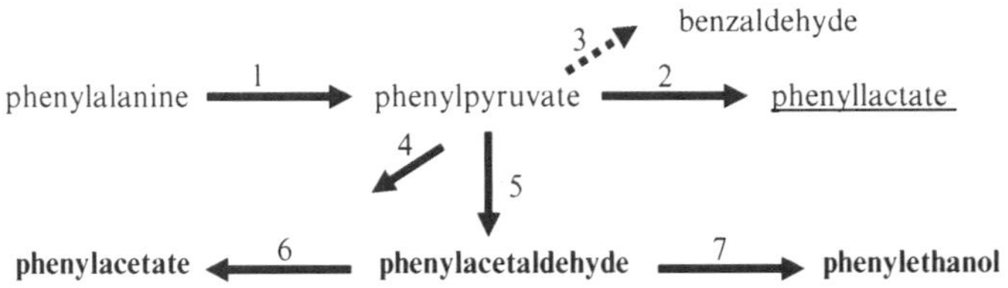

Fig. 7.10 Phenylalanine catabolism in *Lactobacillus plantarum* and *Lactobacillus sanfranciscensis*. The most abundant metabolic products are *underlined*, those responsible for the aroma compounds are in *bold* (Adapted from [3]). *1* Transaminase, *2* dehydrogenase, *3* chemical oxidation, *4* multienzyme complex, *5* decarboxylase, *6* and *7*, dehydrogenase

conversion of amino acids. Glutamate dehydrogenase activity is variable depending on the biotype of lactic acid bacteria [73]. This enzyme is also dependent on NADH + H^+, which links the catabolism of glutamate of *L. sanfranciscensis* DSM20451 to the regeneration of NADH during the central metabolism of carbohydrates (Fig. 7.9) [74]. Glutamate decarboxylase (GAD) activity was also found in sourdough lactic acid bacteria [75, 76]. GAD converts L-glutamate to GABA, through a single-step α-decarboxylation. GABA, a four-carbon nonprotein amino acid, has several functional activities. Sourdough fermentations with *L. plantarum*, and *Lc. lactis* or *L. reuteri* allowed the synthesis of GABA in concentrations of up to 9 g/kg, comparable to those found in functional preparations [75, 76]. Glutamate decarboxylase-mediated acid resistance was shown to contribute to the persistence of *L. reuteri* in type II sourdough fermentations [64].

The catabolism of phenylalanine by *L. sanfranciscensis* and *L. plantarum* is shown in Fig. 7.10. Leucine, isoleucine, and valine undergo a similar mechanism. Alternative pathways for leucine metabolism were proposed but remain to be validated by enzymatic and genetic analyses [77]. Phenyl-lactate or phenyl-acetate are the main end products of the catabolism of phenylalanine. For *L. plantarum* TMW1.468, a glutamate-dehydrogenase negative strain, the synthesis of phenyl-lactate is increased through addition of αKG but not with glutamate and citrate added. Conversely, for *L. sanfranciscensis* DSM20451, a glutamate-dehydrogenase positive strain, the synthesis of phenyl-lactate is increased through addition of αKG, glutamate and peptides containing glutamate. However, αKG and glutamate favor the conversion of phenylalanine only when fructose or citrate are also present, probably owing to the co-factor dependence of glutamate dehydrogenase [74]. During sourdough fermentation, *L. plantarum* TMW1.468 and *L. sanfranciscensis* DSM20451 synthesize ca. 0.35–1.1 mM of phenyl-lactate; the concentration of 1.5 mM inhibits the growth of *L. sanfranciscensis* DSM20451 during growth in mMRS at pH 5.3 [3].

Cystathionine lyase activities are also diffuse within sourdough lactic acid bacteria (Fig. 7.11) [78]. A homotetrameric 160-kDa cystathionine γ-lyase was purified and characterized from *L. reuteri* DSM 20016 [79]. The enzyme is active towards a large number of amino acids and amino acid derivatives, including methionine, and allows the synthesis of ammonia, α-ketobutyrate, and low-molecular-weight volatile sulfur compounds.

Fig. 7.11 Cystathionine metabolism (Adapted from [169]). *1* Cystathionine-γ-lyase, *2* cystathionine-β-lyase, *3* cystathionine-β-synthetases

7.5 Synthesis of Exopolysaccharides

The formation of exopolysaccharide by lactic acid bacteria during sourdough fermentation improves culture survival and stress resistance, impacts dough rheology and bread texture, and allows production of bread with specific nutritional functionality. EPS from lactic acid bacteria can be categorized on the basis of their composition or their biosynthesis. EPS composed of only one type of constituting monosaccharide are classified as homopolysaccharides (HoPS). Examples include levan, which is composed of fructose, or dextran, which is composed of glucose. Heteropolysaccharides (HePS) are composed of two or more constituting monosaccharides. Lactic acid bacteria employ two alternative routes for EPS synthesis, intracellular synthesis from sugar nucleotides by glycosyltransferases, and extracellular synthesis from sucrose by glucansucrases or fructansucrases. EPS produced by glucansucrase or fructansucrase activity are invariably HoPS composed of glucose or fructose, respectively [80, 81]. EPS produced by intracellular glycosyltransferases predominantly are HePS with glucose, galactose, N-acetylglucosamine, N-acetylgalactosamine, rhamnose, or fucose as constituting monosaccharides [82–84]. Noticeable exceptions of HoPS produced by intracellular glycosyltransferases include a galactan from *Lc. lactis* [85] and β-glucan produced by *Pediococcus parvulus* and *Oenococcus oeni* [86].

Large-scale screening of strain collections of sourdough lactic acid bacteria revealed that very few isolates are capable of HePS formation [39, 87–89]; only one report documents formation of HePS in sourdough [90]. In contrast, HoPS formation is a frequent trait of sourdough lactic acid bacteria and any sourdough is likely to contain at least one HoPS-forming strain [39, 88]. Moreover, the in situ formation of HoPS during sourdough fermentation as well as the effect of HoPS on bread quality is well documented ([91, 92]), for a review see [93] and Chap. 8). Subsequent paragraphs will hence only briefly discuss HePS formation and focus on HoPS formation and applications.

7.5.1 *EPS Biosynthesis and HoPS Structure*

HePS formation by lactic acid bacteria is mediated by large gene clusters that are encoded on plasmids or the chromosome (for reviews, see [82–84, 94]). EPS gene clusters generally code for proteins regulating EPS biosynthesis (EpsA in *Streptococcus thermophilus* Sfi6), polymerization, export, and chain length determination (EpsB, EpsC, and EpsD in *S. thermophilus* Sfi6), and one or several

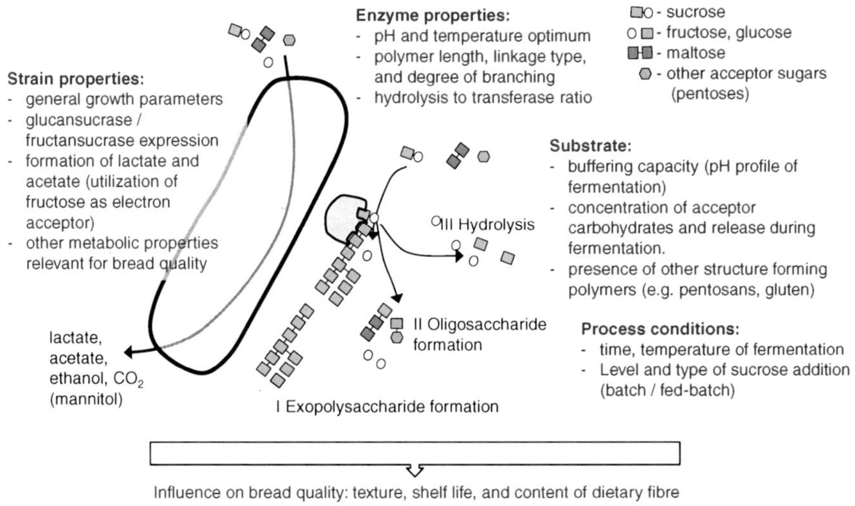

Fig. 7.12 Factors influencing yield of exopolysaccharides and the influence of exopolysaccharides on bread quality. Glucansucrases catalyze three alternative reactions: (I) polymerization of glucose or fructose to exopolysaccharides, (II) glycosyl transfer to acceptor carbohydrates to form oligosaccharides, (III) sucrose hydrolysis to glucose and fructose. EPS properties and yield are additionally dependent on the concentration of substrate-derived acceptor carbohydrates, and the ambient pH and temperature

phosphor-glycosyltransferases or glycosyltransferases (EpsE, EpsF, EpsG, and EpsI in *S. thermophilus* Sfi6). EPS is synthesized by glycosyltransferases with sugar nucleotides derived from glucose-6-phosphate as substrates. The repeating unit of the polysaccharide, consisting of two to eight monosaccharides, is assembled by sequential addition of monosaccharides to the membrane-bound lipid carrier undecadeprenyl pyrophosphate. The repeating unit is exported and polymerized extracellularly. HePS biosynthesis requires energy-rich sugar nucleotides as precursors, and the assembly and export of the repeating units competes with the peptidoglycan biosynthesis; therefore, HePS yields are relatively low, typically well below 1 g/L.

HoPS synthesis is mediated by extracellular cell-wall associated or soluble glucansucrases or fructansucrases [80, 81]. Both groups of enzymes are classified as retaining glycosyl hydrolases and use sucrose as substrate. Fructansucrases but not glucansucrases also employ raffinose, stachyose, or verbascose as substrate [40]. Catalysis by glucansucrases and fructansucrases proceeds through a covalently linked glucosyl-enzyme or fructosyl-enzyme intermediate, respectively, with subsequent transfer of the glycosyl moiety to a growing polymer chain, a suitable acceptor carbohydrate, or water (Fig. 7.12). Sucrose hydrolysis or oligosaccharide formation are thus alternative reactions to polysaccharide synthesis [80, 95–97]. The ratio of sucrose hydrolysis to oligo- or polysaccharide synthesis depends on the enzyme structure as well as the substrate concentrations [98, 99]. Because the catalytic mechanism retains the chemical energy of the glycosidic bond of the substrate, HoPS synthesis does not require energy-rich substrates or co-factors. Several recent reviews provide an excellent overview on structure-function relationships of glucansucrases and fructansucrase [80, 81, 100].

HoPS produced from sucrose are composed only of fructose or glucose but nevertheless have a large diversity with respect to linkage type, or degree of branching, and molecular weight. Dextran is mainly composed of α–(1→6) linked glucose molecules with a varying degree of α–(1→3) or α–(1→2) linkages and α–(1→3,6) branching points. Dextran is produced by many strains of *Leuconostoc* spp. and *Weissella* spp. but dextran-forming *Lactobacillus* spp. were also described [80, 81, 88]. Mutan is mainly composed of α–(1→3) linked glucose moieties and is produced by *Streptococcus* spp. and *L. reuteri* [80, 81]. Glucans with alternating α–(1→6) and α–(1→3) or alternating α–(1→4) and α–(1→6) linkages are referred to as alternan and reuteran, respectively. Reuteran formation has to date exclusively been observed in strains of *L. reuteri* [81]. The current knowledge on structure-function relationships of glucansucrases allows the manipulation of the linkage type [101] as well as the molecular weight and degree of branching [102]. Information on structure-function relationships of glucansucrases has also been used for identification of wild-type glucansucrases producing polymers with desired properties [103]. The linkage type in the oligosaccharides formed by glucansucrases generally matches the linkage types found in the corresponding polysaccharide [81, 97]. For example, *L. reuteri* 121 produces α–(1→4) and α–(1→6) linked reuteran with a relative molecular weight of 8×10^7; in the presence of glucose or maltose as acceptor carbohydrates, maltose and isomaltose or isomaltotriose and panose are produced. Fructans produced by lactic acid bacteria include the predominantly β–(2→1) linked inulin and the predominantly α–(2→6) linked levan; both polymers contain β–(2→1→6) branching points [80, 81, 98, 104]. A majority of fructan-producing sourdough isolates form levan. Remarkably, levansucrases form predominantly inulin-type fructo-oligosaccharides and hetero-oligosaccharides from sucrose [95, 98].

7.5.2 Ecological Function of HoPS Production and HoPS Formation in Dough

The ecological function of glucansucrases and fructansucrases for cereal-associated lactic acid bacteria was related to biofilm formation, sucrose and raffinose utilization, and stress resistance. In oral streptococci, glucansucrases and fructansucrases are primarily responsible for sucrose-dependent biofilm formation on the tooth enamel, and are considered major virulence factors of these organisms [105]. Analogous to the biofilm formation by oral streptococci, colonization and biofilm formation by *L. reuteri* in nonsecretory epithelia of the proximal intestinal tract of animals was found to be dependent on HoPS formation. The reuteransucrase of *L. reuteri* TMW1.106 was a major contributor to formation of the extracellular biofilm matrix, the levansucrase in the same organism functions as a glucan-binding protein. Both enzymes are required for colonization of reconstituted lactobacilli-free mice by *L. reuteri* TMW 1.106 [106].

Glucansucrases and fructansucrases use one of the two constituting monosaccharides of sucrose for poly- or oligosaccharide synthesis, the other monosaccharide, fructose and glucose, respectively, accumulates in the fermentation substrate. Levansucrases also release the α-galactosides melibiose, manninotriose, and manninotetraose from raffinose, stachyose, and verbascose, respectively [40]. In *L. sanfranciscensis* LTH2590, levansucrase is the only enzyme capable of sucrose and raffinose hydrolysis [40, 98]. In *L. reuteri* and *Lc. mesenteroides*, levansucrase, glucansucrase and sucrose phosphorylase constitute alternative pathways for sucrose (and raffinose) metabolism. Metabolism by extracellular levansucrases is the preferred metabolic route for raffinose, stachyose, and verbascose, presumably because the resulting α-galactosides have a smaller degree of polymerization, which facilitates transport across the cytoplasmic membrane [40]. Glucansucrase activity invariably accumulates fructose and thus supports acetate formation by heterofermentative lactic acid bacteria [98]. Because high quantities of acetate have detrimental effects on the structure of wheat bread [107], dextran-producing *Weissella* spp. that do not convert fructose to mannitol with concomitant acetate formation exhibited superior performance in baking applications when compared to *L. reuteri* or *L. sanfranciscensis* [91, 92, 108].

A contribution of HoPS formation to stress resistance of lactic acid bacteria has particularly been demonstrated for *L. reuteri*. Here, HoPS or fructo-oligosaccharides increase survival at acid conditions during the stationary phase [103], and improves survival of freeze-dried *L. reuteri* during storage [109]. The protective effects of levan and fructo-oligosaccharides are partially attributable to their specific interaction with phospholipids of biological membranes [109, 110]. However, protective effects are not limited to levan and fructo-oligosaccharides; dextran and β-glucan also improved acid resistance and stationary phase survival [86, 111].

Lactic acid bacteria producing HoPS from sucrose in laboratory medium generally also produce HoPS during growth in sourdough [39]. HoPS concentrations typically range from 1 to 8 g/kg dough [39, 91, 92, 98, 112]; more than 10 g/kg were reported in optimized, pH-controlled fermentations [113]. Glucansucrases and fructansucrases are extracellular enzymes, thus, their activity is governed by ambient conditions rather than intracellular pH and substrate concentrations (Fig. 7.12). The optimum pH of levansucrases and glucansucrases of sourdough lactobacilli ranges from 4.5 to 5.5, matching the pH profile of sourdough fermentations [98, 99]. HoPS yields in sourdough were maximized by pH-controlled fermentations at a pH of 4.7 [113]. Because sucrose acts both as glycosyl-donor and glycosyl-acceptor in HoPS formation, the sucrose concentration has a decisive influence on the ratio of sucrose hydrolysis, oligosaccharide formation, and polysaccharide formation. Below 10 g/kg sucrose, sucrose hydrolysis is the predominant reaction [98, 114]. Increasing sucrose concentrations initially increase the formation of levan over hydrolysis. After a further increase of the sucrose concentration, sucrose increasingly acts as a glycosylacceptor and oligosaccharide formation is the predominant reaction catalyzed by glucansucrases or fructansucrases [98, 114, 115]. In sourdough fermentation with 100 g/kg sucrose, *L. sanfranciscensis* LTH2590 produced 11 g/kg fructo-oligosaccharides but only 5 g/kg levan [98]. High reuteran yields from *L. reuteri* TMW1.106 were achieved in pH static fed-batch fermentations that maintained sucrose concentrations at 10% throughout fermentation [113]. The addition of acceptor carbohydrates other than sucrose also influences HoPS yields in sourdough fermentations. Maltose is a stronger acceptor for glucansucrases compared to glucose [97, 115]. Correspondingly, dextran yields in wheat sourdoughs, characterized by high maltose concentrations throughout fermentation, were substantially lower compared to yields by the same strains in sorghum fermentations, which are characterized by low maltose and high glucose concentrations [15]. Maltose, xylose, and arabinose also act as alternative acceptor carbohydrates for levansucrase activity [95, 96].

Exopolysaccharide-producing sourdough starter cultures have found industrial application to improve the textural properties of bread [116]. In addition to the function of HoPS as hydrocolloids in baking applications, specific HoPS were found to prevent pathogen adhesion to eukaryotic cells [117] and to exhibit prebiotic activity [92, 118]. HoPS-producing lactic acid bacteria thus enable the formulation of baked goods with specific health properties.

7.6 Antimicrobial Compounds from Sourdough Lactic Acid Bacteria

Leavened baked goods can become contaminated by spores of the genus *Bacillus*, which survive baking, yeasts, mainly belonging to the genera *Pichia* and *Zygosaccharomyces*, which colonize the surface and negatively affect the sensory properties, and, especially, by moulds, mainly belonging to the genera *Penicillium*, *Aspergillus* and *Cladosporium* which alter the color and the sensory properties, and, in some cases, synthesize mycotoxins. More than 40 species of fungi were described as contaminants of baked goods. Although chemical preservatives (e.g., sorbate and propionate, ethanol) are routinely used for preventing the contamination of leavened baked goods, sourdough lactic acid bacteria show a number of natural bio-preservative features that are complementary to chemical preservatives, or can even substitute their use. Antimicrobial compounds from sourdough lactic acid bacteria were additionally shown to influence sourdough microbiota, and to contribute to the stability of individual strains.

The inhibitory activity of sourdough lactic acid bacteria is generally attributable to rapid consumption of oxygen and fermentable carbohydrates, and the formation of lactate with concomitant reduction of the pH. Additional metabolites with specific antimicrobial activity include diacetyl, hydrogen peroxide, acetate and other short-chain fatty acids, and reuterin. Acetate formation by heterofermentative lactobacilli in sourdough is readily adjusted by addition of sucrose or pentoses [39], and contributes to shelf-life extension of bread (see below). The odor threshold of diacetyl, butyrate, and caproate is substantially lower than the concentrations required for antimicrobial activity; these compounds can thus not be accumulated in bread without adverse effects on the sensory bread quality. Although individual sourdough lactic acid bacteria are capable of reuterin synthesis, reuterin formation has not been achieved in cereal fermentations [119].

7.6.1 Antifungal Compounds from Sourdough Lactic Acid Bacteria

Antifungal compounds synthesized by sourdough lactic acid bacteria include diacetyl, hydrogen peroxide, acetate, propionate, caproate, 3-hydroxy fatty acids, phenyllactate, cyclic dipeptides, reuterin, and fungicidal peptides [120]. Numerous reports describe a substantial increase of the mould-free shelf life of bread owing to antifungal activities of sourdough lactic acid bacteria [121, 122]. However, metabolites of lactic acid bacteria responsible for the antifungal effect remain in most cases unknown. Acetate inhibits fungal growth only at concentrations that significantly impair the sensory and textural quality of bread [30, 107, 123]. Phenyllactate and 4-hydroxyphenyllactate were initially characterized as antifungal metabolites from *L. plantarum* ITM21B [124]. Lactobacilli capable of producing phenyllactate and hydroxyphenyllactate delayed the growth of *Aspergillus niger* and *Penicillium roqueforti* for up to seven days of bread storage [125]. The concentration of phenyllactate accumulated by lactobacilli in sourdough fermentation, however, is about 100 fold lower than its minimal inhibitory concentration [74, 126] and the antifungal effect of the sourdough is thus not attributable to phenyllactate accumulation. Likewise, the levels of antifungal cyclic dipeptides in bread are 1,000 fold lower than their MIC, but above the threshold imparting a metallic and bitter taste [127]. Accordingly, the antifungal effect of sourdough lactic acid bacteria is attributed to a synergistic activity of several compounds. The antifungal effect of *L. reuteri*, *L. plantarum*, and *L. brevis* in the presence of Ca propionate (0.2%, w/w) [128] delayed fungal growth by 8 days compared to the bread started with baker's yeast alone. It was hypothesized that a synergistic activity between acetic acid and phenyllactate with chemical preservatives was responsible for this effect. The antifungal effect of *L. buchneri* and *L. diolivorans* against growth of four moulds on bread was attributed to a combination of acetate and propionate [30]. The preservative effect of *L. amylovorus* [121, 122] was attributed to the synergistic activity of more than ten antifungal compounds, including phenyllactate, phenolic acids, fatty acids, and cyclic dipeptides [129].

In addition to the antifungal metabolites of lactic acid bacteria, inhibitory peptides derived from the substrate may also contribute to the preservative effect. A water-extract from beans in combination with sourdough fermented with *L. brevis* AM7 contained three natural inhibitory compounds, two phaseolins (NCBI n. gi 130169, gi 403594) and one lectin (NCBI n. gi 130007) [130]. Antifungal peptides were also identified from the water-extract of sourdough. The combined activity of the above compounds determined a delay in fungal growth of up to 21 days, leading to a shelf life for the bread that was comparable to that found when using Ca propionate (0.3% w/w). A very extensive bread shelf life was also achieved by using sourdough lactic acid bacteria and amaranth flour or wheat germ [131, 132]. Sourdough fermented with the nonconventional yeast *Wickeramomices anomalus* LCF1695 and *L. plantarum* was shown to exhibit strong antifungal activity based on synergistic activities between the inhibitory peptides liberated by *L. plantarum* and ethyl-acetate produced by *W. anomalus* [133]. Fungal contamination was delayed by 28 days under pilot plant bakery conditions. Although indubitable progress has been achieved in the making and storage of baked goods, fungal contamination remains one of the main problems in terms of long-term shelf life of bakery products. The search for novel methods of bio-conservation appears to be one of the most promising tools in this regard, even though a combination of various inhibitory compounds appears to be inevitable to markedly extend the storage of leavened baked goods.

7.6.2 Antibacterial Compounds from Sourdough Lactic Acid Bacteria

Antibacterial metabolites from lactic acid bacteria include reuterin and the organic acids described above. However, the emphasis of research related to antibacterial activities was placed on bacteriocins, ribosomally synthesized peptides with antibacterial activity against closely related organisms, and reutericyclin. The prevention of bread spoilage by rope-forming bacilli does not require the selection of specific protective cultures. Growth of endospores of *Bacillus* spp. is readily inhibited by modest acidification as is characteristic for sourdough bread [134–136]. Acetate and propionate are more effective against rope-forming bacilli than lactic acid [134]. Remarkably, the bacteriocins nisin and pedicoin, either included as additives, or generated in situ by bacteriocin-producing lactic acid bacteria, were ineffective [134].

Bacteriocin formation by sourdough lactic acid bacteria was described particularly for strains of *L. sakei*, *L. plantarum* and *L. amylovorus* (for a review, see [137, 138]). Bacteriocins appear not to be suitable for extending the shelf life of bread but formation in sourdough fermentation was shown to enhance the stability of sourdough microbiota. *Lactococcus lactis* M30, a lacticin 3147-producing strain [139], produced a bacteriocin during fermentation. The inhibitory activity persisted under the low values of pH of sourdough and after thermal treatments that corresponded to baking temperatures. Similarly, the bacteriocin-producing strain *L. amylovorus* DCE471 persisted for a long time during sourdough propagation [140]. The competitiveness of *L. pentosus* 2MF8, which synthesized a bacteriocin-like inhibitory substance [141] and *Lc. lactis* subsp. *lactis* M30, which synthesized lacticin 3147 [139] were studied during sourdough propagation. *Lactococcus lactis* subsp. *lactis* M30 showed a larger spectrum of inhibition compared to *L. pentosus* 2MF8, and did not inhibit the growth of *L. sanfranciscensis*. After 20 days of back-slopping, the persistence of *Lc. lactis* subsp. *lactis* M30 inhibited the indicator strain *L. plantarum* 20, without interference in the growth of *L. sanfranciscensis* CB1. The above-described features of bacteriocins, together with the demonstration of the in situ inhibitory activity, encourage the use of antimicrobial compounds to facilitate the persistence of the starter cultures and the conditioning of the microbial interactions that occur during sourdough fermentation.

Reutericyclin is a tetramic acid derivative with a broad spectrum of activity against Gram-positive bacteria, including rope-forming bacilli [142, 143]. To date, all reutericyclin-producing strains were isolated from an industrial rye sourdough. Reutericyclin is produced to active concentrations in sourdough and the persistence of *L. reuteri* strains during long-term propagation of a rye sourdough was attributed to the synthesis of reutericyclin (for a review, see [144]). The use of reutericyclin-producing *L. reuteri* as a sourdough starter also caused a delay in the growth of *Bacillus* sp. during bread storage [144].

7.7 Metabolism of Phenolic Compounds and Lipids

The major phenolic compound in wheat and rye is ferulic acid bound to cell wall polysaccharides [145]. Changes in the ferulic acid content during sourdough fermentation are predominantly the result of oxidation reactions and cereal enzymes [146, 147]. In contrast, cereal grains used in African cereal fermentations, particularly sorghum and millet, are rich in phenolic compounds, including phenolic acids, phenolic acid esters, desoxyanthocyanidins, and tannins [148, 149]. In these cereal grains, metabolism of phenolic compounds influences product properties, and the content of antimicrobial phenolic compounds influences the microbial ecology of sorghum sourdough fermentations [150]. A review on the metabolism of food phenolics by lactic acid bacteria, based predominantly on isolates from wine and vegetable fermentations, is provided by Rodriguez et al. [151].

Conversion of phenolic compounds is based on glycosyl hydrolases, for example β-glucosidase and α-rhamnosidase, which convert flavonoid glycosides to the corresponding aglycones [152, 153], esterase degrading methyl gallate, tannins, or phenolic acid esters [154], and decarboxylases and reductases with activity on phenolic acids [151]. The strain-specific metabolism of phenolic compounds is particularly well described for *L. plantarum* [151, 155]. Decarboxylation of hydroxyl cinnamic acids generates the corresponding vinyl derivatives [156]. Reductases hydrogenate the double bond of hydroxyl cinnamic acids or their decarboxylated vinyl derivatives [151, 156]. The spectrum of activities is strain specific; organisms capable of conversion of hydroxy cinnamic acids harbor reductase activity, decarboxylase activity, or both [150, 151, 156]. Analysis of phenolic compounds during sorghum sourdough fermentation revealed that strain-specific glycosyl hydrolase, decarboxylase, and phenolic acid reductase contributed to the conversion of phenolic compounds [151].

Phenolic compounds, including phenolic compounds from sorghum and millet, exhibit strong antimicrobial activity [148, 150]. Particularly the activity of hydroxy benzoic and hydroxy cinnamic acids is well characterized [157]. The resistance of lactic acid bacteria towards phenolic acids is highly strain specific and the strain-specific capability for phenolic acid conversion corresponds to resistance [157, 158]. Because the conversion of hydroxy cinnamic acids by reductase- or decarboxylase activities reduced the antimicrobial activity of caffeic acid two to fivefold, metabolism was recently identified as a mechanism of detoxification [157]. To date, lactic acid bacteria capable of conversion of phenolic compounds were predominantly isolated from fermented beverages (wine, whisky, beer), or African sorghum fermentations. These organisms are likely a suitable source for starter cultures of phenolic-rich cereals and pseudocereals used in gluten-free baking [159].

Lipid metabolism in sourdoughs is poorly described and particularly lipase activity of sourdough lactic acid bacteria appears not to be relevant. However, sourdough fermentation influences lipid oxidation and the influence of lipid oxidation products on bread flavor. During flour storage and during dough mixing, chemical oxidation and cereal-derived lipoxygenase, respectively, convert free linoleic acid to lipid peroxides. Peroxides are chemically converted to lipid aldehydes, i.e., (E)-2-nonenal and (E,E)-2,4-decadienal, with a strong influence on the flavor of the bread crumb [21]. Heterofermentative lactobacilli reduce these aldehydes to the corresponding alcohols with a much lower flavor threshold through alcohol reductase activity. Homofermentative lactobacilli may increase lipid oxidation through formation of hydrogen peroxide [21]. Baker's yeast also reduced flavor-active aldehydes resulting from lipid oxidation, but slower and through different metabolic pathways [21]. Thiol accumulation by sourdough lactic acid bacteria [56] may additionally influence lipid oxidation. Cysteine and related thiol compounds reduce linoleic acid peroxides to the corresponding hydroxy fatty acids [160], and thus interrupts the reaction cascade leading to flavor-active aldehydes.

7.8 Cell-to-Cell Communication

Bacteria synthesize, release, sense, and respond to small signaling molecules that are defined as auto-inducers. The signaling molecules accumulate in the environment and lead to a series of physiological and biochemical responses when the quorum (threshold concentration) is reached. The term quorum sensing is derived from this consideration, and is mostly used to describe cell-to-cell communication [161]. Overall, the mechanisms of intraspecies communication include the use of acyl-homoserine lactones (AHL) and auto-inducing peptides (AIP) for Gram-negative and -positive bacteria, respectively [162, 163]. The interspecies communication is mainly based on signaling molecules such as furanone derivatives. The mechanisms involve the LuxS protein or AIP molecules and the three-component regulatory system (3CRS) [161]. Several studies considered the microbial dynamics during sourdough fermentation [164]. Within this complex food ecosystem, an understanding of the mechanisms of interspecies communication should give new insights into the physiological response of sourdough lactic acid bacteria and the interactions in an heterogeneous microbial community that govern growth and metabolism. The mechanism of cell communication was studied in *L. sanfranciscensis* CB1 co-cultivated with other sourdough lactic acid bacteria [165]. The highest number of dead and/or damaged cells of *L. sanfranciscensis* CB1 was found when co-cultured with *L. plantarum* DC400 or *L. brevis* CR13. The co-cultivation with *L. rossiae* A7 did not interfere with survival compared to the mono-culture. The analysis of the proteome revealed the induction of several cytoplasm proteins of *L. sanfranciscensis* CB1, especially when associated with *L. plantarum* DC400. The majority of the induced proteins had a key role in the mechanisms of response to environmental stresses, leading to the hypothesis that co-cultivation with some species of lactic acid bacteria could be considered like an environmental stress which promoted cell communication. A central role for the LuxS protein was established, while also furanone derivatives were identified as presumptive signaling molecules. The synthesis of volatile compounds and the activity of peptidases also decreased when *L. sanfranciscensis* CB1 was cultivated together with *L. plantarum* DC400. The mechanisms of cell communication were also studied in *L. plantarum* DC400 [166]. The co-cultivation of this strain with other sourdough lactic acid bacteria did not interfere with survival compared to the mono-culture. Nevertheless, the analysis of the proteome revealed the induction of proteins involved in the response to environmental stresses and cell communication. *Lactobacillus plantarum* DC400 synthesized the pheromone plantaricin A (PlnA) at variable concentrations that depended on the associated microbial species [167]. The co-cultivation with *L. pentosus*, *L. brevis* and other sourdough lactic acid bacteria species did not modify the synthesis of PlnA compared to the mono-culture conditions and did not show effects on the viability and survival of the other species. On the contrary, the co-cultivation with *L. sanfranciscensis* increased the synthesis of PlnA and caused a decrease of the viability of this species. The same effect was found during growth of *L. sanfranciscensis* in the presence of the chemically synthesized PlnA. Under these environmental conditions, the analysis of the proteome revealed the over-expression of proteins that had a central role in the energy metabolism, catabolism and biosynthesis of proteins and amino acids, stress responses, homeostasis of the redox potential, and programmed cell death. Notwithstanding antimicrobial activities due to substrate competition and synthesis of inhibitory compounds (e.g., phenyl lactic acid), and the elevated capacity to adapt to changing environments, it could be hypothesized that the dominance of *L. plantarum* during sourdough fermentation [168] may relate to the synthesis of the pheromone/bacteriocin PlnA as stimulated by mechanisms of quorum sensing.

References

1. Gobbetti M, De Angelis M, Corsetti A, Di Cagno R (2005) Biochemistry and physiology of sourdough lactic acid bacteria. Trends Food Sci Technol 16:57–69
2. Gobbetti M (1998) The sourdough microflora: interactions of lactic acid bacteria and yeast. Trends Food Sci Technol 9:267–274
3. Gänzle MG, Vermeulen N, Vogel RF (2007) Carbohydrate, peptide and lipid metabolism of lactic acid bacteria in sourdough. Food Microbiol 24:128–138
4. Hammes WP, Gänzle MG, Hammes WP, Gänzle MG (1998) Sourdough bread and related products. In: Wood BJB (ed) Microbiology of fermented foods, vol 199. Blackie Academic and Professional, London

5. Ehrmann MA, Vogel RF (2005) Molecular taxonomy and genetics of sourdough lactic acid bacteria. Trends Food Sci Technol 16:31–42
6. De Vuyst L, Neysens P (2005) The sourdough microflora: biodiversity and metabolic interactions. Trends Food Sci Technol 16:43–56
7. Vogel RF, Knorr R, Müller MRA, Steudel U, Gänzle MG, Ehrmann MA, Vogel RF, Knorr R, Müller MRA, Steudel U, Gänzle MG, Ehrmann MA (1999) Non-dairy lactic fermentations. The cereal world. Antonie van Leeuwenhoek 76:403–411
8. Meroth CB, Walter J, Hertel C, Brandt MJ, Hammes WP (2003) Monitoring the bacterial population dynamics in sourdough fermentation processes by using PCR- denaturing gradient gel electrophoresis. Appl Environ Microbiol 69:475–482
9. Gänzle MG, Ehmann MA, Hammes WP (1998) Modelling of growth of *Lactobacillus sanfranciscensis* and *Candida milleri* in response to process parameters of the sourdough fermentation. Appl Environ Microbiol 64:2616–2623
10. Wolfrum G (2003) Wachstum und Physiologie der Mikroflora in Getreidefermentationen. Dissertation, Fakultät Wissenschaftszentrum Weihenstephan für Ernährung, Landnutzung und Umwelt der Technischen Universität München
11. Messens W, Neysens P, Vansieleghem W, Vanderhoeven J, De Vuyst L (2002) Modeling growth and bacteriocin production by *Lactobacillus amylovorus* DCE 471 in response to temperature and pH values used for sourdough fermentation. Appl Environ Microbiol 68:1431–1435
12. Passos FV, Fleming HP, Ollis DF, Felder RM, McFeeters RF (1994) Kinetics and modeling of lactic acid production by *Lactobacillus plantarum*. Appl Environ Microbiol 60:2627–2636
13. Hammes WP, Vogel RF (1995) The genus *Lactobacillus*. In: Wood BJB, Holzapfel WH (eds) The genera of lactic acid bacteria. Blackie Academic and Professional, London, p 19
14. Vrancken G, Rimaux T, De Vuyst L, Leroy F (2008) Kinetic analysis of growth and sugar consumption by *Lactobacillus fermentum* IMDO 130101 reveals adaptation to the acidic sourdough ecosystem. Int J Food Microbiol 128:58–66
15. Galle S, Schwab C, Arendt E, Gänzle M (2010) Exopolysaccharide forming *Weissella* strains as starter cultures for sorghum and wheat sourdoughs. J Agric Food Chem 58:5834–5841
16. Korakli M, Vogel RF (2003) Purification and characterization of mannitol dehydrogenase from *Lactobacillus sanfranciscensis*. FEMS Microbiol Lett 220:281–286
17. Liu SQ (2003) Practical implication of lactose and pyruvate metabolism by lactic acid bacteria in food and beverage fermentation. Int J Food Microbiol 83:115–131
18. De Angelis M, Gobbetti M (1999) *Lactobacillus sanfranciscensis* CB1: manganese, oxigen, superoxide dismutase and metabolism. Appl Microbiol Biotechnol 51:358–363
19. De Angelis M, Gobbetti M (2004) Environmental stress response in *Lactobacillus*: a review. Proteomics 4:106–122
20. Zhang C, Gänzle MG (2010) Metabolic pathway of α-ketoglutarate in *Lactobacillus sanfranciscensis* and *Lactobacillus reuteri* during sourdough fermentation. J Appl Microbiol 109:1301–1310
21. Vermeulen N, Czerny M, Gänzle MG, Schieberle P, Vogel RF (2007) Reduction of (E)-2-nonenal and (E, E)-2,4-decadienal during sourdough fermentation. J Cereal Sci 45:78–87
22. Vermeulen N, Kretzer J, Machalitza H, Vogel RF, Gänzle MG (2006) Influence of redox-reactions catalysed by homo- and heterofermentative lactobacilli on gluten in wheat sourdoughs. J Cereal Sci 43:137–143
23. Ferain T, Schanck AN, Delcour J (1996) 13 C nuclear magnetic resonance analysis of glucose and citrate end products in an ldhL-ldhD double-knockout strain of *Lactobacillus plantarum*. J Bacteriol 178:7311–7315
24. Hansen Å, Schieberle P (2005) Generation of aroma compounds during sourdough fermentation: applied and fundamental aspects. Trends Food Sci Technol 16:85–94
25. Damiani P, Gobbetti M, Cossignani L, Corsetti A, Simonetti MS, Rossi J (1996) The sourdough microflora. Characterization of hetero- and homofermentative lactic acid bacteria, yeast and their interactions on the basis of the volatile compounds produced. Z Lebensm Wiss Technol 29:63–70

26. Stolz P, Böcker G, Hammes WP, Vogel RF (1995) Utilization of electron acceptors by lactobacilli isolated from sourdough I *Lactobacillus sanfrancisco*. Z Lebensm Unters Forsch 201:91–96
27. Gobbetti M, Corsetti A (1996) Co-metabolism of citrate and maltose by *Lactobacillus brevis* subsp. *lindneri* CB1 citrate-negative strain: effect on growth, end-products and sourdough fermentation. Z Lebensm Unters Forsch 203:82–87
28. Sekwati-Monang B, Gänzle M (2011) Microbiological and chemical characterisation of ting, a sorghum-based sourdough product from Botswana. Int J Food Microbiol 150:115–121
29. Oude Elferink SJWH, Krooneman J, Gottschal JC, Spoelstra SF, Faber F, Driehuis F (2001) Anaerobic conversion of lactic acid to acetic acid and 1,2-propanediol by *Lactobacillus buchneri*. Appl Environ Microbiol 67:125–132
30. Zhang C, Brandt MJ, Schwab C, Gänzle MG (2010) Propionic acid production by cofermentation of *Lactobacillus buchneri* and *Lactobacillus diolivorans* in sourdough. Food Microbiol 27:390–395
31. Titgemeyer F, Hillen W (2002) Global control of sugar metabolism: a gram-positive solution. Antonie Van Leeuwenhoek 82:59–71
32. Leroy F, De Winter T, Adriany T, Neysens P, De Vuyst L (2006) Sugars relevant for sourdough fermentation stimulate growth of and bacteriocin production by *Lactobacillus amylovorus* DCE 471. Int J Food Microbiol 112:102–111
33. Paramithiotis S, Sofou A, Tskalidou E, Kalantzopulos G (2007) Flour carbohydrate catabolism and metabolite production by sourdough lactic acid bacteria. World J Microbiol Biotechnol 23:1417–1423
34. Stolz P, Böcker G, Vogel RF, Hammes WP (1993) Utilization of maltose and glucose by lactobacilli isolated from sourdough. FEMS Microbiol Lett 109:237–242
35. Schwab C, Walter J, Tannock GW, Vogel RF, Gänzle MG (2007) Sucrose utilization and impact of sucrose on glycosyltransferase expression in *Lactobacillus reuteri*. Syst Appl Microbiol 30:433–443
36. Gobbetti M, Corsetti A, Rossi J (1994) The sourdough microflora. Interactions between lactic acid bacteria and yeasts: metabolism of carbohydrates. Appl Microbiol Biotechnol 41:456–460
37. Neubauer H, Glaasker E, Hammes WP, Poolman B, Konings WN (1994) Mechanisms of maltose uptake and glucose excretion in *Lactobacillus sanfrancisco*. J Bacteriol 176:3007–3012
38. De Vuyst L, Schrijvers V, Paramithiotis S, Hoste B, Vancanneyt M, Swingis J, Kalantzopoulos G, Tsakalidou E, Messens W (2002) The biodiversity of lactic acid bacteria in greek traditional wheat sourdoughs is reflected in both composition and metabolite formation. Appl Environ Microbiol 68:6059–6069
39. Tieking M, Korakli M, Ehrmann MA, Gänzle MG, Vogel RF (2003) In situ production of EPS by intestinal and cereal isolates of lactic acid bacteria during sourdough fermentation. Appl Environ Microbiol 69:945–952
40. Teixeira JS, McNeill V, Gänzle MG (2012) Levansucrase and sucrose phoshorylase contribute to raffinose, stachyose, and verbascose metabolism by lactobacilli. Food Microbiol 31:278–284
41. De Angelis M, Gallo G, Settanni L, Corbo MR, McSweeney PLH, Gobbetti M (2005) Purification and characterization of an intracellular family 3 β-glucosidase from *Lactobacillus sanfranciscensis* CB1. Ital J Food Sci 17:131–142
42. Gobbetti M, Lavermicocca P, Minervini F, De Angelis M, Corsetti A (2000) Arabinose fermentation by *Lactobacillus plantarum* in sourdough with added pentosans and α-L-arabinofunarosidase: a tool to increase the production of acetic acid. J Appl Microbiol 88:317–324
43. Gobbetti M, De Angelis M, Arnault P, Tossut P, Corsetti A, Lavermicocca P (1999) Added pentosans in breadmaking: fermentations of derived pentoses by sourdough lactic acid bacteria. Food Microbiol 16:409–418
44. Kunji ER, Mierau I, Hagting A, Poolman B, Konings WN (1996) The proteolytic system of lactic acid bacteria. Antonie Van Leeuwenhoek 70:187–221

45. Guèdon E, Renault P, Ehrlich SD, Delorme C (2001) Transcriptional pattern of genes coding for the proteolytic system of *Lactococcus lactis* and evidence for coordinated regulation of key enzymes by peptide supply. J Bacteriol 183:3614–3622
46. Vermeulen N, Pavlovic M, Ehrmann MA, Gänzle MG, Vogel RF (2005) Functional characterization of the proteolytic system of *Lactobacillus sanfranciscensis* DSM20451T during growth in sourdough. Appl Environ Microbiol 71:6260–6266
47. Wieser H, Vermeulen N, Gaertner F, Vogel RF (2007) Effect of different *Lactobacillus* and *Enterococcus* strains and chemical acidification regarding degradation of gluten proteins during sourdough fermentation. Eur Food Res Technol 226:14
48. Thiele C, Gänzle MG, Vogel RF (2002) Contribution of sourdough lactobacilli, yeast, and cereal enzymes to the generation of amino acids in dough relevant for bread flavour. Cereal Chem 79:45–51
49. Loponen J, Mikola M, Katina K, Sontag-Strohm T, Salovaara H (2004) Degradation of HMW glutenins during wheat sourdough fermentation. Cereal Chem 81:87–93
50. Gänzle MG, Loponen J, Gobbetti M (2008) Proteolysis in sourdough fermentations: mechanisms and potential for improved bread quality. Trends Food Sci Technol 19:513–521
51. Gobbetti M, Smacchi E, Corsetti A (1996) The proteolytic system of *Lactobacillus sanfranciscensis* CB1: purification and characterization of a proteinase, a dipeptidase, and an aminopeptidase. Appl Environ Microbiol 62:3220–3226
52. Di Cagno R, De Angelis M, Lavermicocca P, De Vincenzi M, Giovannini C, Faccia M, Gobbetti M (2002) Proteolysis by sourdough lactic acid bacteria: effects on wheat flour protein fractions and gliadin peptides involved in human cereal intolerance. Appl Environ Microbiol 68:623–633
53. Pepe O, Villani F, Oliviero D, Greco T, Coppola S (2003) Effect of proteolytic starter cultures as leavening agents of pizza dough. Int J Food Microbiol 84:319–326
54. Gallo G, De Angelis M, McSweeney PLH, Corbo MR, Gobbetti M (2005) Partial purification and characterization of an X-prolyl dipeptidyl aminopeptidase from *Lactobacillus sanfranciscensis* CB1. Food Chem 9:535–544
55. Hu Y, Stromeck A, Loponen J, Lopes-Lutz D, Schieber A, Gänzle MG (2011) LC-MS/MS quantification of bioactive antiotensin I-converting enzyme inhibitory peptides in rye malt sourdoughs. J Agric Food Chem 59:11983–11989
56. Jänsch A, Korakli M, Vogel RF, Gänzle MG (2007) Glutathione reductase from *Lactobacillus sanfranciscensis* DSM20451T: contribution to oxygen tolerance and thiol-exchange reactions in wheat sourdoughs. Appl Environ Microbiol 73:4469–4476
57. Coda R, Rizzello CG, Pinto D, Gobbetti M (2012) Selected lactic acid bacteria synthesize antioxidant peptides during sourdough fermentation of cereal flours. Appl Environ Microbiol 78:1087–1096
58. Rizzello CG, De Angelis M, Di Cagno R, Camarca A, Silano M, Losito I, De Vincenzi M, De Bari MD, Palmisano F, Maurano F, Gianfrani C, Gobbetti M (2007) Highly efficient gluten degradation by lactobacilli and fungal proteases during food processing: new perspectives for celiac disease. Appl Environ Microbiol 73:4499–4507
59. Di Cagno R, Barbato M, Di Camillo C, Rizzello CG, De Angelis M, Giuliani G, De Vincenzi M, Gobbetti M, Cucchiara S (2010) Gluten-free sourdough wheat baked goods appear safe for young celiac patients: a pilot study. J Ped Gastroent Nutr 51:777–783
60. Greco L, Gobbetti M, Auricchio R, Di Mase R, Landolfo F, Paparo F, Di Cagno R, De Angelis M, Rizzello CG, Cassone A, Terrone G, Timpone L, D'Aniello M, Maglio M, Troncone R, Auricchio S (2011) Safety for patients with celiac disease of baked goods made of wheat flour hydrolyzed during food processing. Clin Gastroenterol Pathol 9:24–29
61. De Angelis M, Cassone A, Rizzello CG, Gagliardi F, Minervini F, Calasso M, Di Cagno R, Francavilla R, Gobbetti M (2010) Gluten-free pasta made of *Triticum turgidum* L. var. *durum*: mechanisms of epitopes hydrolysis by peptidases of sourdough lactobacilli. Appl Environ Microbiol 75:50–518
62. De Angelis M, Di Cagno R, Gallo G, Curci M, Siragusa S, Crecchio C, Parente E, Gobbetti M (2007) Molecular and functional characterization of *Lactobacillus sanfranciscensis* strains isolated from sourdoughs. Int J Food Microbiol 114:69–82

63. Christensen JF, Dudley EG, Pederson JA, Steele JL (1999) Peptidases and amino acid catabolism in lactic acid bacteria. Antonie Van Leeuwenhoek 76:217–246
64. Su MSW, Schlicht S, Gänzle MG (2011) Contribution of glutamate decarboxylase in *Lactobacillus reuteri* to acid resistance and persistence in sourdough fermentation. Microb Cell Factories 10(suppl1):S8
65. Fernandez M, Zuniga M (2006) Amino acid catabolic pathoways of lactic acid bacteria. Crit Rev Microbiol 32:155–183
66. Kieronczyk A, Skeie S, Olsen K, Langsrud T (2001) Metabolism of amino acids by resting cells of non-starter lactobacilli in relation to flavour development in cheese. Int Dairy J 11:217–224
67. Tonon T, Bourdineaud JP, Lonvaud-Funel A (2001) The arcABC gene cluster encoding the arginine deiminase pathway of *Oenococcus oeni*, and arginine induction of a CRP-like gene. Res Microbiol 152:653–661
68. Hammes WP, Hertel C (2006) The genera *Lactobacillus* and *Carnobacterium*. Prokaryotes 4:320–403
69. De Angelis M, Mariotti L, Rossi J, Servili M, Fox PF, Rollàn G, Gobbetti M (2002) Arginine catabolism by sourdough lactic acid bacteria: Purification and characterization of the arginine deiminase (ADI) pathway enzymes from *Lactobacillus sanfranciscensis* CB1. Appl Environ Microbiol 68:6193–6201
70. Vrancken G, Rimaux T, Wouters D, Leroy F, De Vuyst L (2009) The arginine deiminase pathway of *Lactobacillus fermentum* IMDO 130101 responds to growth under stress conditions of both temperature and salt. Food Microbiol 26:720–727
71. Vermeulen N, Gänzle MG, Vogel RF (2007) Glutamine deamidation by cereal-associated lactic acid bacteria. J Appl Microbiol 103:1197–1205
72. Weingand-Ziadé A, Gerber-Dé Combay C, Affolter M (2003) Functional characterization of a salt- and thermotolerant glutaminase from *Lactobacillus rhamnosus*. Enzyme Microb Technol 32:862–867
73. Tanous C, Kieronczyk A, Helinck S, Chambellon E, Yvon M (2002) Glutamate dehydrogenase activity: a major criterion for the selection of flavour-producing lactic acid bacteria strains. Antonie Van Leeuwenhoek 82:271–278
74. Vermeulen N, Gänzle MG, Vogel RF (2006) nfluence of peptide supply and co-substrates on phenylalanine metabolism of *Lactobacillus sanfranciscensis* DSM20451T and *Lactobacillus plantarum* TMW1.468. J Agric Food Chem 54:3832–3839
75. Coda R, Rizzello CG, Gobbetti M (2010) Use of sourdough fermentation and pseudo-cereals and leguminous flours for the making of a functional bread enriched of γ-aminobutyric acid (GABA). Int J Food Microbiol 37:236–245
76. Stromeck A, Hu Y, Chen L, Gänzle MG (2011) Proteolysis and bioconversion of cereal proteins to glutamate and γ aminobutyrate in rye malt sourdoughs. J Agric Food Chem 59:1392–1399
77. Serrazanetti DI, Ndagijimana M, Sado-Kamdem SL, Corsetti A, Vogel RF, Ehrmann M, Guerzoni ME (2011) Acid-stress mediated metabolic shift in *Lactobacillus sanfranciscensis* LSCE1. Appl Environ Microbiol 77:2659–2666
78. Curtin ÁC, De Angelis M, Cipriani M, Corbo MR, McSweeney PLH, Gobbetti M (2001) Amino acid catabolism in cheese-related bacteria: selection and study of the effects of pH, temperature and NaCl by quadratic response surface methodology. J Appl Microbiol 91:312–321
79. De Angelis M, Curtin ÁC, McSweeney PLH, Faccia M, Gobbetti M (2002) *Lactobacillus reuteri* DSM 20016: purification and characterization of a cystathionine γ-lyase and use as adjunct starter in cheese-making. J Dairy Res 69:255–267
80. Korakli M, Vogel RF (2006) Structure/function relationship of homopolysaccharide producing glycansucrases and therapeutic potential of their synthesised glycans. Appl Microbiol Biotechnol 71:790–803
81. van Hijum SAFT, Kralj S, Ozimek LK, Kijkhuizen L, van Geel-Schutten IGH (2006) Structure-function relationships of glucansucrase and fructansucrase enzymes from lactic acid bacteria. Microbiol Molec Biol Rev 70:157–176

82. de Vuyst L, de Vin F, Vaningelem F, Degeest B (2001) Recent development in the biosynthesis and applications of heteropolysaccharides from lactic acid bacteria. Int Dairy J 11:687–708
83. Boels IC, van Kranenburgt R, Hugenholtz J, Kleerebezem M, de Vos WM (2001) Sugar catabolism and its impact on the biosynthesis and engineering of exopolysaccharide production in lactic acid bacteria. Int Dairy J 11:723–732
84. Broadbent JR, McMahon DJ, Welker DL, Oberg CJ, Moineau S (2003) Biochemistry, k genetics, and applications of exopolysaccharide production from Streptococcus thermophilus: a review. J Dairy Sci 86:407–423
85. Gruter M, Leeflang BR, Kuiper J, Kamerling JP, Vliegenthart JF (1992) Structure of the exopolysaccharide produced by *Lactococcus lactis* subspecies *cremoris* H414 grown in a defined medium or skimmed milk. Carbohydr Res 231:273–291
86. Dols-Lafargue M, Lee HY, le Marrec C, Heyraud A, Chambat GP, Lonvaud-Funel A (2008) Characgterization of *gtf*, a glucosyltransferase gene in the genomes of *Pediococcus parvulus* and *Oenococcus oeni*, two bacterial species commonly found in wine. Appl Environ Microbiol 74:4079–4090
87. Van der Meulen R, Grosu-Tudor S, Mozzi F, Vaningelgem F, Zamfir M, Font de Valdez G, De Vuyst L (2007) Screening of lactic acid bacteria isolates from dairy and cereal products for exoplysaccharide production and genes involved. Int J Food Microbiol 118:250–258
88. Bunaix M-S, Gabriel V, Morel S, Robert H, Rabier P, Remaud-Siméon M, Gabriel B, Fontagné-Faucher C (2009) Biodiversity of exopolysaccharides produced from sucrose by sourdough lactic acid bacteria. J Agric Food Chem 57:10889–10897
89. Palomba S, Cavella S, Torrieri E, Piccolo A, Mazzei P, Blaiotta G, Ventorino V, Pepe O (2012) Polyphasic screening, homopolysaccharide composition, and viscoelastic behavior of wheat sourdough from a *Leuconostoc lactis* and *Lactobacillus curvatus* exopolysaccharide-producing starter culture. Appl Environ Microbiol 78:2737–2747
90. Galle S, Schwab C, Arendt EK, Gänzle MG (2011) Structural and rheological characterisation of heteropolysaccharides produced by lactic acid bacteria in wheat and sorghum sourdough. Food Microbiol 28:547–553
91. Di Cagno R, De Angelis M, Limitone A, Minervini F, Carnevali P, Corsetti A, Gänzle M, Ciati R, Gobbetti M (2006) Glucan and fructan production by sourdough *Weisella cibaria* and *Lactobacillus plantarum* and their effect on bread texture. J Agric Food Chem 54:9873–9881
92. Schwab C, Mastrangelo M, Corsetti A, Gänzle MG (2008) Formation of oligosaccharides and polysaccharides by *Lactobacillus reuteri* LTH5448 and *Weissella cibaria* 10 M in sorghum sourdoughs. Cereal Chem 85:679–684
93. Galle S, Arendt EK Exopolysaccharides from sourdough lactic acid bacteria. A review. Crit Rev Food Sci Nutr (in press)
94. Jolly L, Stingele F (2001) Molecular organization and functionality of exopolysaccharide gene clusters in lactic acid bacteria. Int Dairy J 11:733–746
95. Tieking M, Kühnl W, Gänzle MG (2005) Evidence for formation of heterooligosaccharides by *Lactobacillus sanfranciscensis* during growth in wheat sourdough. J Agric Food Chem 53:2456–2461
96. Beine R, Moraru R, Nimtz M, Na'amineh S, Pawlowski A, Buchholz K, Seibel J (2008) Synthesis of novel fructooligosaccharides by substrate and enzyme engineering. J Biotechnol 138:33–41
97. Dols M, Remaud Simeon M, Willemot R-M, Vignon MR, Monsan PF (1998) Structural characterization of the maltose acceptor-products synthesized by *Leuconostoc mesenteroides* NRRL B-1299 dextransucrase. Carbohydr Res 305:549–559
98. Tieking M, Ehrmann MA, Vogel RF, Gänzle MG (2005) Molecular and functional characterization of a levansucrase from *Lactobacillus sanfranciscensis*. Appl Microbiol Biotechnol 66:655–663
99. Kralj S, Stripling E, Sanders P, van Geel-Schutten GH, Dijkhuizen L (2005) Highly hydrolytic reuteransucrase from probiotic *Lactobacillus reuteri* strain ATCC 55730. Appl Environ Microbiol 71:3942–3950

100. Seibel J, Buchholz K (2010) Tools in oligosaccharide synthesis: current research and application. Adv Carbohydr Chem Biochem 63:101–138
101. Kralj S, van Leeuwen SS, Valk V, Eeuwema W, Kamerling JP, Dijkhuizen L (2008) Hybrid reuteransurase enzymes reveal regions important for glucosidic linkage specificity and the transglucosylation/hydrolysis ratio. FEBS J 275:6002–6010
102. Irague R, Rolland-Sabaté A, Laurence Tarpuis L, Doublier JLO, Moulis C, Monsan P, Remeaud-Siméon M, Potocki-Véronèse G, Buléon A (2011) Structure and property engineering of α–D–glucans synthesized by dextransucrase mutants. Biomacromolecules 13:187–195
103. Kaditzki SJ, Behr J, Stocker A, Kaden P, Gänzle MG, Vogel RF (2008) Influence of pH on the formation of glucan by *Lactobacillus reuteri* TMW 1.106 exerting a protective function against extreme pH values. Food Biotechnol 22:398–418
104. Anwar MA, Kralj S, van der Maarel MJEC, Dijkhuizen L (2008) The probiotic *Lactobacillus johnsonii* NCC533 produces high-molecular-mass inulin from sucrose by using an inulosucrase enzyme. Appl Environ Microbiol 74:3426–3433
105. Cvitkovitch DG, LI YH, Ellen RP (2003) Quorum sensing and biofilm formation in streptococcal infections. J Clin Invest 112:1626–1632
106. Walter J, Schwab C, Loach DM, Gänzle MG, Tannock GW (2008) Glucosyltransferase A (GtfA) and inulosucrase (Inu) of *Lactobacillus reuteri* TMW1.106 contribute to cell aggregation, in vitro biofilm formation, and colonization of the mouse gastrointestinal tract. Microbiology 154:72–80
107. Kaditzky S, Seitter M, Hertel C, Vogel RF (2008) Performance of *Lactobacillus sanfranciscensis* TMW1.392 and its levansucrase deletion mutant in wheat dough and comparison of their impact on bread quality. Eur Food Res Technol 227:433–442
108. Galle S, Schwab C, Dal Bello F, Coffey A, Gänzle M, Arendt E (2012) Influence of in situ synthesised exopolysaccharides on the quality of gluten-free sorghum sourdough bread. Int J Food Microbiol 155:105–112
109. Schwab C, Vogel RF, Gänzle MG (2007) nfluence of oligosaccharides on the viability and membrane properties of *Lactobacillus reuteri* TMW1.106 during freeze-drying. Cryobiology 55:108–114
110. Vereyken IJ, Chupin V, Demel RA, Smeekens SCM, De Kruijff B (2003) Fructans insert between the headgroups of phospholipids. Biochim Biophys Acta 1510:307–320
111. Kim D-S, Thomas S, Fogler HS (2000) Effects of pH and trace minerals on long-term starvation of *Leuconostoc mesenteroides*. Appl Environ Microbiol 66:976–981
112. Katina K, Maina NH, Juvonen R, Flander F, Johansson L, Virkki L, Genkanen M, Laitila A (2009) In situ production and analysis of *Weissella confusa* dextran in wheat sourdough. Food Microbiol 26:734–743
113. Kaditzky S, Vogel RF (2008) Optimization of exopolysaccharide yields in sourdoughs fermented by lactobacilli. Eur Food Res Technol 228:291–299
114. Waldherr FW, Meissner D, Vogel RF (2008) Genetic and functional characterization of *Lactobacillus panis* levansucrase. Arch Microbiol 190:497–505
115. Rodrigues S, Lona LMF, Franco TT (2005) The effect of maltose on dextran yield and molecular weight distribution. Bioprocess Biosyst Eng 28:9–14
116. Decock P, Capelle S (2005) Bread Technology and sourdough technology. Trends Food Sci 16:113–120
117. Wang Y, Gänzle MG, Schwab C (2010) EPS synthesized by *Lactobacillus reuteri* decreases binding ability of enterotoxigenic Escherichia coli to porcine erythrocytes. Appl Environ Microbiol 76:4863–4866
118. Bello FD, Walter J, Hertel C, Hammes WP (2001) In vitro study of prebiotic properties of levan-type exopolysaccharides from lactobacilli and non-digestible carbohydrates using denaturing gradient gel electrophoresis. Syst Appl Microbiol 24:232–237
119. Gänzle MG, Zhang C, Sekwati-Monang B, Lee V, Schwab C (2009) Novel metabolites from cereal-associated lactobacilli – Novel functionalities for cereal products? Food Microbiol 26:712–719
120. Schnürer J, Magnusson J (2005) Antifungal lactic acid bacteria as preservatives. Trends Food Sci Technol 16:70–78

121. Dal Bello F, Clarke CI, Ryan LAM, Ulmer H, Schober TJ, Ström K, Sjörgen J, van Sinderen D, Schnürer J, Arendt EK (2007) Improvement of the quality and shelf life of wheat bread by fermentation with the antifungal strain *Lactobacillus plantarum* FST1.7. J Cereal Sci 45:309–218
122. Ryan LAM, Dal Bello F, Arendt EK (2008) The use of sourdough fermented by anifungal LAB to reduce the amount of calcium propionate in bread. Int J Food Microbiol 125:274–278
123. Drews E (1959) Der Einfluß gesteigerter Essigsäurebildung auf die Haltbarkeit des Schrotbrotes. Brot Gebäck 13:113–114
124. Lavermicocca P, Valerio F, Evidente A, Lazzaroni S, Corsetti A, Gobbetti M (2000) Purification and characterization of novel antifungal compounds from the sourdough *Lactobacillus plantarum* strain 21B. Appl Environ Microbiol 66:4084–4090
125. Lavermicocca P, Valerio F, Visconti A (2003) Antifungal activity of phenyllactic acid against moulds isolated from bakery products. Appl Environ Microbiol 69:634–640
126. Ryan LAM, Dal Bello F, Czerny M, Koehler P, Arendt EK (2009) Quantification of phenyllactic acid in wheat sourdough using high resolution gas chromatography – mass spectrometry. J Agric Food Chem 57:1060–1064
127. Ryan LAM, Dal Bello F, Arendt EK, Koehler P (2009) Detection and quantitation of 2,5-diketopeperazines in wheat sourdough bread. J Agric Food Chem 57:9563–9568
128. Gerez CL, Torino MI, Rollán G, de Valdez GF (2009) Prevention of bread mould spoilage by using lactic acid bacteria with antifungal properties. Food Control 20:144–148
129. Ryan LAM, Zannini E, Dal Bello F, Pawlosksa A, Koehler P, Arendt EK (2011) *Lactobacillus amylovorus* DSM19280 as a novel food-grade antifungal agent for bakery products. Int J Food Microbiol 146:276–283
130. Coda R, Rizzello CG, Nigro F, De Angelis M, Arnault P, Gobbetti M (2008) Long-term fungal inhibitory Activity of water-soluble extracts of *Phaseolus vulgaris* cv. Pinto and sourdough lactic acid bacteria during bread storage. Appl Environ Microbiol 74:7391–7398
131. Rizzello CG, Cassone A, Coda R, Gobbetti M (2011) Antifungal activity of sourdough fermented wheat germ used as an ingredient for bread making. Food Chem 127:952–959
132. Rizzello CG, Coda R, De Angelis M, Di Cagno R, Carnevali P, Gobbetti M (2009) Long-term fungal inhibitory activity of water-soluble extract from *Amaranthus* spp. seeds during storage of gluten-free and wheat flour breads. Int J Food Microbiol 131:189–196
133. Coda R, Cassone A, Rizzello CG, Nionelli L, Cardinali G, Gobbetti M (2011) Antifungal activity of *Wickerhamomyces anomalus* and *Lactobacillus plantarum* during sourdough fermentation: identification of novel compounds and long-term effect during storage of wheat bread. Appl Environ Microbiol 77:3484–3492
134. Rosenquist H, Hansen A (1998) The antimicrobial effect of organic acids, sour dough and nisin against *Bacillus subtilis* and *B. Licheniformis* isolated from wheat bread. J Appl Microbiol 85:621–631
135. Katina K, Sauri M, Alakomi H-L, Mattila-Sandholm T (2002) Potential of lactic acid bacgteria to inhibit rope spoilage in wheat sourdough bread. Lebensm Wiss u-Technol 35:38–45
136. Pepe O, Blaiotta G, Moschetti G, Greco T, Villani F (2003) Rope-producing strains of *Bacillus* spp. from wheat bread and strategy for their control by lactic acid bacteria. Appl Environ Microbiol 69:2321–2329
137. Messens W, De Vuyst L (2002) Inhibitory substances produced by *Lactobacilli* isolated from sourdoughs – a review. Int J Food Microbiol 72:31–43
138. Settanni L, Corsetti A (2008) Application of bacteriocins in vegetable food biopreservation. Int J Food Microbiol 121:123–138
139. Hartnett DJ, Vaughan A, van Sinderen D (2002) Antimicrobial-producing lactic acid bacteria isolated from raw barley and sorghum. J Inst Brew 108:169–177
140. Leroy F, De Winter T, Foulquié Moreno MR, De Vuyst L (2007) The bacteriocin producer *Lactobacillus amylovorus* DCE 471 is a competitive starter culture for type II sourdough fermentations. J Sci Food Agric 87:1726–1736

141. Corsetti A, Settanni L, Van Sinderen D (2004) Characterization of bacteriocin-like inhibitory substances (BLIS) from sourdough lactic acid bacteria and evaluation of their in vitro and in situ activity. J Appl Microbiol 96:521–534
142. Gänzle MG, Höltzel A, Walter J, Jung G, Hammes WP (2000) Characterization of reutericyclin produced by *Lactobacillus reuteri* LTH2584. Appl Environ Microbiol 66:4325–4333
143. Hurdle JG, Heathcott AE, Yan B, Lee RE (2011) Reuter;icyclin and related analogues kill stationary phase *Clostridium difficile* at achievable colonic concentrations. J Antimicrob Chemother 66:1773–1776
144. Gänzle MG (2004) Reutericyclin: biological activity, mode of action, and potential application. Appl Microbiol Biotechnol 64:326–332
145. Andreasen MF, Christensen LP, Meyer AS, Hansen A (2000) Content of phenolic acids and ferulic acid dehydrodimers in 17 rye (*Secale cereale* L.) varieties. J Agric Food Chem 48:2837–2842
146. Piber M, Koehler P (2005) Identification of dehydro-ferulic acid-tyrosine in rye and wheat: evidence for a covalent cross-link between arabinoxylans and proteins. J Agric Food Chem 53:5276–5284
147. Boskov Hansen H, Andreasen MS, Nielsen MM, Larsen LM, Bach Knudsen KE, Meyer AS, Cristensen LP, Hansen A (2002) Changes in dietary fibre, phenolic acids, and activity of endodenous enzymes during rye bread-making. Eur Food Res Technol 214:33–42
148. Taylor JR, Schober TJ, Bean S (2006) Novel and non-food uses for sorghum and millets. J Cereal Sci 44:252–271
149. Dykes L, Rooney LW (2006) Sorghum and millet phenols and antioxidants. J Cereal Sci 44:236–251
150. Svensson L, Sekwati Monang B, Lopez-Lutz D, Schieber A, Gänzle MG (2010) Phenolic acids and flavonoids in non-fermented and fermented red sorghum (*Sorghum bicolor* (L.) Moench). J Agric Food Chem 58:9214–9220
151. Rodriguez H, Curiel JA, Landete JM, de las Rivas B, de Felipe FL, Gómez-Cordovés C, Mancheno JM, Munoz R (2009) Food phenolics and lactic acid bacteria. Int J Food Microbiol 132:79–90
152. Marazza JA, Garro MS, de Giori GS (2009) Aglycone production by *Lactobacillus rhamnosus* CRL981 during soymilk fermentation. Food Microbiol 26:333–339
153. Avila M, Jaquet M, Moine D, Requena T, Pelaez C, Arigoni F, Jankovic J (2009) Physiological and biochemical characterization of the two α-L-rhamnosidases of *Lactobacillus plantarum* NCC245. Microbiology 155:2739–2749
154. Curiel JA, Rodriguez H, Acebron I, Mancheno JM, delas Rivas B, Munoz R (2009) Pdoruction and physiochemical properties of recombinant *Lactobacillus plantarum* tannase. J Agric Food Chem 57:6224–6230
155. De las Rivas B, Rodriguez H, Curiel JA, Landete JM, Munoz R (2009) Meolcular screening of wine lactic acid bacteria degrading hydroxycinnamic acids. J Agric Food Chem 57:490–494
156. Van Beek S, Priest FG (2000) Decarboxylation of substituted cinnamic acids by lactic acid bacteria isolated during malt whisky fermentation. Appl Environ Microbiol 66:5322–5328
157. Sanchez-Maldonado AF, Schieber A, Gänzle MG (2011) Structure-function relationships of the antibacterial activity of phenolic acids and their metabolism by lactic acid bacteria. J Appl Microbiol 111:1176–1184
158. Campos FM, Couto JA, Figueiredo AR, Toth IV, Rangel AOSS, Hogg T (2009) Cell membrane damage induced by phenolic acids on wine lactic acid bacteria. Int J Food Microbiol 135:144–151
159. Moroni AV, Dal Bello F, Arendt EK (2009) Sourdough in gluten-free bread-making: an acient technology to solve a novel issue? Food Microbiol 26:676–684
160. Elshof MBW, Veldink GA, Vliegenthart JFG (1998) Biocatalytic hydroxylation of linoleic acid in a double-fed batch system with lipoxygenase and cysteine. Fett-Lipid 246–251
161. Fuqua C, Winans SC, Greenberg EP (1996) Census and consensus in bacterial ecosystems: the LuxR-LuxI family of quorum-sensing transcriptional regulators. Ann Rev Microbiol 50:727–751

162. Fuqua C, Parsek MR, Greenberg EP (2001) Regulation of gene expression by cell-to-cell communication: acyl-homoserine lactone quorum sensing. Ann Rev Genet 35:439–68
163. Nakayama J, Cao Y, Horii T, Sakuda S, Akkermans AD, de Vos WM, Nagasawa H (2001) Gelatinase biosynthesis-activating pheromone: a peptide lactone that mediates a quorum sensing in *Enterococcus faecalis*. Mol Microbiol 41:145–154
164. Gobbetti M, De Angelis M, Di Cagno R, Minervini F, Limitone A (2007) Cell–cell communication in food related bacteria. Int J Food Microbiol 120:34–45
165. Di Cagno R, De Angelis M, Limitone A, Minervini F, Simonetti MC, Buchin S, Gobbetti M (2007) Cell-cell communication in sourdough lactic acid bacteria: a protomic study in *Lactobacillus sanfranciscensis* CB1. Proteomics 7:2430–2446
166. Di Cagno R, De Angelis M, Coda R, Gobbetti M (2009) Molecular adaptation of sourdough *Lactobacillus plantarum* DC400 under co-cultivation with other lactobacilli. Res Microbiol 20:1–9
167. Di Cagno R, De Angelis M, Calasso M, Vicentini O, Vernocchi P, Ndagijimana M, De Vincenzi M, Dessì MR, Guerzoni ME, Gobbetti M (2010) Quorum sensing in sourdough *Lactobacillus plantarum* DC400: induction of plantaricin A (PlnA) under co-cultivation with other lactic acid bacteria and effect of PlnA on bacterial and Caco-2 cells. Proteomics 10:2175–2190
168. Siragusa S, Di Cagno R, Ercolini D, Minervini F, Gobbetti M (2009) Taxonomic structure and monitoring of the dominant population of lactic acid bacteria during wheat flour sourdough type I propagation using *Lactobacillus sanfranciscensis* starters. Appl Environ Microbiol 75:1099–1109
169. Gobbetti M, De Angelis M, Di Cagno R, Rizzello CG (2007) The relative contributions of starter cultures and non-starter bacteria to the flavour of chees. In: Weimer BC (ed) Improving the flavour of cheese. Woodhead publishing limited, Cambridge, England, pp 121–156

Chapter 8
Sourdough: A Tool to Improve Bread Structure

Sandra Galle

8.1 Introduction

The quality of bread is characterized by its flavor, nutritional value, texture, and shelf life [1]. In the baking industry, these characteristics are improved by addition of bread improvers or enzymes. Alternatively, the addition of sourdough influences all aspects of bread quality and thus meets the consumer demands for a reduced use of additives. As sourdough is an intermediate but not an end product the microbiological activity has to be determined on the bases of their impact on bread quality. Biochemical changes during sourdough fermentation occur in protein and carbohydrate components of the flour. The rate and extent of these changes greatly influence the properties of the sourdough and consequently the quality of the bread dough and bread structure. The effects are associated with the metabolites produced by lactic acid bacteria (LAB) and yeast during fermentation, including organic acids, exopolysaccharides (EPS), enzymes, and CO_2. The following chapter presents the impact of sourdough fermentation on structure-forming components of bread, and bread texture.

8.2 Components Determining Dough and Bread Structure

Gluten proteins of wheat create unique viscoelastic properties of dough, which allow dough to expand in response to formation of carbon dioxide and to retain most of this gas in the dough. Polymeric glutenins give strength and elasticity to wheat dough, whereas the monomeric gliadins are responsible for the viscous properties of the dough [2, 3]. Embedded in the protein network are starch granules that form

S. Galle (✉)
Department of Agricultural, Food and Nutritional Science, Edmonton, University of Alberta, Alberta, Canada
e-mail: galle@ualberta.ca

M. Gobbetti and M. Gänzle (eds.), *Handbook on Sourdough Biotechnology*,
DOI 10.1007/978-1-4614-5425-0_8,

a continuous structure with gluten during kneading. In freshly baked bread, the main networks are the continuous gluten network which forms a matrix between the swollen, gelatinized starch granules and the transient starch network, consisting of entangled, gelatinized starch polymers [4]. During cooling the main changes occur in the starch fraction, where amylose forms a partially crystalline amylose network, with crystalline amylose and amylose-lipid complex. The amylose network together with the gluten network formed during baking determines the resilience of the fresh bread. Additionally pentosans (arabinoxylans) and transient gelatinized amylopectin networks also contribute to the structure formation of freshly baked wheat bread. During storage bread texture becomes harder largely because of physical changes that occur in the starch-protein matrix of the bread crumb [5]. Retrogradation is the process by which starch reverts to a more crystalline form after gelatinization. Because amylose is already completely retrograded in fresh bread, the amylopectin retrograding over time is primarily responsible for bread staling [6]. In addition, water migrates within the crumb and from crumb to crust. This water loss is another important factor contributing to the bread staling process [4].

Compared to wheat flour, rye flour differs with regard to gluten composition, starch gelatinization, and α-amylases activity [7]. Consequently, factors determining structure formation in rye dough and rye bread structure differ from wheat dough and bread. Whereas sourdough was an essential ingredient for ensuring baking properties of dough containing more than 20% of rye flour, its addition to wheat dough remains optional. The use of sourdough has gained increased interest in the last decade as a means to improve the quality and flavor of wheat bread [1]. However, there is much disparity in the results concerning the effect of sourdough on the final wheat product and the use of sourdough has been shown to either decrease [8–10] or increase final bread volume [10–15].

8.3 Organic Acids

The production of organic acids and consequently the pH drop in sourdough have a major effect on structure-forming components like starch, gluten, and arabinoxylans (Table 8.1). The primary effect of acids on the protein fraction is the increased swelling and solubility of gluten proteins [16, 17]. This effect is explained by a positive net charge of the proteins in an acidic environment. Increased intramolecular electrostatic repulsion causes gluten proteins to unfold and expose more hydrophobic groups. The presence of strong intermolecular electrostatic repulsive forces prevents the formation of new bonds. This results in softer dough with less stability and shorter mixing time [18, 19]. Furthermore, softness of the gluten promotes swelling and increased water uptake [20]. Partial acid hydrolysis of starch also exerts positive effects on the starch granules leading to an increased water binding capacity [8, 21]. In addition to the impact of low pH on dough components, secondary effects of acidification and fermentation time include changes in the activity of cereal enzymes. Flour proteases have their pH optima in the acidic range and increased proteolysis

Table 8.1 Effects of sourdough metabolites on bread structure

Metabolites	Effect on flour components	Effect on dough/bread structure
Organic acids	Increased swelling and solubility of the gluten	Shorter mixing time, less stability of the dough
	Increased water uptake of gluten and starch	Increase of elasticity and softness of the dough
	Increased proteolysis of gluten proteins through flour endogenous proteinases	Both increase and reduction of bread volume
Organic acids	Increased solubility of pentosans through acid hydrolysis and endogenous pentosanases	Improved volume and crumb structure of rye and wheat bread
Organic acids	Inhibition of endogenous α-amylases	Bakeability of rye bread
Enzymes	Proteolysis Glutathione-reductase	Weakening of the gluten structure, dough softening
Exopolysaccharides	Increased water absorption	Increased softness of dough and bread texture, increased volume
	Interaction with gluten-starch network Inhibiting retrogradation of starch	Increased shelf life (anti-staling)
CO_2	Expansion of gas cells	Leavening of dough and bread

occurs in doughs at pH 4 in comparison to nonacidified systems [22]. In comparison to straight dough processes, the increased activity of cereal proteases is also attributable to longer fermentation times. The rheological consequence of gluten degradation is a major reduction of elasticity and firmness of the sourdough and subsequent bread dough [10, 23, 24]. Whether this has positive or negative effects on bread volume and staling depends on the acidity profile and gluten network. Physicochemical changes in the protein network resulting from sourdough fermentation enhance gas retention and allow greater expansion due to softer and more extensible doughs [11, 13, 25]. Confocal laser-scanning microscopy visualized the effect of incorporation of sourdough on the dough microstructure: Relative to the fine well-oriented network of the control, the gluten in the dough with added sourdough had a more amorphous nature and there were greater areas of aggregated material composed of thicker protein strands [23]. The presence of thicker strands could also account for a greater increase in loaf volume [26]. Increased volume correlates with the increased softness of the crumb, and is associated with a reduced rate of staling [13, 27]. However, if the acidity of sourdough is further increased, bread volume decreases [8–10]. Quantitative analysis of gluten in sourdough and chemically acidified doughs showed that gliadins and glutenins are hydrolyzed [28, 29]. Especially high molecular glutenins are completely degraded, which leads to a strong gluten softening. Weaker gluten increases the expansion of dough, but also decreases gas retention. Accordingly, the acidity level of sourdough and subsequent bread dough must be carefully controlled to attain increased volume [30].

The drop in pH during fermentation not only affects cereal proteases but modulates amylase activity. Acid conditions partially inactivate amylases. This is an important aspect in rye baking since excessive α-amylase activity results in a sticky crumb, a very open grain, and a reduction in loaf volume [7, 21]. In wheat flour, α-amylases are virtually absent, while β-amylases are abundantly present, but the latter have little if any activity on starch granules and are inactivated before starch gelatinization [4].

In rye flour proteins are not capable of forming a network similar to wheat gluten, here water-binding pentosans take over the function in the structure-forming process [21]. Pentosans mainly consist of arabinoxylans (AX), which can be subdivided into water extractable (WEAX) and water unextractable arabinoxylans (WUAX). WUAX have a deleterious influence on bread quality, in contrast WEAX increase bread volume due to the high water-binding capacity and the ability to undergo oxidative gelation [31] (see Chap. 2). Acidification has a pronounced effect on the solubility of AX due to acid hydrolyzation of WUAX, therefore increasing the proportion of WEAX [32]. Furthermore, during sourdough fermentation pH optima for endogenous rye enzymes like L-arabinofuanosidase, endo-xylanase, and xylosidase are reached [33]. Also in wheat endoxylanases cause a reduction of WUAX and increase the level of WEAX [6]. Thus, beneficial WEAX can increase during sourdough fermentation. In rye doughs, this improves physical properties by increasing the elasticity of the dough and the subsequent bread crumb. The bread crumb is characterized by an improved mouthfeel, improved crumb structure, and volume [7, 31, 33]. In wheat baking the increased solubility of pentosans in sourdough also contributes to enhanced bread volume and improved softness [6, 13, 34].

8.4 Enzymes from LAB Contributing to Dough and Bread Structure

Proteolysis in sourdoughs is mainly based on the pH-mediated activation of endogenous flour proteases [28, 29, 35] (Table 8.1). Lactobacilli used in sourdough fermentation may also exhibit strain-specific proteolytic activity [36, 37], but appear to play a minor role in the overall proteolysis. Although LAB do not influence overall proteolysis when compared to aseptically acidified doughs, they affect the pattern of hydrolyzed products, increasing the amount of dipeptides and amino acids. In fact, the proteolysis by LAB induced softening of the dough in comparison to chemically acidified doughs [37]. Furthermore, the choice of a highly proteolytic starter culture, such as *Enterococcus faecalis*, substantially contributed to the gluten proteolysis [38].

The significant role of enzymes produced by LAB has been proposed to explain observed differences in the staling of sourdough breads [12, 13]. Sourdough breads with comparable acidity levels had varying staling rates in terms of firmness and starch retrogradation. LAB strains possessing proteolytic and amylolytic activities were most effective in delaying staling [12].

In addition to the pH-dependent cereal proteases and LAB-liberated proteases, glutathione reductase expressed by heterofermentative lactobacilli contributes to depolymerization of gluten protein [39] (Table 8.1). Glutathione reductase reduces extracellular oxidized glutathione (GSSG) to the reducing agent glutathione (GSH) leading to increased SH groups in the gluten proteins. GSH undergoes thiol-exchange reactions with gluten proteins and decreases intermolecular disulfide cross-linking, resulting in a decreased molecular weight of the glutenin macropolymer (GMP). Only small amounts of reducing agents are necessary to strongly affect the disulfide network of glutenins [40].

8.5 Exopolysaccharides

Some strains of LAB synthesize EPS from sucrose. EPS exhibit a positive effect on the texture, mouthfeel, taste perception, and stability of fermented food. Moreover, prebiotic effects have been described for specific EPS [41–43].

EPS can be classified in homopolysaccharides (HoPS) and heteropolysaccharides (HePS) on the basis of their biosynthetic pathways. HoPS are synthesized from sucrose by extracellular glucansucrases or fructansucrases and contain only one type of monosaccharide, glucose or fructose, respectively. HePS are synthesized by intracellular glycosyltransferases and consist of one or several monosaccharides (see Chap. 7). HePS formation by LAB is used in dairy fermentation to improve the texture of yoghurt and other fermented dairy products, but was recently also employed in sorghum sourdough fermentation [44]. Glucan and fructan synthesis was observed for *Leuconostoc* spp. and *Weissella* spp., which synthesize dextrans [45]. *Lactobacillus reuteri*, *L. panis*, *L. pontis*, *L. frumenti*, *and L. sanfranciscensis* produce fructans (levan or inulin) and glucans (dextran, reuteran, or mutan) [43]. It was reported that any sourdough is likely to contain at least one EPS-producing strain [43, 46]. In addition to EPS, glucan and fructansucrases can synthesize gluco- and fructooligosaccharides, respectively. In sourdough *L. reuteri*, *L. sanfranciscensis* and *Weissella cibaria* were shown to synthesize FOS, 1-kestose, and isomaltooligosacchrides, respectively [47–49]. Together with EPS, in particular levan from *L. sanfranciscensis*, oligosaccharides have been described for their beneficial health effects. The production of EPS by sourdough LAB exerts two different functions: it improves the nutritional value (Chap. 9) of the bread and improves bread texture.

Bacterial HoPS in sourdough act as hydrocolloids. The effect of bacterial EPS and hydrocolloids on dough and bread is based on two functions: (1) water binding at the dough stage and (2) the interaction with other dough components such as proteins and starch. This interaction influences structural networks during fermentation and the baking process [50]. During sourdough fermentations, LAB can produce EPS in amounts that are sufficient for improving the dough structure [49]. Dextran from *Leuconostoc mesenteroides* was reported to contribute to the long storage stability of Panettone [51]. The application of in situ formed EPS can therefore reduce the requirement for hydrocolloids to meet the consumer demands for clean labels and reduced costs. Furthermore, levan produced in situ was more effective compared to externally added levan [52]. The technological benefits of in situ formed EPS depend on EPS properties and polymer yield. These are in turn dependent on the carbohydrate supply of the fermentation substrate. Depending on the strain, flour, dough yield, and fermentation conditions can be selected to obtain optimal EPS production [53].

Fig. 8.1 Wheat bread prepared without EPS containing sourdough (**a**) compared to bread prepared with sourdough containing EPS (**b**) (Data from Galle [10])

EPS formation in dough reduced elasticity and strength, demonstrating the interaction of EPS with the structure-forming components [10]. The addition of sourdough fermented with HoPS-producing strains in wheat and rye dough induced increased softening and freshness of the crumb and improved specific volume of the bread (Fig. 8.1) [10, 14, 51]. The baking industry currently employs in situ formed EPS and their potential as bread improvers is of growing importance. A process was developed to adapt sourdough microbiota to high sucrose, and to reach a content of 25% dextran [54]. Dextran interacts with the gluten network and builds structure resulting in improved dough stability and gas retention. Its high water-binding capacity accounts for the freshness of the baked product. Linear, high molecular weight dextran has a greater effect on bread volume compared to more branched, high molecular weight dextran [51, 54, 55]. When the same amounts of reuteran, levan, or dextran were added to wheat dough, dextran exhibited the best effect on viscoelastic properties of dough and the volume of the final bread [43].

Sucrose metabolism of LAB not only yields EPS but also glucose and fructose. Glucose supports gas production by yeast and contributes to dough leavening [10, 25, 56]. Fructose stimulates mannitol and acetate formation by heterofermentative LAB. Acetate has antimicrobial properties, but high acetate concentrations negatively influence flavor and bread structure. Beneficial effects on bread volume of levan and reuteran from *L. sanfranciscensis* and *L. reuteri*, respectively, were mitigated by the detrimental effects of acetate (Fig. 8.2) [9, 10]. Different from heterofermentative lactobacilli, *Weissella* spp. form only minor amounts of acetate and at the same time significant amounts of dextran [15, 47, 48]. The application of dextran-containing *Weissella*-sourdoughs provided mildly acidic bread with improved structure and shelf life. Incorporation of higher amounts of *Weissella*-sourdough leads to a higher concentration of dextran in the final bread dough without negative impacts on the flavor and texture of the bread [10, 15].

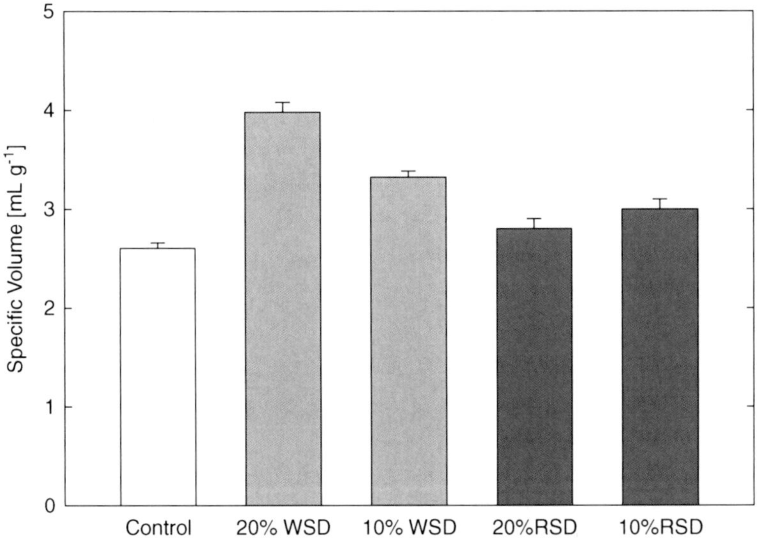

Fig. 8.2 Specific volume of bread prepared with 10 and 20% EPS-containing *Weissella*-sourdough (*WSD*) compared to bread made with 10 and 20% EPS- and acetate-containing *L. reuteri*-sourdough (*RSD*) (Data from Galle [10])

The findings of the influence of EPS in wheat and rye baking can be transferred into gluten-free baking (Table 8.2) and are discussed in detail in Chap. 10. In summary, EPS formed during sourdough fermentation from sucrose can be successfully applied to improve bread-making performance, however, it is necessary to select an EPS-producing starter culture not solely based on its polymer properties and polymer yield but also according to its by-products and fermentation performance.

8.6 CO_2 Formation

Sourdough typically contains yeasts and lactic acid bacteria. Interactions of yeasts and lactobacilli are important for microbial activity and CO_2 formation in sourdough [57]. Gas formation by microorganisms is necessary in order to leaven bread. In sourdough, carbon dioxide is produced by both heterofermentative LAB and yeasts. The contribution of each group to the overall gas volume depends on the starter culture and the dough technology applied [21, 25]. An increase in loaf volume was observed when sourdough with increased fermentation time was added to the dough, but when baker's yeast was applied the gas production by yeasts superseded the gas production by LAB [11, 21, 30]. Unless sourdough is used as the sole leavening agent, it is generally assumed that sourdough improves gas retention but not gas production in bread dough [11, 21].

Table 8.2 In situ formed EPS and their effect on bread quality

EPS	Strains	Amounts [g kg^{-1} SD[a]]	Effects on bread structure	References
Dextran	*Ln. mesenteroides*	0.6–16 g/kg	Increases volume (wheat)	[10, 15, 48, 51, 56]
	W. cibaria		Decreases firmness (wheat and sorghum)	
	W. confusa		Improves freshness (wheat, rye, sorghum)	
Glucan	*W. cibaria*	2.5 g/kg	Increases dough viscosity (wheat)	[14]
	L. plantarum		Increases volume (wheat)	
			Decreased firmness (wheat)	
Levan	*L. reuteri*	1.5–5.2 g kg^{-1}	No improvement (wheat, sorghum)	[9, 48]
	L. sanfranciscensis			
Fructan	*L. reuteri*	3.3 g kg^{-1}	Decreases firmness (sorghum)	[56]
			Decreases freshness (sorghum)	
Reuteran	*L. reuteri*	0.6–5.1 g kg^{-1}	Decreases firmness (sorghum, wheat)	[10, 56]
			Improves freshness (sorghum, wheat)	
			Decrease in dough strength and elasticity (wheat, sorghum)	
HePS	*L. buchneri*	n.d.	Decreases dough strength and elasticity (sorghum)	[44]

[a]SD … sourdough; n.d. … HePS amount not detectable. HePS formation in sourdough was confirmed by gene expression

8.7 Alternative Fermentations and Synergistic Affects of Sourdough and Dough Additives

The nutritional importance of dietary fiber has been demonstrated in many studies. Acceptable loaf volume with high-fiber, whole-grain bread is difficult to obtain. The fermentation of bran allows enhanced water absorption and textural modification of bran particles. The use of fermented bran in combination with an enzyme mixture (α-amylases, xylanase and lipase) improved the volume, texture, and shelf life of high-fiber wheat bread [58]. Combined use of exogenous enzymes and sourdough in wheat baking also enhanced the rate of acidification, improved bread volume, and retarded staling [13, 59, 60]. Recently, sourdough in combination with cryoprotectants and/or conventional additives (e.g., honey, hydrocolloids) in frozen dough technology were used to overcome problems such as prolonged final leavening time, lower loaf volume, and poor bread characteristics [61].

8.8 Conclusion

The addition of sourdough improves the quality of bread and has a long tradition in production of wheat and rye bread. Organic acid, EPS, enzymes, and CO_2 synthesized during sourdough fermentation by LAB and yeast are responsible for the positive changes in dough and bread quality. Organic acids positively influence structure-forming compounds of wheat and rye dough, particularly gluten, starch, and arabinoxylans. Furthermore, the production of organic acids and consequently the drop in pH induces/inhibits endogenous enzymes like proteinases, pentosanases, and amylases, also interfering with structural components. EPS formed from sucrose during fermentation can replace hydrocolloids, currently used as dough and bread improvers. However, it has to be considered that the metabolism of the sourdough microflora can also have negative impacts on the bread quality, i.e., increased acidification compromises crumb structure and masks beneficial effects of EPS. An improved knowledge of metabolites formed during sourdough fermentation and their interaction with the dough components enables a more direct optimization of the resulting bread quality. The studies presented in this review demonstrate that the use of an optimized sourdough process in the production of bread provides a feasible technology for producing breads with improved texture, volume, and shelf life. Sourdough technology can be useful to reduce or eliminate the level of additives often used in baked products, and furthermore helps to meet consumer demands for clean labels, natural products, and reduced costs.

References

1. Arendt EK, Ryan LAM, Dal Bello F (2007) Impact of sourdough on the texture of bread. Food Microbiol 24:165–174
2. Shewry PR, Halford NG, Tatham AS (1992) High molecular weight subunits of wheat glutenin. J Cereal Sci 15:105–120

3. Belton PS (1999) On the elasticity of wheat gluten. J Cereal Sci 29:103–107
4. Goesaert H, Slade H, Levine H, Delcour JA (2009) Amylases and bread firming – an integrated view. J Cereal Sci 50:345–352
5. Gray JA, Bemiller JN (2003) Bread staling: molecular basis and control. Compr Rev Food Sci Food Safety 2:1–20
6. Goesaert H, Brijs K, Veraverbeke WS, Courtin CM, Gebruers K, Delcour JA (2005) Wheat flour constituents: how they impact bread quality, and how to impact their functionality. Trends Food Sci Tech 16:12–30
7. Brandt MJ (2006) Bedeutung von Sauerteig für die Brotqualität. In: Brandt MJ, Gänzle MG (eds) Handbuch Sauerteig. B. Behr's Verlag, Hamburg, pp 21–40
8. Barber B, Ortolá C, Barber S, Fernández F (1992) Storage of packaged white bread. Z Lebensm Forsch 194:442–449
9. Kaditzky S, Seitter M, Hertel C, Vogel RF (2008) Performance of *Lactobacillus sanfranciscensis* TMW 1.392 and its levansucrase deletion mutant in wheat dough and comparison of their impact on bread quality. Eur Food Res Tech 227:433–442
10. Galle S (2011) Isolation, characterization and application of exopolysaccharides from lactic acid bacteria to improve the quality of wheat and gluten-free bread. Doctoral thesis, University College Cork, Ireland
11. Clarke CI, Schober TJ, Arendt EK (2002) The effect of single strain and traditional mixed strain starter cultures on rheological properties of wheat dough and bread quality. Cereal Chem 79:640–647
12. Corsetti A, Gobbetti M, Balestrieri F, Paoletti F, Russi L, Rossi J (1998) Sourdough lactic acid bacteria effects on bread firmness and staling. J Food Sci 63:347–351
13. Corsetti A, Gobbetti B, De Marco B, Balestrieri F, Paoletti F, Rossi J (2000) Combined effect of sourdough lactic acid bacteria and additives on bread firmness and staling. J Agric Food Chem 48:3044–3051
14. Di Cagno R, De Angelis M, Limitone A, Minervini F, Carnevali P, Corsetti A, Gänzle M, Ciati R, Gobbetti M (2006) Glucan and fructan production by sourdough *Weissella cibaria* and *Lactobacillus plantarum*. J Agric Food Chem 54:9873–9881
15. Katina K, Maina NH, Juvonen R, Flander L, Johansson L, Virkki L, Tenkanen M, Laitila A (2009) In situ production and analysis of *Weissella confusa* dextran in wheat sourdough. Food Microbiol 26:734–743
16. Axford DWE, McDermott EE, Redman DG (1979) Note on the sodium dodecyl sulfate test of breadmaking quality: comparison with Pelshenke and Zeleny tests. Cereal Chem 56:582–583
17. Takeda K, Matsumura Y, Shimizu M (2001) Emulsifying and surface properties of wheat gluten under acidic conditions. J Food Sci 66:393–399
18. Hoseney C (1994) Principles of cereals science and technology, 2nd edn. American Association of Cereal Chemists, St. Paul
19. Wehrle K, Grau H, Arendt EK (1997) Effects of lactic acid, acetic acid, and table salt on fundamental rheological properties of wheat dough. Cereal Chem 74:739–744
20. Schober TJ, Dockery P, Arendt EK (2003) Model studies for wheat sourdough systems using gluten, lactate buffer and sodium chloride. Eur Food Res Tech 217:235–243
21. Hammes WP, Ganzle MG (1998) Sourdough breads and related products. In: Woods BJB (ed) Microbiology of fermented foods, vol 1. Blackie Academic/Professional, London, pp 199–216
22. Thiele C, Gänzle MG, Vogel RF (2002) Contribution of sourdough lactobacilli, yeast and cereal enzymes to the generation of amino acids in dough relevant for bread flavour. Cereal Chem 79:45–51
23. Clarke C, Schober T, Dockery P, O'Sullican K, Arendt EK (2004) Wheat sourdough fermentation: effects of time and acidification on fundamental rheological properties. Cereal Chem 81:409–417
24. Thiele C, Grassl S, Gänzle M (2004) Gluten hydrolysis and depolymerization during sourdough fermentation. J Agric Food Chem 52:1307–1314
25. Gobbetti M, Corsetti A, Rossi J (1995) Interaction between lactic acid bacteria and yeasts in sour-dough using a rheofermentometer. World J Microbiol Biotech 11:625–630
26. Kieffer R, Stein N (1999) Demixing in wheat doughs-its influence on dough and gluten rheology. Cereal Chem 76:688–693

27. Crowley P, Schober T, Clarke C, Arendt E (2002) The effect of storage time on textural and crumb grain characteristics of sourdough wheat bread. Eur Food Res Tech 214:489–496
28. Thiele C, Gänzle G, Vogel RF (2003) Fluorescence labeling of wheat proteins for determination of gluten hydrolysis and depolymerization during dough processing and sourdough fermentation. J Agri Food Chem 51:2745–2752
29. Loponen J, Mikola M, Katina K, Sontag-Strohm T, Salovaara H (2004) Degradation of HMW glutenins during wheat sourdough fermentations. Cereal Chem 81:87–90
30. Clarke CI, Schober TJ, Angst E, Arendt EK (2003) Use of response surface methodology to investigate the effects of processing conditions on sourdough wheat bread quality. Eur Food Res Tech 217:23–33
31. Vinkx CJA, Delcour JA (1996) Rye (*Secale cereale* L.) arabinoxylans: a critical review. J Cereal Sci 24:1–14
32. Boskov Hansen H, Andreasen M, Nielsen M, Larsen L, Knudsen BK, Meyer A, Christensen L, Hansen A (2002) Changes in dietary fibre, phenolic acids and activity of endogenous enzymes during rye bread-making. Eur Food Res Tech 214:33–42
33. Brandt MJ (2006) Bedeutung von Rohwarenkompontenten. In: Brandt MJ, Gänzle MG (eds) Handbuch Sauerteig. B. Behr's Verlag, Hamburg, pp 41–55
34. Courtin C, Delcour J (2002) Arabinoxylans and endoxylanases in wheat flour bread-making. J Cereal Sci 35:225.243
35. Gänzle MG, Loponen J, Gobbetti M (2008) Proteolysis in sourdough fermentations: mechanisms and potential for improved bread quality. Trends Food Sci Tech 19:513–521
36. Gobbetti M, Smacchi E, Fox P, Stepaniak L, Corsetti A (1996) The sourdough microflora. Cellular localisation and charcterization of proteolytic enzymes in lactic acid bacteria. Lebensm Wissensch Tech 29:561–569
37. Di Cagno R, de Angelis M, Lavermicocca P, de Vincenzi M, GiovanniniC FM, Gobbetti M (2002) Proteolysis by sourdough lactic acid bacteria: effects on wheat flour protein fractions and gliadin peptides involved in human cereal intolerance. Appl Environ Microbiol 68:623–633
38. Wieser H, Vermeulen N, Gaertner F, Vogel RF (2008) Effect of different *Lactobacillus* and *Enterococcus* strains and chemical acidification regarding degradation of gluten proteins during sourdough fermentation. Eur Food Res Tech 226:1495–1502
39. Vermeulen N, Kretzer J, Machalitza H, Vogel RF, Gänzle MG (2006) Influence of redox-reactions catalysed by homo- and heterofermentative lactobacilli on gluten in wheat sourdoughs. J Cereal Sci 43:137–143
40. Grosch W, Wieser H (1999) Redox reactions in wheat dough as affected by ascorbic acid. J Cereal Sci 29:1–16
41. Korakli M, Rossmann A, Gänzle MG, Vogel RF (2001) Sucrose metabolism and exopolysaccharide production in wheat and rye sourdoughs by *Lactobacillus sanfranciscensis*. J Agri Food Chem 49:5194–5200
42. Korakli M, Gänzle MG, Vogel RF (2002) Metabolism by bifidobacteria and lactic acid bacteria of polysaccharides from wheat and rye, and exopolysaccharides produced by *Lactobacillus sanfranciscensis*. J Appl Microbiol 92:958–965
43. Tieking M, Gänzle MG (2005) Exopolysaccharides from cereal-associated lactobacilli. Trends Food Sci Tech 16:79–84
44. Galle S, Schwab C, Arendt EK, Gänzle MG (2011) Structural and rheological characterisation of heteropolysaccharides produced by lactic acid bacteria in wheat and sorghum sourdough. Food Microbiol 26:547–553
45. Bounaix MS, Robert H, Gabriel V, Morel S, Remaud-Simeon M, Gabriel B, Fontagne-Faucher C (2010) Characterization of dextran-producing *Weissella* strains isolated from sourdoughs and evidence of constitutive dextransucrase expression. FEMS Microbiol Lett 311:18–26
46. Bounaix MS, Gabriel V, Morel S, Robert H, Rabier P, Remaud-Simeon M, Gabriel B, Fontagne-Faucher C (2009) Biodiversity of exopolysaccharides produced from sucrose by sourdough lactic acid bacteria. J Agri Food Chem 57:10889–10897
47. Galle S, Schwab C, Arendt E, Gänzle MG (2010) Exopolysaccharide-forming *Weissella* strains as starter cultures for sorghum and wheat sourdoughs. J Agri Food Chem 58:5834–5841

48. Schwab C, Mastrangelo M, Corsetti A, Gänzle MG (2008) Formation of oligosaccharides and polysaccharides by *Lactobacillus reuteri* LTH5448 and *Weissella cibaria* 10M in sorghum sourdoughs. Cereal Chem 85:679–684
49. Tieking M, Korakli M, Ehrmann MA, Gänzle MG, Vogel RF (2003) In situ production of exopolysaccharides during sourdough fermentation by cereal and intestinal isolates of lactic acid bacteria. Appl Environ Microbiol 69:945–952
50. Waldherr FW, Vogel RF (2009) Commercial exploitation of homo-exopolysaccharides in non-dairy food systems. In: Ullrich M (ed) Bacterial polysaccharides: Current innovations and future trends. Caister Academic Press, Norfolk, UK, pp 313–330
51. Lacaze G, Wick M, Cappelle S (2007) Emerging fermentation technologies: development of novel sourdoughs. Food Microbiol 24:155–160
52. Brandt MJ, Roth K, Hammes WP (2003) Effect of an exopolysaccharide produced by *Lactoabacillus sanfranciscensis* LTH1729 on dough and bread quality. In: Vyust L, Vrije (eds) Sourdough from fundamentals to application, de Universiteit Brussels (VUB), Brussels, Hamburg, p 80
53. Kaditzky S, Vogel R (2008) Optimization of exopolysaccharide yields in sourdoughs fermented by lactobacilli. Eur Food Res Tech 228:291–299
54. Decock P, Cappelle S (2005) Bread technology and sourdough technology. Trends Food Sci Tech 16:113–120
55. Ross AS, McMaster GJ, David Tomlinson J, Cheetham NWH (1992) Effect of dextrans of differing molecular weights on the rheology of wheat flour doughs and the quality characteristics of pan and arabic breads. J Sci Food Agri 60:91–98
56. Galle S, Schwab C, Dal Bello F, Coffey A, Gänzle MG, Arendt EK (2012) Influence of in-situ synthesized exopolysaccharides on the quality of gluten-free sorghum sourdough bread. Int J Food Microbiol (in press), Corrected proof
57. Martínez-Anaya MA (2003) Associations and interactions of micro-organisms in dough fermentations: effects on dough and bread characteristics. In: Kulp K, Lorenz K (eds) Handbook of dough fermentations. Marcel Dekker, New York, pp 63–195
58. Katina K, Salmenkallio-Marttila M, Partanen R, Forssell P, Autio K (2006) Effects of sourdough and enzymes on staling of high-fibre wheat bread. Food Sci Tech 39:479–491
59. Di Cagno R, De Angelis M, Corsetti A, Lavermicocca P, Arnault P, Tossut P, Gallo G, Gobbetti M (2003) Interactions between sourdough lactic acid bacteria and exogenous enzymes: effects on the microbial kinetics of acidification and dough textural properties. Food Microbiol 20:67–75
60. Martínez-Anaya MA, Devesa A, Andreu P, Escrivá C, Collar C (1998) Effects of the combination of starters and enzymes in regulating bread quality and self-life. Food Sci Tech Int 4:425–435
61. Minervini F, Pinto D, Di Cagno R, De Angelis M, Gobbetti M (2011) Scouting the application of sourdough to frozen dough bread technology. J Cereal Sci 54:296–304

Chapter 9
Nutritional Aspects of Cereal Fermentation with Lactic Acid Bacteria and Yeast

Kati Katina and Kaisa Poutanen

9.1 Introduction

Sourdough fermentation is best known and most studied for its effects on the sensory quality and shelf life of baked goods. Acidification, activation of enzymes and their effects on the cereal matrix as well as production of microbial metabolites all produce changes in the dough and bread matrix that also influence the nutritional quality of the products. The nutritional quality is formed through the chemical composition and structure of the fermented foods, i.e. content and bioavailability of nutrients and non-nutrients. Sourdough fermentation can change all of these, as previously reviewed by Poutanen et al. [1] and Katina et al. [2].

Sourdough fermentation has been traditionally applied to whole grain foods, and it is a good means of making whole grain bread more palatable. Rye bread is an extreme example of this, as most of the whole-grain rye bread is made through sourdough fermentation [3]. Sourdough fermentation, also in the form of pre-treating raw materials, is again gaining interest also in mixed flour and dietary-fibre-enriched baking [4], where it also can change the properties of the dietary fibre complex. Fermentation has been studied for reducing the glycaemic response of bread [5, 6], and for increasing the uptake of minerals [7]. Microbial metabolism during sourdough fermentation may also produce new nutritionally active compounds, such as vitamins [8] and potentially prebiotic exopolysaccharides [9].

This chapter will deal with nutritionally relevant changes in cereal starch, protein, dietary fibre, vitamins, minerals and some phytochemicals, and discuss the potential of microorganisms to produce new compounds.

K. Katina (✉) • K. Poutanen
VTT Technical Research Centre of Finland,
PB 1000, 02044 VTT Espoo, Finland
e-mail: kati.katina@vtt.fi

M. Gobbetti and M. Gänzle (eds.), *Handbook on Sourdough Biotechnology*,
DOI 10.1007/978-1-4614-5425-0_9, © Springer Science+Business Media New York 2013

9.2 Effects on Cereal Biopolymers

9.2.1 *Starch*

Dietary carbohydrate is the major source of plasma glucose. An increase in the amount of rapidly digestible carbohydrate in the diet causes a rapid increase in blood glucose levels and a large demand for insulin in the postprandial period. The major carbohydrate sources in the Western diet contain rapidly digestible starch, and many common starchy foods like bakery goods, breakfast cereals, potato products and snacks produce high glycaemic responses. There are strong indications that the large amounts of rapidly available glucose derived from starch and free sugars in the modern diet [foods with high glycaemic index (GI) and high insulin index (II)] lead to periodic elevated plasma glucose and insulin concentrations that may be a risk factor to health [10].

Most processed starchy foods have low to medium moisture contents, thus their digestion is basically a solid–liquid two phase reaction, and the enzyme (particularly α-amylase) needs first to diffuse into the hydrated solid food matrix, bind to the substrate, and then cleave the glycosidic linkages of the starch molecules [11]. Factors affecting the binding of α-amylase to substrates [e.g. inhibition by the hydrolysis products (maltose and maltotriose)] will slow down the enzymatic reaction and thus digestion of starch. Other physiological factors affecting starch digestibility include gastric emptying, enzyme inhibitors and viscosity in the digestive tract [12].

Macro- and microstructure of cereal foods has a profound influence on the digestibility of starch, as reviewed by Singh et al. [13]. Especially, the characteristics of starch per se are of crucial importance for glucose response. Amylose-rich starches are more resistant to amylolysis than waxy or normal starches. The major intrinsic factors affecting raw starch digestibility include the supramolecular structure (packing of crystallites inside the starch granule), the ratio of amylase and amylopectin, the amylopectin fine structure, and the surface characteristics of starch granules [14]. In vitro, native starches are hydrolysed very slowly, and to a limited extent, by amylases [15–17]. When starch is used in food processing, starch gelatinisation, i.e. the process of disrupting starch crystalline structure with heat and moisture, usually results in a decrease or loss of the slow digestion property of native cereal starches [18]. Gelatinised starch will exist for example in bakery products in a partially or completely amorphous state. Thus, the more gelatinised starch is, the more rapidly it will be digested [19]. In many common starchy foods, such as in regular white wheat bread, the starch is highly gelatinised and product structure very porous, resulting in rapid degradation of starch in the small intestine and a very rapid rise of blood glucose level (high GI).

There are several mechanisms leading to slow digestion of gelatinised starch [20]. The first group of important factors is related to the state of starch in the food matrix. Starch retrogradation, which is the reassociation of amylose and amylopectin to form double helices and possible crystalline structures, promotes slow digestibility. The molecular structure of amylopectin is also an important factor, as high branch density has been shown to be linked to slow digestibility. Lowering the degree of starch gelatinisation and partially retaining the A-type crystalline structure of related starches is one effective way to increase the content of slowly digestible starch in food products. The second group of factors include impact of chewing on food structure, gastric emptying rate, transit time in the small intestine and the properties of digestive enzymes [16].

The means to slower starch digestibility in wheat flour-based products such as bread, biscuits and breakfast cereals are rare, if the addition of a high amount of intact kernels is excluded due to the resulting inferior product quality and consumer preferences. For wheat bread, the use of pre-fermentation technology (sourdough) or the addition of soluble fibres were identified in a recent review as the only suggested means to reduce GI [21].

The fermentation of the wheat and rye flour matrix with lactic acid bacteria (sourdough process) has been shown to lower GI of wholemeal barley bread [19, 22] and wheat bread [5, 23–25], and insulin index (II) of rye breads with varying dietary fibre (DF) content [26]. Several mechanisms have been proposed to be involved in sourdough processing contributing to reduced starch digestibility. Formation of organic acids, especially lactic acid, during fermentation has been suggested to be a main reason. The physiological mechanisms for the acute effects of acids appear to vary. Whereas lactic acid lowers the rate of starch digestion in bread [22], acetic and propionic acids appear instead to prolong the gastric emptying rate [27]. Chemical changes taking place during sourdough fermentation have been postulated to diminish the degree of starch gelatinisation [19], which would partly explain the lower digestibility of sourdough-fermented cereal foods. Sourdough fermentation has been also shown to promote the formation of resistant starch, which has slower digestibility [28].

At the product level, tissue integrity, porosity and structure of starch are important characteristics influencing glycaemic response. Rye breads baked from wholemeal or white rye flour with very different fibre contents produced lower insulin responses than white wheat bread, when the food portion size was standardised to provide 50 g of starch [26]. The breads were baked with a sourdough process and with 40% of a total amount of rye flour being pre-fermented before incorporation into the dough. The results suggested that with all rye breads, regardless of bran content, less insulin was needed to regulate blood sugar from the same amount of starch in comparison to normal wheat bread. The influence is probably due to the more rigid and less porous structure of rye bread, and because of the presence of organic acid formed during sourdough fermentation [29].

There also may be other mechanisms for the sourdough to regulate GI/II of the products. For example, pH-dependent proteolysis generally occurs during sourdough fermentation [30] producing significant amounts of peptides and amino acids in the sourdough. These may have a role in regulating glucose metabolism [31]. Furthermore, the results of Katina et al. [32] demonstrate that sourdough fermentation increases the amount of free phenolic compounds, which may also have an impact on lowering the GI/II [6, 33].

However, not all the sourdough breads automatically have low GI/II [34]. In general, a rather low pH of sourdough and subsequent bread is required to obtain lowered GI or II; typical values being 3.5–4 for sourdoughs and 3.8–5.1 for sourdough breads [5, 24, 25, 35, 36]. The efficacy of individual acids reducing GI is not completely clarified [6], and may vary between different bread types. In addition, such a low pH will in many cases reduce bread volume and increase density, which have been shown to promote low GI per se also in regular wheat breads [37]. Furthermore, the sensory quality of highly acidic breads may be a limiting factor for consumer acceptability of such breads, and means for enhancing the efficacy of fermentation while maintaining higher pH levels would be desirable. Further studies will be needed to clarify the direct influence of sourdough metabolites (acids, peptides, and exopolysaccharides) on starch digestibility, and the indirect impact of sourdough fermentation on cereal matrix properties (density, liberation of phenolic compounds, state of protein, and formation of resistant starch), which all influence digestibility.

9.2.2 *Protein*

Protein degradation that occurs during sourdough fermentation is among the key phenomena that affect the overall quality of sourdough bread as reviewed by Gänzle et al. [30]. Proteolysis by sourdough fermentation has been found to be higher than in just yeasted doughs. During dough fermentation, the proteolysis by LAB releases small peptides and free amino acids, which are important for rapid microbial growth and acidification and as precursors for the flavour development of leavened baked products [38]. Furthermore, this proteolytic activity might be used as a tool to reduce certain allergen compounds. Cereal proteins are one of the most frequent causes of food allergies. Wheat proteins may induce a classical allergy affecting the skin, gut or respiratory tract, exercise-induced anaphylaxis, occupational rhinitis or asthma [36, 39], and protein modification with fermentation offers possibilities to reduce their allergy-causing properties. For example, De Angelis et al. [36] demonstrated the capacity of probiotic VSL#3 to hydrolyse wheat flour allergens. Albumins, globulins, and gliadins extracted from wheat flour, a chemically acidified and started doughs, and total proteins extracted from breads were analysed by immunoblotting with pooled sera from patients with an allergy to wheat. Several IgE-binding proteins persisted after treatment of baker's yeast bread with pepsin and pancreatin. The signal of all these IgE-binding proteins disappeared after further treatment by VSL#3. Utilisation of the VSL#3 strain as a starter for bread making, caused a marked degradation of wheat proteins, including some IgE-binding proteins. De Angelis et al. [36] showed that the IgE-binding profile of the bread manufactured by VSL#3 was largely different from that of baker's yeast bread. The IgE-binding proteins that persisted in the bread made with VSL#3 were completely degraded by pepsin and pancreatin.

Intensive degradation of prolamin of wheat and rye has also opened new possibilities to use these cereals even as part of gluten-free diets [23, 40, 41]. Controlled proteolysis in wheat and rye doughs was suggested to reduce gluten levels to such an extent that the products were tolerated by celiac patients [42]. While such sourdoughs with extended fermentation time are not suitable for bread production as such, they can be incorporated as baking improvers into gluten-free recipes. It was shown in a 60-day clinical trial that biscuits and cakes produced using a hydrolysed wheat product made using sourdough lactobacilli and fungal proteases were not toxic to patients with celiac disease [43].

The quality of gluten-free bread is often inferior when compared to conventional (wheat) products [2]. However, by degrading prolamins of wheat or rye with a proteolysis-intensive sourdough process, it is possible to produce good quality gluten-free bread with sourdough technology [40, 42]. The concept of complete elimination of gluten, however, is controversial. Gluten is considered essential for wheat baking and the complete elimination of gluten from wheat and rye, albeit possible, is technically challenging in industrial baking operations. The use of germinated rye in sourdoughs may avoid, in part, such controversy because the water binding as well as gas retention in rye doughs are mediated by pentosans which remain unaffected by proteolysis [30]. De Angelis et al. [23] demonstrated that fermentation by selected sourdough lactic acid bacteria to decrease celiac intolerance to rye flour [44, 45] used flour from germinated wheat and rye grains to enhance the proteolysis and efficient degradation of wheat and rye prolamins.

Recently, it has been demonstrated that sourdough fermentation can promote the formation of bioactive peptides [46–48]. Bioactive peptides are defined as specific protein fragments that have positive effects on body functions or conditions and that may influence human health. Usually, bioactive peptides correspond to specific sequences from native proteins, which are released through hydrolysis by digestive, microbial, and plant proteolytic enzymes, and their levels generally increase during food fermentation. Coda et al. [46] summarised that bioactive peptides, on the basis of in vitro and in vivo studies, have demonstrated a large spectrum of biological functions, such as opioid-like, mineral-binding, immunomodulatory, antimicrobial, antioxidative, antithrombotic, hypocholesterolemic, and antihypertensive activities. The ability of selected lactic acid bacteria to produce antioxidant peptides during sourdough fermentation by using various cereal flours as substrates was demonstrated [46]. The radical-scavenging activity of water/salt-soluble extracts (WSE) from sourdoughs was shown to be significantly ($P<0.05$) higher than that of chemically acidified doughs. Twenty-five peptides of 8–57 amino acid residues were identified in their study and nearly all sequences shared compositional features that are typical of antioxidant peptides. All of the purified fractions showed ex vivo antioxidant activity on mouse fibroblasts artificially subjected to oxidative stress. Recently, interest in antioxidant peptides derived from food proteins has increased, and evidence that bioactive peptides prevent oxidative stresses associated with numerous degenerative aging diseases (e.g. cancer and arteriosclerosis) is accumulating [49].

Rizzello et al. [47] exploited the potential of sourdough lactic acid bacteria to release lunasin, an anticarcinogenic peptide, during fermentation of cereal and non-conventional flours. They used selected lactic acid bacteria as sourdough starters to ferment wholemeal wheat, soybean, barley, amaranth, and rye flours. Sourdough-originated lunasin was identified in their study and the concentration of lunasin was shown to increase up to two to four times during fermentation.

From a practical standpoint, baked cereal goods are currently manufactured by highly accelerated processes. Long-term fermentations by sourdough, characterised by a cocktail of acidifying and proteolytic LAB and yeasts, have been almost totally replaced by the indiscriminate use of chemical and/or baker's yeast leavening agents. In these technological circumstances, cereal components (e.g. proteins) are subjected to very mild or no degradation during manufacture, resulting in less easily digestible foods compared to traditional and ancient sourdough baked goods [41].

9.2.3 Dietary Fibre

Dietary fibre consists of the plant polysaccharides and lignin that are resistant to hydrolysis by the digestive enzymes of man. A high consumption of dietary fibre may lower the risk of cardiovascular disease, diabetes, hypertension, obesity, and gastrointestinal disorders [50, 51]. Cereal foods are an important source of dietary fibre, and because of their role as a staple food provide an important food group to increase the currently too low intake of dietary fibre. Sourdough fermentation provides two main options for enhancing utilisation of fibre-enriched products: (1) It is important technology in the manufacture of whole grain bread, especially rye bread, and (2) it may be used to modify fibre-rich cereal ingredients such as bran and germ for improved technological functionality.

Wholemeal rye and wheat are very good sources of dietary fibre. However, a high content of fibre poses technological challenges for baking. For whole-grain rye baking, sourdough fermentation is an essential part of the process [2]. Without sourdough wholemeal rye or wheat-rye flour mixes are very difficult to process, and sourdough improves the overall quality and shelf life of whole-grain rye breads. The rye sourdough process not only improves flavour and texture of rye bread but enables consumption of wholemeal rye, which is well known for its high nutritional quality and health-promoting properties.

Bran sourdough (or bran pre-ferment) is a potential means to improve the quality of high fibre bread [4, 52–54]. The use of bran sourdough improves loaf volume and crumb softness of high-fibre wheat breads [4, 52, 55] and bread with 10-% fermented bran has been reported to provide the best sensory properties of bread [53]. The impact of fermentation is assumed to be related to control of endogenous microbiota of bran, endogenous xylanase activity and subsequent solubilisation of arabinoxylans in bran fermentation [4]. Enzyme activity and gluten characteristics of dough containing fermented bran will be modified by the acidity produced during fermentation, and subsequently decreased pH. The fibre content of the bran does not change significantly in a short fermentation time but can decrease slightly during prolonged fermentation due to hydrolysis of cell wall structures (Katina, unpublished data).

Use of enzymes (α-amylase, xylanase, lipase) in combination with yeast fermentation of bran has been shown to increase the volume of the subsequent bread, and soften its texture significantly [55]. Use of fermented bran improves carbon dioxide retention of the dough, and the use of enzymes strengthens that effect. Also, addition of insoluble arabinoxylans and xylanase enzymes has been shown to increase the volume of the sour dough bread [56]. Arabinoxylans function as the source material of xylose and arabinose, which accelerate the acidification rate and positively interfere with the metabolism of sourdough microflora.

Sourdough fermentation improves the technological functionality of bran as a baking ingredient, but it most probably also changes the quality of dietary fibre. The physiological effects of dietary fibre depend on the chemical but also physical characteristics, including degree of polymerisation of the polysaccharides, presence of side chains and degree of cross-linking, particle size and cell wall integrity [51]. Because of solubilisation of arabinoxylan, sourdough fermentation may influence its fermentation pattern and also produce prebiotic oligosaccharides [57]. It may also influence the bioaccessibility of phytochemicals associated with the dietary fibre complex, as shown below.

Wheat germ, in addition to vitamins and lipids, contains a significant amount of dietary fibre. Fermentation of wheat germ has recently been noticed to enhance the volume of the bread and decrease the rate of firmness [58]. The use of wheat germ as a source of dietary fibre for bread is still moderate because of its poor shelf-life stability. The high lipase and lipoxygenase activities cause sensitivity to oxidation which leads to the release of free fatty acids and, consequently, to the appearance of rancidity in baked goods. Sourdough fermentation stabilised and enhanced some nutritional and chemical properties of the wheat germ. Because of lactic acidification, the lipase activity of the sourdough fermented wheat germ has been shown to be lower than that found in the raw wheat germ [58].

9.3 Micronutrients

9.3.1 Vitamins

Whole-grain cereal foods are an important source of vitamins, such as thiamine, vitamin E and folates. Yeast fermentation increases the folate content during the pre-fermentation process of both wheat flour and bran [32, 59] and rye [32, 59, 60], causing over a doubling of folate in rye fermentation [60]. The presence of yeast has been shown to be a crucial factor for increased folate production in rye sourdough as sourdough bacteria had only slight effects on the synthesis of folates [61]. Yeast strains have been shown to be different in their capability to produce folate, and thus a high folate-producing strain could be used as an alternative to folate fortification [62, 63]. The folate content in fermented cereal foods can be further increased by the use of malted or germinated grains, as reviewed by Jagerstad et al. [64]. Conversely, 25–38% reduction of folate content in yeast and LAB fermented breads have been reported by Gujska et al. [65].

Thiamine content has been reported to increase especially in elongated yeast fermentation [8, 66], but also to decrease in the actual baking process [67]. Prolonged yeast or sourdough fermentation maintained the original content of vitamin B_1 in whole wheat baking in contradiction to a short process, which reduced its amount. Whole wheat breadmaking with yeast (from kneading to final bread), with long fermentations time resulted in a 30% enrichment in riboflavin. The fermentation step can thus improve the retention of vitamins in the baking process. The use of both yeast and sourdough did not have a synergistic effect on B-vitamin levels [8]. Production of the B_2 vitamin with strain selection for enrichment of pasta and bread has also been recently demonstrated by Capozzi et al. [68]. The applied approaches resulted in a considerable increase of vitamin B_2 content (about two- and threefold increases in pasta and bread, respectively), thus representing a convenient and efficient food-grade biotechnological application for the production of vitamin B_2-enriched bread and pasta. This methodology may be extended to a wide range of cereal-based foods, feed, and beverages. However, sourdough or yeast fermentation do not automatically increase the levels of all vitamins; decreased levels have been observed for vitamin E during sourdough preparation and dough making [69], and for levels of tocopherol and tocotrienol in rye sourdough baking [60].

9.3.2 *Minerals*

Whole grains are a good source of minerals, including calcium, potassium, magnesium, iron, zinc and phosphorus. As the bran fraction of the grain also contains phytate (myo-inositol hexaphosphate), the bioavailability of minerals may be limited. This has a large impact especially in developing countries, where iron deficiency is a common nutritional disorder, especially among children and women. Grains contain 3–22-mg phytic acid per gram [70], concentrated in the aleurone layers. Phytate has strong chelating capacity and forms insoluble complexes with dietary cations, thus impairing mineral absorption. Phytases are able to dephosphorylate phytate, forming free inorganic phosphate and inositol phosphate esters, which have less capacity to influence mineral solubility and bioavailability. It has been shown that iron was more bioavailable in mice when fed in sourdough bread vs. straight dough bread [71], and absorption of zinc, magnesium, and iron was higher in rats when bread was baked using sourdough [72].

Grain endogenous phytase activity is accelerated in the acidic environment produced in sourdough fermentation. Lactic acid bacteria and yeasts may also possess some phytase activity. The pH optimum of wheat phytase is pH 5.0, and that of yeast is somewhat lower, i.e. pH 3.5 [73]. A moderate decrease of pH to 5.5 in fermentation reduces phytate content of whole wheat flour by 70% due to enhanced action of endogenous phytase present in the flour [74]. It was suggested that the endogenous flour phytase activity was much more influential than the microbial phytase of the sourdough. No major phytase activity was found in screening of 50 lactic acid bacteria strains isolated from sourdoughs [75], even though in studies with phytic acid as the only carbon source sourdough-originated lactic acid bacteria have been reported to utilise it [76, 77]. Phytase activity has been detected in commercial baker's yeasts [78], and variable activities were detected in traditional sourdough starters containing both yeast and lactic acid bacteria [79, 80]. Yeast strains high in phytase activity have also been suggested to be potential phytase carriers in the gastrointestinal tract [81].

Phytase action is dependent on the fermentation conditions: flour particle size, acidity, temperature, time and water content [82, 83]. Sourdough fermentation has been shown to be more effective in solubilising minerals in whole-wheat flours than its bran fraction. Bran particle size influenced calcium and iron solubilisation, which only happened if the bran was finely milled [7]. Pre-fermentation of bran with lactic acid bacteria increased phytate breakdown (up to 90%) and increased magnesium and phosphorus solubility [84].

Selenium-enriched rye and wheat seeds have been used to produce fermented sourdough bread, and studied in human volunteers for bioavailability of selenium [85, 86]. The selenium enrichment was made by incubating the seeds in selenium solution. The high content of selenium in raw material was reflected in high contents in the sourdough bread and further in humans having consumed the bread.

9.3.3 *Phytochemicals*

Phytochemicals are biologically active compounds in the cereal grain and they have been suggested to be among the factors contributing to the protective properties of whole grain foods [87]. The outer layers of grains, such as bran, contain much higher levels of phytochemicals, such as phenolic acids, alkylresorcinols, lignans, phytosterols, tocols and folate, than the inner parts [60, 88]. Processing may decrease or increase the levels, and also modify the bioavailability of these compounds as reviewed by Slavin et al. [89], and for the phenolic compounds of rye as reviewed by Bondia-Pons et al. [90].

Wheat bread containing a sourdough-fermented wheat bran-flour mixture was recently shown to provide higher antioxidant potential as compared to regular wheat bread [91]. Traditional rye sourdough has been shown to increase the antioxidant activity (DPPH radical scavenging activity) in the methanol-extracted fraction of rye sourdough, concurrently with increased levels of easily extractable phenolic compounds [60]. Accordingly, the antioxidant capacity of traditional rye breads baked with sourdough has been shown to be higher than that of common white wheat bread, the highest values reported for breads made with whole meal flour [67, 92].

Fermentation of rye or wheat bran with yeast and especially with added cell wall-degrading enzymes was able to increase the level of free ferulic acid [4, 32, 93]. Ferulic acid is a structural component in cell walls, cross-linked to arabinoxylan. Since most of the ferulic acid is covalently bound to the cell wall structures, its bioaccessibility in physiological conditions is low, and bioprocessing can be used as an effective means to increase the bioaccessibility of ferulic acid. Wheat bread supplemented with bioprocessed bran increased the in vitro and in vivo bioaccessibility of phenolic compounds as well as the colonic end metabolite 3-phenylpropionic in breads, and exerted anti-inflammatory effects ex-vivo [93, 94].

9.4 Microbial Exopolysaccharides

Dietary non-digestible oligosaccharides (NDO) have been shown to modulate the composition and activity of intestinal microbiota, and they may also exert health benefits in humans by improving bowel function, prevention of overgrowth of pathogenic bacteria through selective stimulation of non-pathogenic members of intestinal microbiota and by increased production of short-chain fatty acids (SCFA) [95]. Intestinal fermentation and health benefits of fructo-oligosaccharides, galacto-oligosaccharides and xylo-oligosaccharides have been well documented in animal and human studies [96, 97]. Recently, stimulation by isomalto-oligosaccharides (IMO) of the growth of intestinal lactic acid bacteria in a rat model was also shown by Ketabi et al. [95]. The relationship between diet, intestinal microbiota and host nutrition is currently under active investigation, and the integration of the functional analyses of gut microbiota and sourdough genomes and metagenomes may allow for design of prebiotic molecules with specific functional properties [98].

Microbes are able to produce a variety of polysaccharides. Exopolysaccharides (EPS) are sugar biopolymers that are secreted by bacteria, microalgae and by some yeasts and filamentous fungi. They may protect cells from external stress factors such as desiccation and antimicrobial substances, and mediate interactions of cells with surfaces and other cells, thus playing an important role, for instance, in biofilm formation. EPS can be divided into capsular polysaccharides that are more or less tightly bound on cells, and extracellular slime which cells excrete to their surrounding medium. EPS production can usually be detected on solid and liquid medium, respectively, from a slimy or ropy colony appearance and from an increase in medium viscosity. Microbial EPS vary greatly in mass; from ~10 kDa to 1–2 mDa. On the basis of their chemical composition, all microbial EPS can be broadly divided into homopolysaccharides (=Hops), consisting of only one monosaccharide type, and heteropolysaccharides (=Heps), made of two or more different monosaccharide units. Additionally, various inorganic or organic constituents may be attached. The possible complexities of polysaccharide structures are almost infinite as, for instance, even a disaccharide may be linked in eight different ways [99, 100].

Lactobacilli from wheat and rye sourdoughs have been shown to produce EPS [9, 101], and especially gluco-oligosaccharides [102] and fructo-oligosaccharides, which have prebiotic properties [9]. For example, *Lactobacillus sanfranciscensis* LTH2590 produced 0.5–1% levan (flour basis) during 24-h fermentation in wheat and rye doughs [101]. Tieking et al. [103] studied the ability of seven fructan- or glucan-positive LAB (*Lb. sanfranciscensis* LTH 2581 and 2590, *Lb. frumenti* TMW 1.103, 1.660, 1.669, *Lb. pontis* TMW 1.675, *Lb. reuteri* TMW

1.1.06) to produce these EPS during wheat dough fermentation in the presence of 12% sucrose (flour weight). For all the strains the production of the same EPS at a level of 0.5–2 g kg^{-1} was shown. Levans from *Lb. sanfranciscensis* may also exert probiotic effects as they are preferentially degraded by bifidobacteria in the intestinal tract [101]. Formation of oligo- and polysaccharides with prebiotic potential has also been shown by *Lb. reuteri* LTH5448 and *Weissella cibaria* 10 M in sorghum sourdoughs [104].

9.5 Future Prospects

Sourdough fermentation is a food processing method with a long history, traditionally used mainly to improve product quality. During the past 15 years, the use of microbial fermentation has also been proven to intensively modify the nutritional quality of cereal foods. Because of complex microbial and food structure interactions present in a sourdough system, fermentation can be tuned for multi-functional nutritional modifications of both traditional and novel fermentable substrates.

In the future, sourdough technology can provide an effective means to utilise and upgrade side streams from both food and non-food processing, provide novel protein functionalities and produce completely novel oligo- and polysaccharides for new nutritional improvements such as fat or sugar replacement. They also show potential in producing and influencing bioavailability of minor food constituents with high biological activity. Next-generation fermentations with yeast and lactic acid bacteria can thus be considered effective cell factories to modify cereal and also other fermentable materials for nutritionally tailored food or feed.

References

1. Poutanen K, Flander F, Katina K (2009) Sourdough and cereal fermentation in a nutritional perspective. Food Microbiol 26:693–699
2. Katina K, Arendt E, Liukkonen K-H, Autio K, Flander L, Poutanen K (2005) Potential of sourdough for healthier cereal products. Trends Food Sci Technol 16(1–3):104–112
3. Valjakka TT, Kerojoki H, Katina K (2003) Chapter 11: Sourdough bread in Finland and Eastern Citation Information Europe. In: Kulp K, Lorenz K (eds) Handbook of dough fermentations. Marcel Dekker Inc, New York, USA
4. Katina K, Juvonen R, Laitila A, Flander L, Nordlund E, Kariluoto S, Piironen V, Poutanen K (2012) Fermented wheat bran as a functional ingredient in baking. Cereal Chem 189(2): 126–134
5. Lappi J, Selinheimo E, Schwab U, Katina K, Lehtinen P, Mykkänen H, Kolehmainen M, Poutanen K (2010) Sourdough fermentation of wholemeal wheat bread increases solubility of arabinoxylan and protein and decreases postprandial glucose and insulin responses. J Cereal Sci 51(1):152–158
6. Novotni D, Ćurić D, Bituh M, Barić IC, Škevin D, Čukelj N (2011) Glycemic index and phenolics of partially-baked frozen bread with sourdough. Int J Food Sci Nutr 62(1):26–33

7. Lioger D, Leenhardt F, Demigne C, Remesy C (2007) Sourdough fermentation of wheat fractions rich in fibres before their use in processed food. J Sci Food Agric 87:1368–1373
8. Batifoulier F, Verny M-A, Chanliaud E, Rémésy C, Demigne C (2005) Effect of different breadmaking methods on thiamine, riboflavin and pyridoxine contents of wheat bread. J Cereal Sci 42:101–108
9. Tieking M, Gänzle M (2005) Exopolysaccharides from cereals associated lactobacilli. Trends Food Sci Technol 16:79–84
10. Barclay AW, Petocz P, McMillan-Price J, Flood VM, Prvan T, Mitchell P, Brand-Miller JC (2008) Glycemic index, glycemic load, and chronic disease risk – a meta-analysis of observational studies. Am J Clin Nutr 87:627–637
11. Leloup VM, Colonna P, Ring SG (2004) α-Amylase adsorption on starch crystallites. Biotechnol Bioeng 38:127–134
12. Zhang G, Hamaker B (2009) Slowly digestible starch: concept, mechanism, and proposed extended glycemic index. Crit Rev Food Sci Nutr 49:852–867
13. Singh J, Dartois A, Kaur L (2010) Starch digestibility in food matrix: a review. Trends Food Sci Technol 21:168–180
14. Oates CG (1997) Towards an understanding of starch granule structure and hydrolysis. Trends Food Sci Technol 8:375–382
15. Lauro M, Suortti T, Autio K, Linko P, Poutanen K (1993) Accessibility of barley starch granules to α-amylase during different phases of gelatinization. J Cereal Sci 17:125–136
16. Zhang G, Ao Z, Hamaker BR (2008) Nutritional property of endosperm starches from maize mutants: a parabolic relationship between slowly digestible starch and amylopectin fine structure. J Agric Food Chem 56:4686–4694
17. Björk I, Granfeldt Y, Liljeberg H, Tovar J, Asp NG (1994) Food properties affecting the digestion and absorption of carbohydrates. Am J Clin Nutr 59:699S–705S
18. Zhang G, Venkatachalam M, Hamaker BR (2006) Structural basis for the slow digestion property of native cereal starch. Biomacromolecules 7:3259–3266
19. Östman E (2003) Fermentation as a means of optimizing the glycaemic index – food mechanisms and metabolic merits with emphasis on lactic acid in cereal products. Ph.D. thesis, Lund University, Department of Applied Nutrition and Food Chemistry
20. Zhang G, Sofyan M, Hamaker BR (2008) Slowly digestible state of starch: mechanism of slow digestion property of gelatinized maize starch. J Agric Food Chem 56:4695–4702
21. Fardet A, Leenhardt F, Lioger D, Scalbert A, Rémésy C (2006) Parameters controlling the glycaemic response to breads. Nutr Res Rev 19:18–25
22. Liljeberg H, Lönner C, Björck I (1995) Sourdough fermentation or addition of organic acids or corresponding salts to bread improves nutritional properties of starch in healthy humans. J Nutr 125:1503–1511
23. De Angelis M, Coda R, Silano M, Minervini F, Rizzello C, Di Cagno R, Vicentini O, De Vincenzi M, Gobbetti M (2006) Fermentation by selected sourdough lactic acid bacteria to decrease coeliac intolerance to rye flour. J Cereal Sci 43:301–314
24. De Angelis M, Damiano N, Rizzello CG, Cassone A, Di Cagno R, Gobbetti M (2009) Sourdough fermentation as a tool for the manufacture of low-glycemic index white wheat bread enriched in dietary fibre. Eur Food Res Technol 229:593–601
25. Maioli M, Pes GM, Sanna M, Cherchi S, Dettori M, Manca E, Farris GA (2008) Sourdough-leavened bread improves postprandial glucose and insulin plasma levels in subjects with impaired glucose tolerance. Acta Diabetol 45:91–96
26. Juntunen K, Laaksonen D, Autio K, Niskanen L, Holst J, Savolainen K, Liukkonen K-H, Poutanen K, Mykkänen H (2003) Structural differences between rye and wheat bread but not total fiber content may explain the lower postprandial insulin response to rye bread. Am J Clin Nutr 78:957–964
27. Liljeberg H, Björck I (1998) Delayed gastric emptying rate may explain improved glycaemia in healthy subjects to a starchy meal with added vinegar. Eur J Clin Nutr 52:368–371
28. Scazzina F, Del Rio D, Pellegrini N, Brighenti F (2009) Sourdough bread: starch digestibility and postprandial glycemic response. J Cereal Sci 49:419–421

29. Autio K, Liukkonen K-H, Juntunen K, Katina K, Laaksonen D, Mykkänen H, Niskanen L, Poutanen K (2003) Food structure and its relation to starch digestibility and glycaemic response. In: Fischer P, Marti I, Windhab EJ (eds) Third international conference of food rheology and structure, Zurich, 10–13 Feb 2003, pp 7–11
30. Gänzle M, Loponen J, Gobbetti M (2008) Proteolysis in sourdough fermentations: mechanisms and potential for improved bread quality. Trends Food Sci Technol 19:513–521
31. Nilsson M, Holst JJ, Björck IME (2007) Metabolic effects of amino acid mixtures and whey protein in healthy subjects: studies using glucose-equivalent drinks. Am J Clin Nutr 85:996–1004
32. Katina K, Laitila A, Juvonen R, Liukkonen K-H, Kariluoto S, Piironen V, Landberg R, Åman P, Poutanen K (2007) Bran fermentation as a means to enhance technological properties and bioactivity of rye. Food Microbiol 24:175–186
33. Solomon TPJ, Blannin AK (2007) Effects of short-term cinnamon ingestion on in vivo glucose tolerance. Diabetes Obes Metab 9:895–901
34. Hardman Fredensborg M, Perry T, Mann J, Chisholm A, Rose M (2010) Rising methods and leavening agents used in the production of bread do not impact the glycaemic index. Asia Pac J Clin Nutr 19:188–194
35. Najjar AM, Parsons PM, Duncan AM, Robinson LE, Yada RY, Graham TE (2009) The acute impact of ingestion of breads of varying composition on blood glucose, insulin and incretins following first and second meals. Br J Nutr 101:391–398
36. De Angelis M, Rizzello CG, Scala E, De Simone C, Farris GA, Turrini F et al (2007) Probiotic preparation has the capacity to hydrolyze wheat protein responsible for food allergy. J Food Prot 70:135–144
37. Burton P, Lightowler HJ (2006) Influence of bread volume on glycaemic response and satiety. Br J Nutr 96:877–882
38. Rollán G, De Angelis M, Gobbetti M, de Valdez GF (2005) Proteolytic activity and reduction of gliadin-like fractions by sourdough lactobacilli. J Appl Microbiol 99:1495–1502
39. Palosuo K (2003) Update on wheat hypersensitivity. Curr Opin Allergy Clin Immunol 3:205–209
40. Di Cagno R, De Angelis M, Auricchio S, Greco L, Clarke C, De Vincenzi M et al (2004) Sourdough bread made from wheat and nontoxic flours and started with selected lactobacilli is tolerated in celiac sprue patients. Appl Environ Microbiol 70:1088e1096
41. Gobbetti M, Rizzello C, Di Cagno R, De Angelis M (2007) Sourdough lactobacilli and celiac disease. Food Microbiol 24:187–196
42. Rizzello CG, De Angelis M, Di Cagno R, Camarca A, Silano M, Losito I et al (2007) Highly efficient gluten degradation by lactobacilli and fungal proteases during food processing: new perspectives for celiac disease. Appl Environ Microbiol 73:4499–4507
43. Greco L, Gobbetti M, Auricchio R, Di Mase R, Landolfo F, Paparo F, Di Cagno R, De Angelis M, Rizzello CG, Cassone A, Terrone G, Timpone L, D'Aniello M, Maglio M, Troncone R, Auricchio S (2011) Safety for patients with celiac disease of baked goods made of wheat flour hydrolyzed during food processing. Clin Gastroenterol Hepatol 9(1):24–29
44. Loponen J, Kanerva P, Zhang C, Sontag-Strohm T, Salovaara H, Gänzle M (2009) Prolamin hydrolysis and pentosan solubilization in germinated-rye sourdoughs determined by chromatographic and immunological methods. J Agric Food Chem 57:746–753
45. Loponen J, Sontag-Strohm T, Venäläinen J, Salovaara H (2007) Prolamin hydrolysis in wheat sourdoughs with differing proteolytic activities. J Agric Food Chem 55:978–984
46. Coda R, Rizzello CG, Pinto D, Gobbetti M (2012) Selected lactic acid bacteria synthesize antioxidant peptides during sourdough fermentation of cereal flours. Appl Environ Microbiol 78(4):1087–1096
47. Rizzello CG, Nionelli L, Coda R, Gobbetti M (2012) Synthesis of the cancer preventive peptide lunasin by lactic acid bacteria during sourdough fermentation. Nutr Cancer 64:111–120
48. Rizzello CG, Cassone A, Di Cagno R, Gobbetti M (2008) Synthesis of angiotensin I-converting enzyme (ACE)-inhibitory peptides and gamma-aminobutyric acid (GABA) during sourdough fermentation by selected lactic acid bacteria. J Agric Food Chem 56:6936–6943

49. Adebiyi AP, Adebiyi AO, Yamashita J, Ogawa T, Muramoto K (2009) Purification and characterization of antioxidative peptides derived from rice bran protein hydrolysates. Eur Food Res Technol 228:553–563
50. Anderson JW, Baird P, Davis RH Jr, Ferreri S, Knudtson M, Karaym A, Waters V, Williams CL (2009) Health benefits of dietary fiber. Nutr Rev 67:188–205
51. Raninen K, Lappi J, Mykkänen H, Poutanen K (2011) Dietary fiber type reflects physiological functionality: comparison of grain fiber, inulin, and polydextrose. Nutr Rev 69:9–21
52. Salmenkallio-Marttila M, Katina K, Autio K (2001) Effect of bran fermentation on quality and microstructure of high-fiber wheat bread. Cereal Chem 78:429–435
53. Eiman G, Amir M, Alkareem A, Moniem A, Mustafa A (2008) Effect of fermentation and particle size of wheat bran on the antinutritional factors and bread quality. Pak J Nutr 7:521–526
54. Katina K, Liukkonen K-H, Kaukovirta-Norja A, Adlercreutz H, Heinonen S-M, Lampi A-M, Pihlava J-M, Poutanen K (2007) Fermentation-induced changes in the nutritional value of native or germinated rye. J Cereal Sci 46:348–355
55. Katina K, Salmenkallio-Marttila M, Partanen R, Forssell P, Autio K (2006) Effects of sourdough and enzymes on staling of high-fibre wheat bread. LWT- Food Sci Technol 39:479–491
56. Corsetti A, Gobbetti B, De Marco B, Balestrieri F, Paoletti F, Rossi J (2000) Combined effect of sourdough lactic acid bacteria and additives on bread firmness and staling. J Agric Food Chem 48:3044–3051
57. Broekaert WF, Courtin CM, Verbeke K, van de Wiele T, Verstraete W, Delcour JA (2011) Prebiotic and other health-related effects of cereal-derived arabinoxylans, arabinoxylan-oligosaccharides, and xylooligosaccharides. Crit Rev Food Sci Nutr 51:178–194
58. Rizzello CG, Cassone A, Coda R, Gobbetti M (2011) Antifungal activity of sourdough fermented wheat germ used as an ingredient for bread making. Food Chem 127(3):952–959
59. Kariluoto S, Vahteristo L, Salovaara H, Katina K, Liukkonen K-H, Piironen V (2004) Effect of baking method and fermentation on folate content of rye and wheat breads. Cereal Chem 81:134–139
60. Liukkonen K-H, Katina K, Wilhelmson A, Myllymäki O, Lampi A-M, Kariluoto S, Piironen V, Heinonen S-M, Nurmi T, Adlercreutz H, Peltoketo A, Pihlava J-M, Hietaniemi V, Poutanen K (2003) Process-induced changes on bioactive compounds in whole grain rye. Proc Nutr Soc 62:117–122
61. Kariluoto S, Aittamaa M, Korhola M, Salovaara H, Vahteristo L, Piironen V (2006) Effects of yeasts and bacteria on the levels of folates in rye sourdoughs. Int J Food Microbiol 106:137–143
62. Hjortmo S, Patring J, Jastrebova J, Andlid T (2008) Biofortification of folates in white wheat bread by selection of yeast strain and process. Int J Food Microbiol 127:32–36
63. Hjortmo S, Patring J, Jastrebova J, Andlid T (2005) Inherent biodiversity of folate content and composition in yeasts. Trends Food Sci Technol 16:311–316
64. Jägerstad M, Piironen V, Walker C, Ros G, Carnovale E, Holasova M, Nau H (2005) Increasing natural food folates through bioprocessing and biotechnology. Trends Food Sci Technol 16:298–306
65. Gujska E, Michalak J, Klepacka J (2009) Folates stability in two types of rye breads during processing and frozen storage. Plant Foods Hum Nutr 64:129–134
66. Ternes W, Freund W (1988) Effects of different doughmaking techniques on thiamin content of bread. Getreide Mehl Brot 42:293–297
67. Martinez-Villaluenga C, Michalska A, Frias F, Piskula M-K, Vidal-Valverde C, Zielinski H (2009) Effect of flour extraction rate and baking on thiamine and riboflavin content and antioxidant capacity of traditional rye bread. J Food Sci 74:49–55
68. Capozzi V, Menga V, Digesù AM, De Vita P, Van Sinderen D, Cattivelli L, Fares C, Spano G (2011) Biotechnological production of vitamin B_2-enriched bread and pasta. J Agric Food Chem 59:8013–8020

69. Wennermark B, Jägerstad M (1992) Breadmaking and storage of various wheat fractions affect vitamin E. J Food Sci 57:1205–1209
70. García-Estepa R, Guerra-Hernández E, García-Vilanova B (1999) Phytic acid content in milled cereal products and breads. Food Res Int 32:217–221
71. Chaoui A, Faid M, Belahsen R (2006) Making bread with sourdough improves iron bioavailability from reconstituted fortified wheat flour in mice. J Trace Elem Med Biol 20:217–220
72. Lopez H, Duclos V, Coudray C, Krespine V, Feillet-Coudray C, Messager A, Demigné C, Rémésy C (2003) Making bread with sourdough improves mineral bioavailability from reconstituted whole wheat flour in rats. Nutrition 19:524–530
73. Türk M, Carlsson N, Sandberg A-S (1996) Reduction of the levels of phytate during wholemeal bread baking; effects of yeast and wheat phytases. J Cereal Sci 23:257–264
74. Leenhardt F, Levrat-Verny M-A, Chanliaud E, Remesy C (2005) Moderate decrease of pH by sourdough fermentation is sufficient to reduce phytate content of whole wheat flour through endogenous phytase activity. J Agric Food Chem 53:98–102
75. Reale A, Konietzny U, Coppola R, Sorrentino E, Greiner R (2007) The importance of lactic acid bacteria for phytate degradation during cereal dough fermentation. J Agric Food Chem 55:2993–2997
76. Shirai K, Revah-Moiseev S, García-Garibay M, Marshall V (1994) Ability of some strains of lactic acid bacteria to degrade phytic acid. Lett Appl Microbiol 19:366–369
77. Lopez H, Ouvry A, Bervas E, Guy C, Messager A, Demigne C, Remesy C (2000) Strains of lactic acid bacteria isolated from sourdoughs degrade phytic acid and improve calcium and magnesium solubility from whole wheat flours. J Agric Food Chem 48:2281–2285
78. Türk M, Sandberg A-S, Carlsson N, Andlid T (2000) Inositol hexaphosphate hydrolysis by baker's yeast. Capacity, kinetics and degradation products. J Agric Food Chem 48:100–104
79. Chaoui A, Faid M, Belhcen R (2003) Effect of natural starters used for sourdough bread in Morocco on phytate biodegradation. East Mediterr Health J 9:141–147
80. Reale A, Mannina L, Tremonte P, Sobolev AP, Succi M, Sorrentino E, Coppola R (2004) Phytate degradation by lactic acid bacteria and yeast during the wholemeal dough fermentation: a 31P NMR study. J Agric Food Chem 52:6300–6305
81. Haraldsson A-K, Veide J, Andlid T, Larsson Alminger M, Sandberg A-S (2005) Degradation of phytate by high-phytase *Saccharomyces cerevisiae* during simulated gastrointestinal digestion. J Agric Food Chem 53:5438–5444
82. Harinder K, Tiwana AS, Kaur B (1998) Studies on the baking of whole wheat meals; effect of pH, acids, milling and fermentation on phytic acid degradation. Adv Food Sci 20:181–189
83. De Angelis M, Gallo G, Corbo MR, McSweeney PL, Faccia M, Giovine M, Gobbetti M (2003) Phytase activity in sourdough lactic acid bacteria: purification and characterization of a phytase from *Lactobacillus sanfranciscensis* CB1. Int J Food Microbiol 87:259–570
84. Lopez H, Krspine V, Guy C, Messager A, Demigne C, Remesy C (2001) Prolonged fermentation of whole wheat sourdough reduces phytate level and increases soluble magnesium. J Agric Food Chem 49:2657–2662
85. Bryszewska MA, Ambroziak W, Diowksz A, Baxter MJ, Langford NJ, Lewis DJ (2005) Changes in the chemical form of selenium observed during the manufacture of a selenium-enriched sourdough bread for use in a human nutrition study. Food Addit Contam 22(2):135–140
86. Bryszewska MA, Ambroziak W, Langford NJ, Baxter MJ, Colyer A, Lewis DJ (2007) The effect of consumption of selenium enriched rye/wheat sourdough bread on the body's selenium status. Plant Foods Hum Nutr 62:121–126
87. Slavin J (2003) Why whole grains are protective: biological mechanisms. Proc Nutr Soc 62:129–134
88. Mattila P, Pihlava J-M, Hellström J (2005) Contents of phenolic acids, alkyl- and alkenylresorcinols, and avenanthramides in commercial grain products. J Agric Food Chem 53:8290–8295
89. Slavin J, Jacobs D, Marquardt L (2000) Grain processing and nutrition. Crit Rev Food Sci Nutr 40:309–326

90. Bondia-Pons I, Aura A-M, Vuorela S, Kolehmainen M, Mykkänen H, Poutanen K (2009) Rye phenolics in nutrition and health. J Cereal Sci 49:323–336
91. Rizzello CG, Coda R, Mazzacane F, Minervini D, Gobbetti M (2012) Micronized by-products from debranned durum wheat and sourdough fermentation enhanced the nutritional, textural and sensory features of bread. Food Res Intern 46:304–313
92. Michalska A, Ceglinska A, Amarowicz R, Piskula M-K, Szawara-Nowak D, Zielinski H (2007) Antioxidant contents and antioxidative properties of traditional rye breads. J Agric Food Chem 55:734–740
93. Mateo Anson N, Selinheimo E, Havenaar R, Aura A-M, Mattila I, Lehtinen P, van den Berg R, Haenen GRMM, Poutanen K, Bast A (2009) Effect of bioprocessing on in vitro bioaccessibility of phenolic acids and their microbial metabolites from wheat bran. J Agric Food Chem
94. Mateo Anson N, Aura A-M, Selinheimo E, Mattila I, Poutanen K, van den Berg R, Havenaar R, Bast A, Haenen GR (2011) Bioprocessing of wheat bran in whole wheat bread increases the bioavailability of phenolic acids in men and exerts anti-inflammatory effects ex vivo. J Nutr 141(1):137–143
95. Ketabi A, Dieleman LA, Gänzle MG (2011) Influence of isomalto-oligosaccharides on intestinal microbiota in rats. J Appl Microbiol 110:1297–1306
96. Meyer D, Stasse-Wolthuis M (2009) The bifidogenic effect of inulin and ligofructose and its consequences for gut health. Eur J Clin Nutr 63:1277–1289
97. Gibson GR, Scott KP, Rastall RA, Tuohy KM, Hotchkiss A, Dubert-Ferrandon A, Gareau M, Murphy EF (2010) Dietary prebiotics: current status and new definition. Food Sci Technol Bull 7:1–19
98. Candela M, Fiori J, Dipalo S, Brigidi P (2010) Development of a high-performance affinity chromatography-based method to study the biological interaction between whole microorganisms and target proteins. Lett Appl Microbiol 51:678–682
99. De Vuyst L, Degeest B (1999) Heteropolysaccharides from lactic acid bacteria. FEMS Microbiol Rev 23:153–177
100. Sutherland IW (2001) The biofilm matrix – an immobilized but dynamic microbial environment. Trends Microbiol 9:222–227
101. Korakli M, Gänzle M, Vogel R (2002) Metabolism by bifidobacteria and lactic acid bacteria of polysaccharides from wheat and rye, and exopolysaccharides produced by *Lactobacillus sanfranciscensis*. J Appl Microbiol 92:958–965
102. Seo E-C, Nam S-H, Kang H-K, Cho J-Y, Lee H-S, Ryu H-W, Kim D (2007) Synthesis of thermo- and acid-stable novel oligosaccharides by using dextransucrase with high concentration of sucrose. Enzyme Microb Technol 40:1117–1123
103. Tieking M, Korakli M, Ehrmann MA, Gänzle MG, Vogel RF (2003) In situ production of exopolysaccharides during sourdough fermentation by cereal and intestinal isolates of lactic acid bacteria. Appl Environ Microbiol 69:945–952
104. Schwab C, Mastrangelo M, Corsetti A, Gänzle M (2008) Formation of oligosaccharides and polysaccharides by *Lactobacillus reuteri* LTH5448 and *Weissella cibaria* 10M in sorghum sourdoughs. Cereal Chem 85:679–684

Chapter 10
Sourdough and Gluten-Free Products

Elke K. Arendt and Alice V. Moroni

10.1 Introduction: Gluten-Free Cereal Products

Celiac disease is one of the most common food intolerances, with an incidence of 1 in every 100 people worldwide, a number that is set to rise [1]. A lifelong avoidance of gluten-containing cereals and related products is the only effective treatment for people who suffer from celiac disease. Foods that are not allowed in the gluten-free (GF) diet are all the gluten-containing products prepared from barley, kamut, oat, wheat and their derivates, in which the gluten content exceeds 20 mg/kg on a total basis [2]. As the request for GF products is significantly rising, food technologists and manufacturers are called upon to satisfy the increasing requirements of the GF consumers [3]. In particular, people who suffer from celiac disease and those who are allergic to gluten ask for high-quality GF products, with the same textural, sensorial and nutritional properties as their gluten-containing counterparts [4, 5]. Nonetheless, the replacement of gluten with other non-toxic ingredients in conventional products, primarily bread and pasta, constitutes a major technological obstacle for the food industry. In fact, gluten represents the structure-forming protein in flour and it is responsible for the unique viscoelastic properties (extensibility, resistance to deformation, mixing tolerance and gas-holding capacity) of the dough [6]. The proteins present in GF flours do not possess these fundamental structural features, and, upon

E.K. Arendt (✉)
School of Food Science, Food Technology and Nutrition,
National University of Ireland, Cork, Ireland

Department of Food and Nutritional Sciences,
University College Cork, Western Road, Cork, Ireland
e-mail: e.arendt@ucc.ie

A.V. Moroniv
School of Food Science, Food Technology and Nutrition,
National University of Ireland, Cork, Ireland

National Food Biotechnology Centre, National University of Ireland, Cork, Ireland

M. Gobbetti and M. Gänzle (eds.), *Handbook on Sourdough Biotechnology*,
DOI 10.1007/978-1-4614-5425-0_10, © Springer Science+Business Media New York 2013

mixing, a weak batter, resembling a cake dough, is obtained [7]. Because of the impaired rheological properties of the GF batters in comparison to conventional doughs, most of the GF products available on the market are characterised by overall low quality, lacking flavour and showing poor textural characteristics and mouth feel [4, 8]. Furthermore, as GF products are mainly made from starch and are generally not fortified [9], their contribution in terms of different nutrients, such as folate, B vitamins, iron and dietary fibre, is poor [10, 11].

Over the last decade, most of the academic research in the GF field has been focused on the improvement of the quality of GF breads, producing breads that would meet the expectations of the GF consumers in terms of appearance, structure and nutritional benefits. Recent advances have been made in the incorporation of nutrient-dense whole grains in GF bread formulations [12–14]. In particular, increasing attention has been drawn to the utilization of pseudocereals, i.e. amaranth, buckwheat and quinoa, for their excellent protein quality and high fibre, mineral and phytochemical contents [15, 16]. Incorporation of pseudocereals in GF formulations was shown to induce significant improvements in the baking quality of the GF dough and in the nutritional benefits of the GF bread [12, 13, 17, 18]. However, these "healthy" flours are not yet extensively used for the production of GF products. Additionally, incorporation of prebiotics was also shown to be a successful approach for improving the dietary fibre content of GF breads produced from starch [19].

The attempts aimed at the improvement of the textural properties of GF bread are primarily affected by the absence of standardised baking tests for GF flours. For example, a recent study has shown how the physiochemical composition, i.e. starch content, particle size and rate of damaged starch, of oat flour can dramatically influence its bread-making performances [20]. However, no guidelines are currently available for the evaluation of the baking quality of GF flours.

Various additives, such as starches, hydrocolloids, non-toxic proteins, enzymes and combinations thereof have been investigated in an attempt to improve the poor structural and gas-holding capacity of GF batters. Because of their structure-forming properties [21], hydrocolloids have been extensively used to imitate the viscoelastic properties of gluten [22]. In particular, hydroxyl-propril-methyl-cellulose (HPMC), carbossi- or methyl-cellulose (CMC, MC), locust bean and guar gum, xhantan and pectins have been efficiently incorporated in GF bread formulations [9, 14, 23, 24, 25]). Overall, these investigations suggest that the obtained quality improvements, i.e. improved gas retention, crumb texture, specific volume and prolonged shelf life, are correlated to the amount of hydrocolloids used and the interaction between the type of flour and type of hydrocolloid employed.

Non-toxic proteins can also be applied to promote structure formation in GF breads. Incorporation of milk, legume and egg proteins can induce the formation of a gluten-like matrix in the batter and, therefore, improve the volume and the crumb texture of the final bread [14, 18, 26, 27]. As for hydrocolloids, the positive effects of the added proteins strongly depend on the interaction between type of flour and nature of the proteins [25]. However, the application of structuring proteins may represent a matter of concern, as these ingredients can be potential allergens for celiac patients [28].

Enzymatic processing of GF flours has also been extensively investigated. In particular, both cross-linking promoting enzymes, such as transglutaminase (TGase), cyclodextrin glycosyl transferase, glucose oxidase [23, 29, 30, 44] and proteases [31] were shown to improve the viscoelastic properties of batter and the final bread quality. TGase was proved to be particularly efficient when applied in GF breads produced with rice and buckwheat flours, used individually or in combination with other ingredients [32, 33]. In particular, Renzetti et al. [33] showed that the mode of action and the effects that TGase exerts on the pseudoplastic behaviour of the batter and, ultimately, on bread quality vary according to the raw material used for baking. Glucose oxidase was also used as a bread improver in rice and oat GF breads [23, 29]). Interestingly, protein hydrolysis was shown to improve the baking quality of rice flour, by decreasing the resistance to deformation of the batter during the baking process [31]. Thus, the concept that not only structure-promoting treatments can improve the bread-making performances of GF flours brings about new opportunities for GF baking [30, 34].

Overall, even if over the last decade promising improvements in the quality of GF breads have been obtained, the production of superior quality GF bread still represents a challenging task. Considering the lack of standardized baking tests, the high costs of the investigated additives and the great variability of flour-additive interactions, alternative ways to produce high-quality GF breads need to be investigated. This chapter discusses the recent advances in the application of sourdough in GF baking as a low-cost, efficient and natural tool to improve the quality of GF bread.

10.2 Sourdough Bread

Sourdough is a mixture of flour and water fermented with lactic acid bacteria (LAB) and yeasts, which can be added as starter cultures or originate as contaminants in the flour [35, 36]. During sourdough fermentation, the resident LAB are the main factor responsible for the acidification of the dough and for the synthesis of aroma compounds, exopolysaccharides, enzymes and anti-fungal compounds [37–39]. The addition of sourdough can strongly influence the quality of bread, in terms of enhanced texture, prolonged shelf life and improved organoleptic and nutritional profile [5, 40]. A profound knowledge of the metabolic events and the microbiological interactions occurring during sourdough fermentation is of crucial importance for controlling the fermentation and ensuring constant quality of the sourdough bread. However, while extensive research has been done on wheat and rye sourdoughs, for example on their microbial composition and their functional and organoleptic properties [35], only little information is available for GF sourdoughs [41–51]. Despite being limited in number, these studies indicate that the GF flours represent a unique source of novel strains and that GF sourdough can be successfully applied for modifying the rheological properties of the batters and for producing high-quality GF breads (Tables 10.1 and 10.2). In the next paragraphs we will discuss recent findings in the application of sourdough fermentation to GF flours.

Table 10.1 GF sourdough fermentations and their effects on the properties of GF batters and breads

Substrate/starter	Sourdough properties	Effects on GF batter	Effects on GF bread
Sorghum flour – *L. plantarum* [48]	Proteolysis of water soluble proteins	Increased strength lof the starch gel	Improved bread volume and crumb structure
Sorghum flour – *W. cibaria* [52]	Synthesis of dextran and GOS		Softer crumb and presence of undigested GOS
Sorghum flour – *W. kimchii* or *W. cibaria* MG1 [41]	Synthesis of dextran and GOS Production of low amounts of acetate	Improved viscoelastic properties (?)	Improved crumb structure, specific volume and delayed staling (?)
Composite formulation – *L. plantarum* FST 1.7 [44]	Production of anti-fungal compounds Activation of endogenous enzymes	Increased elasticity	Increase in the shelf life Delayed staling
Red sorghum – *L. plantarum*/*L. casei* or *L. reuteri*/*L. fermentum*	Hydrolysis of oligosaccharides		Improved nutritional and sensorial quality (?)
Pulse flour – *L. reuteri* [53]	Hydrolysis of glycol esters of phenolic compounds and flavonoid glucosides		
Amaranth – *L. paralimentarius* lAL28 or *L. plantarum* AL30 [42]		Improved viscoelastic properties	

(?) predicted effects

Table 10.2 Microbiota of GF sourdoughs and traditional GF products

Raw material Product	Dominant microbiota[a]	Starter (S) Spontaneous (Sp)[b]	Reference
Sorghum			
Sudanese kisra	*L. fermentum, L. reuteri, L. vaginalis, L. helveticus, L. pontis, P. pentosaceus, I. orientalis*	Sp	[54–56]
Sudanese khamir	*P. pentosaceus, L. brevis, L. lactis, L. cellobiosus, C. parapsilosis, C. orvegnsis, R. glutinis*	Sp	[57]
Botswana – porridge	*L. plantarum, L. casei/paracasei, L. buchneri, L. reuteri, L. perolens*	Sp	Sekwati-Monang, Gänzle, unpublished
Rice			
Industrial sourdough	*L. paracasei, L. paralimentarius, L. perolens, L. spicheri, S. cerevisiae*	Sp	[43]
	L. fermentum, L. gallinarum, L. pontis, C. krusei, S. cerevisiae	S	
Laboratory-scale sourdough	*L. gallinarum, L. plantarum, L. helveticus, L. fermentum, L. kimchii, L. pontis, I. orientalis, S. cerevisiae*	S	50]
Maize			
Kenkey	*L. fermentum, L. reuteri, P. pentosaceus, C. krusei, S. cerevisiae*	Sp	[58–60]
Mexican pozol	*L. delbrueckii, L. casei, L. fermentum, L. plantarum, Streptococcus* spp., *Leuconostoc* spp., *Weisella* spp.	Sp	61–63]
Laboratory-scale sourdough	*L. brevis, L. casei, L. fermentum, L. plantarum, Lc. mesenteroides, L. dextranicum, P. acidilactici, C. halbicans, S. cerevisiae, S. pombe*	Sp	[64, 65]
Laboratory-scale sourdough	*L. fermentum, L. paralimentarius, L. helveticus, L. pontis, S. cerevisiae, I. orientalis*	S	[50]
Teff			
Injera	*P. cerevisiae, L. brevis, L plantarum, L. fermentum, Saccharomyces, Torulopsis, Candida* spp.	Sp	[66]
Laboratory-scale sourdough	*L. plantarum* (S), *L. paralimentarius* (S), *L. fermentum* (S,Sp), *L. sanfranciscensis* (S), *L. frumenti* (S), *L. pontis* (S,Sp), *L. reuteri* (S), *L. amylovorus* (S), *L. brevis* (S), *L. vaginalis* (Sp), *L. gallinarum* (Sp), *P. acidilactici* (S), *P. pentosaceus* (Sp), *Lc. holzapfelii* (Sp), *K. barnetti* (S), *S. cerevisiae* (S,Sp), *C. glabrata* (Sp)	Sp and S	[46, 47]

(continued)

Table 10.2 (continued)

Raw material Product	Dominant microbiota[a]	Starter (S) Spontaneous (Sp)[b]	Reference
Buckwheat			
Laboratory-scale sourdoughs	*L. plantarum* (S,Sp), *L. paralimentarius* (S), *Lc. argentinum* (S), *L. sanfranciscensis* (S), *W. cibaria* (S,Sp), *L. brevis* (S), *L. fermentum* (S, Sp), *L. amylovorus* (S), *L. helveticus*(S), *P. pentosaceus* (Sp), *Lc. holzapfelii* (Sp), *L. graminis* (Sp), *L. sakei* (Sp), *L. vaginalis* (Sp), *L. crispatus* (Sp), *L. gallinarum* (Sp), *K. barnetti* (Sp)	Sp and S	[46, 47, 50]
Amaranth			
Laboratory-scale sourdoughs	*L. plantarum* (S,Sp), *L. sakei* (Sp), *L. paralimentarius* (S,Sp), *L. fermetum* (S), *L. helveticus* (S), *L. spicheri* (S), *P. pentosaceus* (Sp), *Enterococcus* spp. (Sp), *S. cerevisiae* (S), *C. glabrata* (S)	Sp and S	[49, 50]

Adapted from [5]

[a]I. *Issatchenkia,* L. *Lactobacillus,* Lc. *Leuconostoc,* P. *Pediococcus,* R. *Rhodotorula,* S. *Saccharomyces,* C. *Candida*

[b]The fermentation was either started by addition of starter strains (S) or by the spontaneous biota of the flour (Sp)

10.3 Ecology of GF Fermentations and Development of GF Sourdough Starters

The species diversity of wheat and rye sourdoughs has been intensively investigated by culture dependent and independent approaches. Heterofermentative species belonging to the genus *Lactobacillus* are among the most frequently isolated, but also species of the genera *Leuconostoc*, *Pediococcus* and *Weissella* were retrieved in traditional sourdoughs [67]. It is widely accepted that the selection of the competitive biota in wheat and rye sourdoughs is mainly driven by the fermentation parameters [36, 68], whereas the role played by the flour and its autochthonous microorganisms is under discussion [67].

Most ecological studies have been performed on sourdoughs produced from wheat, rye or spelt, fermented at laboratory scale or previously processed in bakery environments [69–73]. All together, these investigations indicate that the quality and nature of the flour play only a marginal role in the selection of the competitive species in conventional sourdoughs. In this regard, we must consider that wheat, rye and spelt are three closely related cereals, and that the bakery environment has been shown to define the predominant species in sourdoughs [73]. Therefore, a question still remains unsolved: would this principle be applicable for alternative sourdoughs? Recent studies indicate a different trend for GF flours [46, 50]. Vogelmann et al. [50] investigated the adaptability of various LAB and yeast starter strains in continuously propagated sourdoughs prepared from GF cereals and pseudocereals. The authors concluded that some of the starter strains used could not persist over the fermentation and their adaptability to the GF sourdoughs was strongly influenced by the chosen flour. Unfortunately, because of the high number of different substrates used, the authors could not identify the reasons for such variability. Moroni et al. [46] recently investigated the development of buckwheat and teff sourdoughs using commercial starters. The substrate used determined the persistence of specific starter strains; whereas spontaneous species, originating from the flour, either outcompeted or co-dominated with the starter LAB and yeasts. The unique sugar composition of certain GF flours is a key factor in the selection of dominant species in GF sourdoughs [41, 46, 47]. For example, when added as a starter strain, *Lactobacillus sanfranciscensis* failed to grow in sorghum sourdough due to lack of maltose at the beginning of the fermentation [41]. Instead, the high glucose level in sorghum sourdough favoured the growth and metabolic activities of *Weissella* spp. [41]. Similarly, *Weissella cibaria* was found to be highly competitive in buckwheat sourdough, in which the initial content of the monosaccharide was higher than that of maltose [46, 47]. Furthermore, the high ratio glucose/maltose favoured the coexistence of maltose positive yeasts and LAB in teff sourdoughs [46]. All together, these findings suggest that the nature of the GF flours used for sourdough fermentations has a strong impact in the selection of the dominant LAB and yeast strains. Furthermore, these studies clearly indicate that commercial starters as such cannot be efficiently used for the production of GF sourdough. In fact, the main criteria for selecting useful starters are that the starter's strains must be highly adapted to the GF substrate, dominate the fermentation and inhibit the growth of contaminant and/or autochthonous strains [67].

Ecological studies on GF sourdoughs are essential for developing GF starters, but to date only few data are available (Table 10.2). Most of these investigations have been carried out on traditional products, mainly produced from maize, sorghum and teff in tropical countries. These fermentation products were dominated by LAB species that mainly overlapped with those frequently isolated in wheat and rye sourdoughs [68]. In particular, *L. fermentum*, *L. reuteri* and *L. plantarum* were the ones most frequently isolated (Table 10.1). However, an interpretation of the results obtained in these studies is difficult due to the inappropriate techniques used for species identification, and to the non-sterile conditions under which the fermentations were carried out. More recently, the species diversity of laboratory-scale and industrial sourdoughs produced from different GF cereals and pseudocereals has been investigated through integrated approaches of culture dependent and independent techniques (Table 10.1). These sourdoughs have either been developed using starter cultures [43, 46, 50] or by spontaneous fermentation [47, 49, 64, 65]. As observed in traditional products, *L. fermentum*, *L. plantarum*, and also *L. paralimentarius* were present in virtually all the spontaneously and starter fermented GF sourdoughs from rice, maize, buckwheat, teff and amaranth (Table 10.1). Instead, the dominance of the most common sourdough species *L. sanfranciscensis* and *L. pontis* was substrate-specific. Species such as *L. gallinarum*, *L. graminis*, *L. sakei* and *Pediococcus pentosaceus*, which are not frequently isolated in conventional sourdoughs, were present in various GF sourdoughs, in particular when produced by spontaneous fermentation. The pseudocereals buckwheat and amaranth were found to be good substrates for the growth of *L. sakei*, *P. pentosaceus* and *L. paralimentarius*, whereas buckwheat sourdoughs induced inhibition of yeast growth [46, 47, 50, 51]. In conclusion, the ecological studies on GF fermentations suggest that GF flours represent an important reservoir for novel, competitive LAB and yeast strains that can be selected as starters for the production of stable GF sourdoughs. In a second stage, these strains can be screened for their functional properties, such as production of EPS, aroma and/or anti-fungal compounds and rate of acidification [5, 72, 74].

10.4 Proteolysis as a Tool to Improve the Baking Performances of GF Flours

The proteolytic events occurring during sourdough fermentations of wheat and rye flours have been exhaustively reviewed by Gänzle et al. [75]. Protein degradation during sourdough fermentation is among the key phenomena that affect the overall quality of sourdough bread, by inducing formation of precursors for flavour compounds and by modifying the viscoelastic properties of the dough [75]. Because of the gradual acidification of the dough by LAB, endogenous enzymes are activated and exert primary proteolysis on flour proteins. In a second stage, the released peptides are further hydrolysed into amino acids by intracellular peptidases of LAB, in a strain-specific manner [76]. In general, most sourdough LAB do not possess extracellular proteinase activity and prefer peptide uptake over amino acid transport [77]. LAB can affect the pattern of hydrolysed products by increasing the amount of dipeptides and amino acids released in the sourdough ([76, 77] As yeasts consume amino acids during growth, amino acid accumulation in sourdough can occur only after yeast growth has stopped [78–80]. Studies on the proteolytic events occurring during sourdough fermentation of GF flours and their effects on GF bread quality are still limited. Recently, sourdough fermentation was effectively applied for the production of GF bread based on sorghum flour, potato starch and HPMC ([48]; Table 10.1). Superior quality bread could be produced only when the total amount of sorghum flour was replaced by sorghum sourdough, fermented with the starter strain *L. plantarum*. The authors ascribed this quality improvement mainly to the proteolytic events occurring on soluble sorghum proteins during sourdough fermentation. The hydrolysed proteins did not interfere with the starch gel upon gelatinisation and a stronger starch gel with superior structural properties was obtained. As shown by confocal laser scanning microscopy (CLSM), the presence of small peptides in the sourdough bread prevented the formation of protein aggregates in the crumb upon cooking; whereas aggregates were formed in the chemically acidified control. Elkhalifa et al. [81] investigated the molecular and structural changes occurring during fermentation of sorghum flour for the preparation of the traditional Sudanese food *kisra*. As shown by light and CLS microscopy, fermentation resulted in the hydrolysis of the water-soluble proteins that constitute the outer shell of the starch granules. As a result, smaller starch granules were released from the matrix and the pasting properties of sorghum flour were modified. Proteolysis has also been investigated in the GF product *towga*, a traditional Tanzanian fermented food prepared by fermentation of either sorghum, maize, cassava, millet or combinations thereof [82]. Both spontaneous and starter-induced fermentations induced the activation of proteinases and, depending on the fermentation conditions, an increase in the content of specific amino acids, such as glutamic acid, proline, ornithine, methionine and lysine.

Overall, the proteolytic events occurring during sourdough fermentation can positively affect the baking quality of GF flours. Therefore, sourdough fermentation could replace the use of proteolytic enzymes in GF bread formulations [30, 31], which would allow a reduction in the cost of the bread and enable consumers' acceptance issues to be overcome. However, more studies are needed in order to understand which GF flours can be positively treated by sourdough fermentation and what degree of proteolysis is required in order to enhance their baking performances.

10.5 Proteolysis for Reducing the Toxicity of Wheat Flour

During endoluminal digestion, a family of peptides rich in Pro and Gln are released from prolamins of wheat and rye. These toxic peptides are responsible for the autoimmune response in celiac patients [83, 84]. During sourdough fermentation of wheat flour, gliadins are among the most affected proteins, where the extent of hydrolysis of monomeric gliadins (α-, β-, γ-, ω-gliadins) is strain-specific [76, 85]. Di Cagno et al. [76] showed that selected LAB, possessing proteolytic activities, could efficiently hydrolyse the 31–43 fragment of the toxic peptide A-gliadin in wheat sourdough. The same authors further applied the proteolytic LAB for producing non-toxic sourdough from a mixture of toxic and non-toxic flours, in which the highly toxic 33-mer peptide was completely hydrolysed [86]. Breads produced with this sourdough showed acceptable quality and when they were fed to celiac individuals, no alterations to the baseline values of the patients were observed. The same pool of proteolytic LAB was proven to be efficient for detoxifying rye flour [87] and, when used in association with *L. sanfranciscensis*, for producing non-toxic wheat sourdough bread of acceptable quality [88]. Sourdough fermentation can also be efficiently applied for eliminating the gluten that can eventually be present as a contaminant in GF flours [89].

Prolonged sourdough fermentation of wheat and rye using specific LAB may represent a novel technology for baking good-quality breads that can be consumed by celiac individuals. Nonetheless, long-term in vivo tests are needed in order to confirm the suitability of these products for celiac patients. Furthermore, manufacturers will have to face the obstacle of gaining the acceptance of consumers towards GF products containing detoxified wheat and/or rye.

10.6 Exopolysaccharides: A Low-Cost Alternative to Hydrocolloids in GF Breads

Exopolysaccharides (EPS) produced by LAB are alternative biothickeners that act as viscosifying, stabilising, emulsifying or gelling agents in a wide range of food products [90]. EPS are generally classified in two categories: homopolysaccharides (HoPS), glucose or fructose polymers, and heteropolysaccharides, containing (ir) regular repeating units [90]. To date, only HoPS have been shown to be useful in bread making. Glucans and fructans are synthesised by extracellular glucan- or fructansucrases, respectively, by various sourdough-associated LAB. *Lactobacillus reuteri, L. panis, L. pontis, L. frumenti and L. sanfranciscensis* were shown to produce fructans (levan or inulin) and glucans (dextran, reuteran or mutan) [91]. In particular, *Leuconostoc* spp. and *Weisella* spp. were proved to synthesise a large variety of dextrans [92, 93]. The structure, molecular weight and the production yield of HoPS varies among the producing LAB species and also depends on the carbohydrate concentration and composition of the source [92, 94]. During sourdough fermentations, LAB can produce EPS in high amounts, sufficient for improving the structural properties of the dough [95, 96]. In particular, in situ production of EPS was shown to be more effective than external addition of the same polysaccharide in the bread formulation [97]. Addition of sourdough fermented with HoPS-producing strains in wheat dough had a dramatic effect on bread quality by inducing softening of the crumb and increasing specific volume of the bread [92, 93, 95].

In addition to EPS, glucan- and fructansucrase can synthesise gluco- and fructooligosaccharides (FOS), respectively [98]. FOS have been associated with prebiotic effects [94, 99]. In particular, the levan produced by *L. sanfranciscensis* LHT2590, was proved to stimulate bifidobacterial growth in vitro [100]. In sourdough, *L. reuteri, L. acidophilus* and *L. sanfranciscensis* LHT2590 showed the ability to produce the prebiotic FOS 1-kestose [94, 94]. Therefore, EPS-producer strains can be applied not only to improve the bread-making performance of the flour, but also to enhance the nutritional value of the bread. These features render EPS the ideal replacement for hydrocolloids in GF breads. However, to date, only few studies [41, 52] have investigated EPS formation in GF sourdoughs and their feasibility for GF baking.

Schwab et al. [52] recently analysed the applicability of the EPS-producers *L. reuteri* LHT5448 and *W. cibaria* 10M in GF sourdoughs (Table 10.1). Both strains were shown to be suitable starters for sourdough fermentation of quinoa and sorghum, and during fermentation they produced levan/FOS and dextran/GOS (galacto-oligosaccharides), respectively. When applied in baking, sorghum sourdough fermented with *W. cibaria* was shown to be the most effective. In fact, softer sorghum breads were obtained upon addition of this sourdough, and the isomaltooligosaccharides present in the dough were not digested by baker's yeast during proofing [52]. The authors concluded that consumption of 300 g of sorghum GF bread prepared with *W. cibaria* 10M would account for a significant intake of prebiotic GOS.

Galle et al. [41] have also screened EPS-forming *Weissella* strains for their potential as starter strains in sorghum and wheat sourdoughs (Table 10.1). Independent from the strain used, higher amounts of EPS were formed in sorghum sourdough than in wheat, because of the higher concentration of glucose in the GF flour. In particular, the amount of dextrans produced by *W. kimchii* and *W. cibaria* MG1 were high enough for the sourdough to be applicable as a replacement for hydrocolloids in bread. In both sourdoughs, *Weissella* strains also produced oligosaccharides, whose structure was dependant on the chosen flour. Furthermore, both strains released fructose and formed only small amounts of acetate, a characteristic which is favourable for bread production [101]. All together, these studies indicate that sourdough fermented with EPS-forming *Weissella* strains can be applied in GF bread as a substitute for hydrocolloids. In addition, different types of oligosaccharides can be formed in the GF sourdough through manipulation of the carbon sources.

In conclusion, EPS-forming LAB can be successfully applied for improving the baking performances of GF flours. However, screening must be performed in order to identify EPS-producing strains that can be successfully applied as starter cultures in GF sourdoughs. In addition, more investigations are needed to evaluate the applicability of specific starters in GF sourdough breads, as the type of EPS and its interactions with the matrix strongly influence the structural properties of the dough [93].

10.7 Starch Hydrolysis for Delaying the Staling of GF Bread

Retrogradation/recrystallisation of starch is one of the key events involved in bread staling. During storage, the amylopectin network present in fresh bread gradually turns into a semi-crystalline network, which is responsible for crumb firming [102]. Amylases, in particular maltogenic amylases, and malt are commonly applied in bread baking as anti-staling agents [103]. Addition of sourdough to wheat bread can retard staling and sourdoughs with high TTA (total titratable acidity) and low pH are favourable for such purposes [104]. Even if most LAB species do not possess amylolytic activities, amylolytic strains were isolated from cereal fermentations in tropical climates [75, 105, 106]. In this regard, Corsetti et al. [107, 108] showed that the biological acidification together with the proteolytic and amylolytic activities of the selected starter strains delayed staling of sourdough wheat bread.

Staling represents one of the major issues in GF baking, since most GF breads are mainly starch based. Research on the effects of sourdough fermentation in starch hydrolysis and staling in GF bread is extremely limited. Malting and boiling combined with fermentation of sorghum resulted in a decrease of crumb firmness and dryness of sorghum-wheat composite bread [109]. Songré-Ouattara et al. [110] isolated amylolytic strains of *L. plantarum* in pearl millet gruels and efficiently used them as starter cultures for starch hydrolysis. The authors also suggested that more amylolytic strains can be isolated from the traditional product [110]. Schober et al. [48] showed that the combination of α-amylase and sourdough improved the quality of sorghum bread, but it did not have any positive effects on the staling rate. Instead, chemical acidification had rather negative effects on the bread volume and it dramatically increased the firmness over the whole storage period. Incorporation of sourdough into a GF bread mixture was also shown to delay the firming process, whereas chemical acidification increased the staling rate of the bread ([44]; Table 10.1).

Thus, depending on the flour and the starter strains, sourdough fermentation can be used for controlling the firming process of GF bread. However, the mechanism of bread staling is still under discussion and virtually no studies address this complex event in GF formulations. Therefore, more investigations are needed in order to characterise the activity of flour and bacterial amylases in GF sourdoughs. Through in depth investigations, it will be possible to identify the best process conditions for retarding staling of GF bread by application of sourdough, i.e. type and amount of added sourdough and its combination with commercial amylases and/or malts.

10.8 Sourdough as a Natural Tool for Improving the Shelf Life of GF Bread

The distribution of dust and mould spores in the bakery environment is the main cause of bread spoilage [111]. Bread spoilage represents a main issue for the baking enterprises, in so far that it induces economical losses and health risks for the consumers. The most common bread spoilage moulds are *Aspergillus*, *Cladiosporum*, *Endomyces*, *Fusarium*, *Monilia*, *Mucor*, *Penicillium* and *Rhizopus* [111]. An effective and natural way to improve the shelf life of bread consists in the application of sourdough. Sourdough-associated LAB can produce various substances with anti-fungal properties [112, 113]. Heterofermentative LAB release anti-fungal organic acids, among which acetic and propionic acids are more effective than lactic acid [113]. A mixture of organic acids, i.e. caproic, acetic, formic, propionic, butyric, and *n*-valeric acids, produced by *L. sanfranciscensis* CB1 were shown to be the main factors responsible for its anti-mould activity against *Fusarium*, *Penicillium*, *Aspergillus* and *Monilia* [108]. Strains of *L. plantarum* were proven to have broad anti-fungal activity, with 4-hydroxyphenlyllactic and phenyllactic acid being the major inhibiting compounds produced by the sourdough-LAB [114–116]. Ryan et al. [39] also showed that addition of sourdough fermented with the anti-fungal strain of *L. plantarum* allowed a reduction of the calcium propionate level in wheat bread by around 30%, without any negative effects on the shelf life. Reutericyclin is a low molecular weight antibiotic active against Gram-positive LAB and yeasts produced in active concentrations by *L. reuteri* [117]. In addition, *L. reuteri* strains can produce reuterin, an anti-microbial substance active against bacteria, yeasts and fungi (reviewed by [118]). Sourdough LAB are also effective against rope spoilage induced by *Bacillus* spp., because of production of organic acids and other unknown anti-bacterial compounds [119, 120].

Research on the application of sourdough for prolonging the shelf life of GF breads is still in its infancy. Recently, Moore et al. [44] employed sourdough (20%) fermented by the anti-fungal strain *L. plantarum* FST 1.7 in a composite GF bread (Table 10.1). *Lactobacillus plantarum* FST 1.7 retained its inhibitory activity against the test fungus *Fusarium culmorum* in the GF bread and it delayed the growth of the mould for up to 3 days in respect to the non-acidified control. Thus, the potentiality of sourdough as a shelf-life improver is retained in GF breads, but more investigations should evaluate this application. GF fermentations represent an important source of novel LAB strains and their potentiality as anti-fungal starter strains should be evaluated.

10.9 Sourdough Fermentation for Enhancing the Health Benefits of GF Bread

Sourdough has the potential to enhance the nutritional benefits of bread by improving mineral bioavailability, lowering the glycaemic response and regulating the accessibility of bioactive compounds in the flour (reviewed by [38]).

Phytic acid (PA) is the major storage form of phosphorous in grains and it is considered an anti-nutritional factor. In fact, PA impairs mineral absorption by strongly binding dietary cations to form insoluble complexes [121]. For celiac patients, who suffer from micronutrient deficiencies, the presence of PA in GF bread is particularly undesirable. Cereal grains contain phytases which can increase mineral bioavailability by dephosphorylating phytate to free inorganic phosphate and inositol phosphate esters. Sourdough fermentation favours the activity of endogenous phytases by creating optimal pH conditions for their activation [122, 123]. Lopez et al. [123] showed that sourdough fermentation enhanced phytate hydrolysis twice as much in respect to conventional yeast fermentation in wheat and in whole wheat bread. In addition, sourdough-associated LAB and yeasts were found to exert phytase activity in traditional sourdoughs [124]. Accordingly, De Angelis and co-workers [122] showed that fermentation by *L. sanfranciscensis* CB1 induced a decrease of PA by over 50% in wheat dough.

To date no studies have investigated the fate of PA in GF sourdoughs destined for bread production. Yet, phytase activity has been characterised during fermentation of GF cereals. Fermentation induced a decrease in PA content of sorghum and pearl millet [125]. Two phytase-positive strains, i.e. *L. plantarum* and *L. fermentum*, have been recently isolated from the above fermentation products [126].

Sourdough fermentation can also increase the bioavailability of several nutrients, such as folate, thiamine, vitamin B_1 and antioxidants in wheat and rye bread (reviewed by [38]). However, the excessive content of bioactive compounds in GF substrates can hamper the nutritional attributes of the GF flour [53]. Gänzle et al. [53] applied sourdough fermentation to red sorghum and pulse flour for reducing their content of polyphenols and oligosaccharides, respectively (Table 10.1). The α-galactosidase positive strain *L. reuteri* efficiently hydrolysed raffinose, stachyose and verbascose in pulse flour. Additionally, sourdough fermentation of red sorghum with binary combinations of LAB, i.e. *L. plantarum* / *L. casei* or *L. reuteri* / *L. fermentum,* resulted in quantitative hydrolysis of glycol esters of phenolic compounds and flavonoid glucosides. Therefore, LAB-induced fermentation of GF flours has the potential to improve their nutritional and sensorial characteristics [53]. Furthermore, as shown by Galle et al. [41] and Schwab et al. [52] fermentation of GF materials with selected EPS-producer strains can result in the release of significant amounts of GOS with prebiotic activity.

The poor nutritional benefits of GF breads can be enhanced by sourdough fermentation, which can efficiently decrease the content of anti-nutritional factors in the GF flour and increase the content of prebiotic GOS. However, more studies are needed in order to identify the best flour/starter combinations and process conditions for producing GF bread of high nutritional quality.

10.10 Conclusions

Production of GF bread of superior structural and nutritional quality still represents a major issue for food technologists. Sourdough fermentation of wheat and rye has been used for improving the overall quality of bread since ancient times. Recent studies have shown that its main positive effects on bread quality are retained when sourdough fermentation is applied to GF materials. Proteolysis and EPS production occurring during sourdough fermentation of GF substrates were shown to improve the rheological properties of the bread dough and the final bread quality. Furthermore, addition of sourdough has the potential of delaying the staling process through starch hydrolysis. From a nutritional point of view, application of sourdough technology leads to both an increase in certain bioactive compounds and a decrease in the content of anti-nutritional factors in GF flours. However, even if promising results have been obtained, the use of sourdough in GF baking is only in its infancy and more research should focus on this matter. For example, the potential of sourdough as a flavour-carrier has not been investigated in GF sourdoughs yet, even though the lack of flavour is one of the main negative aspects of GF breads. Following the rising interest in the use of sourdough in GF baking, investigations on the ecology of GF fermentations are increasing. Through such studies, it will be possible to select novel starter strains for producing GF sourdoughs with specific properties for achieving the desired quality improvements in GF breads.

With more data available, the industrial production of GF sourdough starters and GF sourdough breads is likely to become a reality in the market of GF products.

References

1. Catassi C, Fasano A (2008) Celiac disease. In: Arendt EK, Dal Bello F (eds) Gluten-free cereals products and beverages. Academic Press (Elsevier), London, pp 1–22
2. Deutsch H (2009) Gluten-free diet and food legislation. In: Arendt EK, Dal Bello F (eds) The science of gluten free foods and beverages. AACC International, St Paul

3. Bogue J, Sorenson D (2008) The marketing of gluten free products. In: Arendt EK, Dal Bello F (eds) Gluten free cereal products and beverages. Academic Press (Elsevier), London, pp 393–408
4. Gallagher E, Gormley TR, Arendt EK (2004) Recent advances in the formulation of gluten-free cereal-based products. Trends Food Sci Technol 15:143–152
5. Moroni AV, Dal Bello F, Arendt EK (2009) Sourdough in gluten-free bread-making: an ancient technology to solve a novel issue? Food Microbiol 26:676–684
6. Don C, Lichtendonk WJ, Plijter JJ, Hamer RJ (2003) Glutenin macropolymer: a gel formed by glutenin particles. J Cereal Sci 37:1–7
7. Arendt EK, Morrissey A, Moore MM, Dal Bello F (2008) Gluten-free breads. In: Arendt EK, Dal Bello F (eds) Gluten-free cereal products and beverages. Academic Press (Elsevier), London, pp 289–319
8. Gallagher E, Gormley TR, Arendt EK (2003) Crust and crumb characteristics of gluten free breads. J Food Eng 56:153–161
9. Ahlborn GJ, Pike OA, Hendrix SB, Hess WM, Huber CS (2005) Sensory, mechanical, and microscopic evaluation of staling in low-protein and gluten-free breads. Cereal Chem 82:328–335
10. Thompson T (2000) Folate, iron, and dietary fiber contents of the gluten-free diet. J Am Diet Assoc 100:1389–1396
11. Yazynina E, Johansson M, Jägerstad M, Jastrebova J (2008) Low folate content in gluten-free cereal products and their main ingredients. Food Chem 111:236–242
12. Alvarez-Jubete L, Holse M, Hansen A, Arendt EK, Gallagher E (2009) Impact of baking on vitamin E content of pseudocereals amaranth, quinoa, and buckwheat. Cereal Chem 86:511–515
13. Kiskini A, Argiri K, Kalogeropoulos M, Komaitis M, Kostaropoulos A, Mandala I, Kapsokefalou M (2007) Sensory characteristics and iron dialyzability of gluten-free bread fortified with iron. Food Chem 102:309–316
14. Moore MM, Schober TJ, Dockery P, Arendt EK (2004) Textural comparisons of gluten-free and wheat-based doughs, batters, and breads. Cereal Chem 81:567
15. Alvarez-Jubete L, Arendt EK, Gallagher E (2010) Nutritive value of pseudocereals and their increasing use as functional gluten-free ingredients. Trends Food Sci Technol 21:106–113
16. Schoenlechner R, Siebenhandl S, Berghofer E (2008) Pseudocereals. In: Arendt EK, Dal Bello F (eds) Gluten-free cereal products and beverages. Academic Press (Elsevier), London
17. Alvarez-Jubete L, Auty M, Arendt E, Gallagher E (2010) Baking properties and microstructure of pseudocereal flours in gluten-free bread formulations. Eur Food Res Technol 230:437–445
18. Mariotti M, Lucisano M, Pagani A, Ng MPKW (2009) The role of corn starch, amaranth flour, pea isolate, and Psyllium flour on the rheological properties and the ultrastructure of gluten-free doughs. Food Res Int 42:963–975
19. Korus J, Grzelak K, Achremowicz K, Sabat R (2006) Influence of prebiotic additions on the quality of gluten-free bread and on the content of inulin and fructooligosaccharides. Food Sci Technol Int 12:489–495
20. Hüttner EK, Dal Bello FD, Arendt EK (2010) Rheological properties and bread making performance of commercial wholegrain oat flours. J Cereal Sci 62:65–71
21. BeMiller JN (2008) Hydrocolloids. In: Arendt EK, Dal Bello F (eds) Gluten-free cereal products and beverages. Academic Press (Elsevier), London, pp 203–215
22. Lazaridou A, Duta D, Papageorgiou M, Belc N, Biliaderis CG (2007) Effects of hydrocolloids on dough rheology and bread quality parameters in gluten-free formulations. J Food Eng 79:1033–1047
23. Gujural HS, Rosell CM (2004) Improvement of the breadmaking quality of rice flour by glucose oxidase. Food Res Int 37:75–81
24. McCarthy DF, Gallagher E, Gormley TR, Schober TJ, Arendt EK (2005) Application of response surface methodology in the development of gluten-free bread. Cereal Chem 82:609–615

25. Schober TJ, Messerschmidt M, Bean SR, Park S-H, Arendt EK (2005) Gluten-free bread from sorghum: quality differences among hybrids. Cereal Chem 82:394–404
26. Gallagher E, Kunkel A, Gormley TR, Arendt EK (2003) The effect of dairy and rice powder addition on loaf and crumb characteristics, and on shelf life (intermediate and long-term) of gluten-free breads stored in a modified atmosphere. Eur Food Res Technol 218:44–48
27. Nunes M, Ryan L, Arendt E (2009) Effect of low lactose dairy powder addition on the properties of gluten-free batters and bread quality. Eur Food Res Technol 229:31–41
28. Ojetti V, Nucera G, Migneco A, Gabrielli M, Lauritano C, Danese S, Assunta Zocco MA, Nista EC, Cammarota G, de Lorenzo A, Gasbarrini G, Gasbarrini A (2005) High prevalence of celiac disease in patients with lactose intolerance. Digestion 71:106–110
29. Gujral HS, Guardiola I, Carbonell JV, Rosell CM (2003) Effect of cyclodextrinase on dough rheology and bread quality from rice flour. J Agric Food Chem 51:3814–3818
30. Renzetti S, Courtin CM, Delcour JA, Arendt EK (2010) Oxidative and proteolytic enzyme preparations as promising improvers for oat bread formulations: rheological, biochemical and microstructural background. Food Chem 119:1465–1473
31. Renzetti S, Arendt EK (2009) Effect of protease treatment on the baking quality of brown rice bread: from textural and rheological properties to biochemistry and microstructure. J Cereal Sci 48:33–45
32. Gujral HS, Rosell CM (2004) Improvement of the breadmaking quality of rice flour by glucose oxidase. Food Res Int 37:75–81
33. Renzetti S, Dal BF, Arendt EK (2008) Microstructure, fundamental rheology and baking characteristics of batters and breads from different gluten-free flours treated with a microbial transglutaminase. J Cereal Sci 48:33–45
34. Schober TJ, Bean SR, Boyle DL, Park SH (2008) Improved viscoelastic zein-starch doughs for leavened gluten-free breads: their rheology and microstructure. J Cereal Sci 48:755–767
35. De Vuyst L, Vancanneyt M (2007) Biodiversity and identification of sourdough lactic acid bacteria. Food Microbiol 24:120–127
36. Hammes WP, Brandt MJ, Francis KL, Rosenheim J, Seitter MFH, Vogelmann SA (2005) Microbial ecology of cereal fermentations. Trends Food Sci Technol 16:4–11
37. Gänzle MG, Vermeulen N, Vogel RF (2007) Carbohydrate, peptide and lipid metabolism of lactic acid bacteria in sourdough. Food Microbiol 24:128–138
38. Poutanen K, Flander L, Katina K (2009) Sourdough and cereal fermentation in a nutritional perspective. Food Microbiol 26:693–699
39. Ryan LAM, Dal BF, Arendt EK (2008) The use of sourdough fermented by antifungal LAB to reduce the amount of calcium propionate in bread. Int J Food Microbiol 125:274–278
40. Arendt EK, Ryan LAM, Dal Bello F (2007) Impact of sourdough on the texture of bread. Food Microbiol 24:165–174
41. Galle S, Schwab C, Arendt E, Gänzle M (2010) Exopolysaccharide-forming *Weissella* strains as starter cultures for sorghum and wheat sourdoughs. J Agric Food Chem 58:5834–5841
42. Houben A, Goetz H, Mitzscherling M, Becker T (2010) Modification of the rheological behaviour of Amaranth (*Amaranthus hypochondriacus*) dough. J Cereal Sci 51:350–356
43. Meroth CB, Hammes WP, Hertel C (2004) Characterisation of the microbiota of rice sourdoughs and description of *Lactobacillus spicheri* sp. nov. Syst Appl Microbiol 27:151–159
44. Moore M, Dal BF, Arendt E (2008) Sourdough fermented by *Lactobacillus plantarum* FST 1.7 improves the quality and shelf life of gluten-free bread. Eur Food Res Technol 226:1309–1316
45. Moore MM, Juga B, Schober TJ, Arendt EK (2007) Effect of lactic acid bacteria on properties of gluten-free sourdoughs, batters, and quality and ultrastructure of gluten-free bread. Cereal Chem 84:357–364
46. Moroni AV, Arendt EK, Dal Bello F (2010a) Biodiversity of lactic acid bacteria and yeasts in spontaneously fermented buckwheat and teff sourdoughs. Food Micr 28, 497–502
47. Moroni AV, Arendt EK, Morrissey JP, Dal Bello F (2010b) Development of buckwheat and teff sourdoughs with the use of commercial starters. In J food Microb 142:142–148
48. Schober TJ, Bean SR, Boyle DL (2007) Gluten-free sorghum bread improved by sourdough fermentation: biochemical, rheological, and microstructural background. J Agric Food Chem 55:5137–5146

49. Sterr Y, Weiss A, Schmidt H (2009) Evaluation of lactic acid bacteria for sourdough fermentation of amaranth. Int J Food Microbiol 136:75–82
50. Vogelmann SA, Seitter M, Singer U, Brandt MJ, Hertel C (2009) Adaptability of lactic acid bacteria and yeasts to sourdoughs prepared from cereals, pseudocereals and cassava and use of competitive strains as starters. Int J Food Microbiol 130:205–212
51. Weiss A, Bertsch D, Struett S, Sterr Y, Schmidt H (2009) Isolierung und Charakterisierung potentieller Starterkulturen aus Amaranth-, Buchweizen-und Hirse-Sauerteigen. Getreidetechnologie 63:68–75
52. Schwab C, Mastrangelo M, Corsetti A, Gänzle M (2008) Formation of oligosaccharides and polysaccharides by *Lactobacillus reuteri* LTH5448 and *Weissella cibaria* 10M in sorghum sourdoughs. Cereal Chem 85:679–684
53. Gänzle MG, Schieber A, Svensson L, Teixeira J, McNeill V (2010) Formation and modification of bioactive compounds in gluten free sourdoughs. In: Second international symposium on gluten-free cereal products and beverages, Tampere, pp 89–90
54. Hamad SH, Böcker G, Vogel RD, Hammes WP (1992) Microbiological and chemical analysis of fermented sorghum dough for Kisra production. Appl Microbiol Biotechnol 37:728–731
55. Hamad SH, Dieng MC, Ehrmann MA, Vogel R (1997) Characterisation of the bacterial flora of Sudanese sorghum flour and sorghum sourdough. J Appl Microbiol 83:764–770
56. Mohammed SI, Steenson LR, Kirleis AW (1991) Isolation and characterization of microorganisms associated with the traditional sorghum fermentation for production of Sudanese kisra. Appl Environ Microbiol 57:2529–2533
57. Gassem MAA (1999) Study of the micro-organisms associated with the fermented bread (khamir) produced from sorghum in Gizan region, Saudi Arabia. J Appl Microbiol 86:221–225
58. Hayford AE, Petersen A, Vogensen FK, Jakobsen M (1999) Use of conserved randomly amplified polymorphic DNA (RAPD) fragments and RAPD pattern for characterization of *Lactobacillus fermentum* in Ghanaian fermented maize dough. Appl Environ Microbiol 65:3213–3221
59. Jespersen L, Halm M, Kpodo K, Jakobsen M (1994) Significance of yeasts and moulds occurring in maize dough fermentation for "kenkey" production. Int J Food Microbiol 24:239–248
60. Olsen A, Halm M, Jakobsen M (1995) The antimicrobial activity of lactic acid bacteria from fermented maize (kenkey) and their interactions during fermentation. J Appl Bacteriol 79:506–512
61. Ampe F, ben Omar N, Moizan C, Wacher C, Guyot JP (1999) Polyphasic study of the spatial distribution of microorganisms in Mexican pozol, a fermented maize dough, demonstrates the need for cultivation-independent methods to investigate traditional fermentations. Appl Environ Microbiol 65:5464–5473
62. Bẹn Omar N, Ampe F (2000) Microbial community dynamics during production of the Mexican fermented maize dough pozol. Appl Environ Microbiol 66:3664–3673
63. Escalante A, Wacher C, Farrés A (2001) Lactic acid bacterial diversity in the traditional Mexican fermented dough pozol as determined by 16S rDNA sequence analysis. Int J Food Microbiol 64:21–31
64. Edema MO, Sanni AI (2008) Functional properties of selected starter cultures for sour maize bread. Food Microbiol 25:616–625
65. Sanni AI, Onilude AA, Fatungase MO (1998) Production of sour-maize bread using starter-cultures. World J Microbiol Biotechnol 14:101–106
66. Ashenafi M (2006) A review on the microbiology of indigenous fermented food and beverages in Ethiopia. Ethiop J Microbiol Sci 5:189–245
67. De Vuyst L, Vrancken G, Ravyts F, Rimaux T, Weckx S (2009) Biodiversity, ecological determinants, and metabolic exploitation of sourdough microbiota. Food Microbiol 26:666–675
68. Gänzle M, Schwab C (2009) Exploitation of the metabolic potential of lactic acid bacteria for improved quality of gluten-free bread. In: Arendt EK, Dal Bello F (eds) The science of gluten-free food and beverages. AACC International, St Paul

69. Meroth CB, Walter J, Hertel C, Brandt MJ, Hammes WP (2003) Monitoring the bacterial population dynamics in sourdough fermentation processes by using PCR-denaturing gradient gel electrophoresis. Appl Environ Microbiol 69:475–482
70. Meroth CB, Hammes WP, Hertel C (2003) Identification and population dynamics of yeasts in sourdough fermentation processes by PCR-denaturing gradient gel electrophoresis. Appl Environ Microbiol 69:7453–7461
71. Rosenquist H, Hansen A (2000) The microbial stability of two bakery sourdoughs made from conventionally and organically grown rye. Food Microbiol 17:241–250
72. Siragusa S, Di Cagno R, Ercolini D, Minervini F, Gobbetti M, De Angelis M (2009) Taxonomic structure and monitoring of the dominant population of lactic acid bacteria during wheat flour sourdough type I propagation using *Lactobacillus sanfranciscensis* starters. Appl Environ Microbiol 75:1099–1109
73. Scheirlinck I, Van der Meulen R, Van Schoor A, Vancanneyt M, De Vuyst L, Vandamme P, Huys G (2008) Taxonomic structure and stability of the bacterial community in Belgian sourdough ecosystems as assessed by culture and population fingerprinting. Appl Environ Microbiol 74:2414–2423
74. Holzapfel WH (2002) Appropriate starter culture technologies for small-scale fermentation in developing countries. Int J Food Microbiol 75:197–212
75. Gänzle MG, Loponen J, Gobbetti M (2008) Proteolysis in sourdough fermentations: mechanisms and potential for improved bread quality. Trends Food Sci Tech 19:513–521
76. Di Cagno R, De Angelis M, Lavermicocca P, De Vincenzi M, Giovannini C, Faccia M, Gobbetti M (2002) Proteolysis by sourdough lactic acid bacteria: effects on wheat flour protein fractions and gliadin peptides involved in human cereal intolerance. Appl Environ Microbiol 68:623–633
77. Thiele C, Gaenzle MG, Vogel RF (2003) Fluorescence labeling of wheat proteins for determination of gluten hydrolysis and depolymerization during dough processing and sourdough fermentation. J Agric Food Chem 51:2745–2752
78. Gobbetti M, Simonetti MS, Rossi J, Cossignani L, Corsetti A, Damiani P (1994) Free D- and L-amino acid evolution during sourdough fermentation and baking. J Food Sci 59:881–884
79. Spicher G, Nierle W (1988) Proteolytic activity of sourdough bacteria. Appl Microbiol Biotechnol 28:487–492
80. Thiele C, Gänzle MG, Vogel RF (2002) Contribution of sourdough lactobacilli, yeast, and cereal enzymes to the generation of amino acids in dough relevant for bread flavor. Cereal Chem 79:45–51
81. Elkhalifa A, Bernhardt R, Bonomi F, Iametti S, Pagani M, Zardi M (2006) Fermentation modifies protein/protein and protein/starch interactions in sorghum dough. Eur Food Res Technol 222:559–564
82. Mugula JK, Nnko SAM, Narvhus JA, Sørhaug T (2003) Microbiological and fermentation characteristics of togwa, a Tanzanian fermented food. Int J Food Microbiol 80:187–199
83. Anderson RP, Degano P, Godkin AJ, Jewell DP, Hill AVS (2000) In vivo antigen challenge in celiac disease identifies a single transglutaminase-modified peptide as the dominant A-gliadin T-cell epitope. Nat Med 6:337–342
84. Kendall M, Schneider R, Cox PS, Hawkins CF (1972) Gluten subfractions in coeliac disease. Lancet 18:1065–1067
85. Wieser H, Vermeulen N, Gaertner F, Vogel R (2008) Effects of different *Lactobacillus* and *Enterococcus* strains and chemical acidification regarding degradation of gluten proteins during sourdough fermentation. Eur Food Res Technol 226:1495–1502
86. Di Cagno R, De Angelis M, Auricchio S, Greco L, Clarke C, De Vincenzi M, Giovannini C, D'Archivio M, Landolfo F, Parrilli G, Minervini F, Arendt E, Gobbetti M (2004) Sourdough bread made from wheat and nontoxic flours and started with selected lactobacilli is tolerated in celiac sprue patients. Appl Environ Microbiol 70:1088–1096
87. De Angelis M, Coda R, Silano M, Minervini F, Rizzello CG, Di Cagno R, Vicentini O, De Vincenzi M, Gobbetti M (2006) Fermentation by selected sourdough lactic acid bacteria to decrease coeliac intolerance to rye flour. J Cereal Sci 43:301–314

88. Rizzello CG, De Angelis M, Di Cagno R, Camarca A, Silano M, Losito I, De Vincenzi M, De Bari MD, Palmisano F, Maurano F, Gianfrani C, Gobbetti M (2007) Highly efficient gluten degradation by lactobacilli and fungal proteases during food processing: new perspectives for celiac disease. Appl Environ Microbiol 73:4499–4507
89. Giuliani GM, Benedusi A, Di Cagno R, De Angelis M, Luisi A, Gobbetti M (2006) Miscela di batteri lattici per la preparazione di prodotti da forno senza glutine, RM2006A000369
90. De Vuyst L, Degeest B (1999) Heteropolysaccharides from lactic acid bacteria. FEMS Microbiol Rev 23:153–177
91. Tieking M, Gänzle MG (2005) Exopolysaccharides from cereal-associated lactobacilli. Trends Food Sci Technol 16:79–84
92. Di Cagno R, De Angelis M, Limitone A, Minervini F, Carnevali P, Corsetti A, Gaenzle M, Ciati R, Gobbetti M (2006) Glucan and fructan production by sourdough *Weissella cibaria* and *Lactobacillus plantarum*. J Agric Food Chem 54:9873–9881
93. Lacaze G, Wick M, Cappelle S (2007) Emerging fermentation technologies: development of novel sourdoughs. Food Microbiol 24:155–160
94. Korakli M, Pavlovic M, Ganzle MG, Vogel RF (2003) Exopolysaccharide and kestose production by *Lactobacillus sanfranciscensis* LTH2590. Appl Environ Microbiol 69:2073–2079
95. Katina K, Maina NH, Juvonen R, Flander L, Johansson L, Virkki L, Tenkanen M, Laitila A (2009) In situ production and analysis of *Weissella confusa* dextran in wheat sourdough. Food Microbiol 26:734–743
96. Tieking M, Korakli M, Ehrmann MA, Ganzle MG, Vogel RF (2003) In situ production of exopolysaccharides during sourdough fermentation by cereal and intestinal isolates of lactic acid bacteria. Appl Environ Microbiol 69:945–952
97. Brandt MJ, Roth K, Hammes WP (2003) Effect of an exopolysaccharides produced by *Lactobacillus sanfranciscensis* LHT 1729 on dough and bread quality. In: De Vuyst L (ed) Sourdough from fundamentals to application. Vrije Universiseit Brussels (VUB), IMDO, Brussels, p 80
98. Monsan P, Bozonnet S, Albenne C, Joucla G, Willemot RM, Remaud-Siméon M (2001) Homopolysaccharides from lactic acid bacteria. Int Dairy J 11:675–685
99. Cummings JH, Macfarlane GT, Englyst HN (2001) Prebiotic digestion and fermentation. Am J Clin Nutr 73:415S–420S
100. Dal Bello F, Walter J, Hertel C, Hammes WP (2001) In vitro study of prebiotic properties of levan-type exopolysaccharides from lactobacilli and non-digestible carbohydrates using denaturing gradient gel electrophoresis. Syst Appl Microbiol 24:232–237
101. Kaditzky S, Vogel R (2008) Optimization of exopolysaccharide yields in sourdoughs fermented by lactobacilli. Eur Food Res Technol 228:291–299
102. Goesaert H, Slade L, Levine H, Delcour JA (2009) Amylases and bread firming – an integrated view. J Cereal Sci 50:345–352
103. Goesaert H, Gebruers K, Courtin CM, Brijs K, Delcour J (2006) Enzymes in breadmaking. In: Hui YH (ed) Bakery products. Science and technology. Blackwell Publishing, Ames, pp 337–364
104. Barber B, Ortola C, Barber S, Fernandez F (1992) Storage of packaged white bread. III: Effects of sourdough and addition of acids on bread characteristics. Z Lebensm Unters Forsch 194:442–449
105. Sanni AI, Morlon-Guyot J, Guyot JP (2002) New efficient amylase-producing strains of *Lactobacillus plantarum* and *L. fermentum* isolated from different Nigerian traditional fermented foods. Int J Food Microbiol 72:53–62
106. Tou EH, Mouquet-Rivier C, Rochette I, Traoré AS, Trèche S, Guyot JP (2007) Effect of different process combinations on the fermentation kinetics, microflora and energy density of ben-saalga, a fermented gruel from Burkina Faso. Food Chem 100:935–943
107. Corsetti A, Gobbetti M, De Marco B, Balestrieri F, Paoletti F, Russi L, Rossi J (2000) Combined effect of sourdough lactic acid bacteria and additives on bread firmness and staling. J Agric Food Chem 48:3044–3051

108. Corsetti A, Gobbetti M, Rossi J, Damiani P (1998) Antimould activity of sourdough lactic acid bacteria: identification of a mixture of organic acids produced by *Lactobacillus sanfrancisco* CB1. Appl Microbiol Biotechnol 50:253–256
109. Hugo LF, Rooney LW, Taylor JRN (2003) Fermented sorghum as a functional ingredient in composite breads. Cereal Chem 80:495–499
110. Songré-Ouattara LT, Mouquet-Rivier C, Icard-Vernière C, Rochette I, Diawara B, Guyot JP (2009) Potential of amylolytic lactic acid bacteria to replace the use of malt for partial starch hydrolysis to produce African fermented pearl millet gruel fortified with groundnut. Int J Food Microbiol 130:258–264
111. Legan JD (1993) Mould spoilage of bread. Int Biodeter Biodegr 32:33–53
112. Messens W, De Vuyst L (2002) Inhibitory substances produced by Lactobacilli isolated from sourdoughs – a review. Int J Food Microbiol 72:31–43
113. Schnürer J, Magnusson J (2005) Antifungal lactic acid bacteria as biopreservatives. Trends Food Sci Technol 16:70–78
114. Dal Bello F, Clarke CI, Ryan LAM, Ulmer H, Schober TJ, Ström K, Sjögren J, van Sinderen D, Schnürer J, Arendt EK (2007) Improvement of the quality and shelf life of wheat bread by fermentation with the antifungal strain *Lactobacillus plantarum* FST 1.7. J Cereal Sci 45:309–318
115. Lavermicocca P, Valerio F, Visconti A (2003) Antifungal activity of phenyllactic acid against molds isolated from bakery products. Appl Environ Microbiol 69:634–640
116. Ryan LAM, Dal Bello F, Czerny M, Koehler P, Arendt EK (2009) Quantification of phenyllactic acid in wheat sourdough using high resolution gas chromatography-mass spectrometry. J Agric Food Chem 57:1060–1064
117. Gänzle MG, Holtzel A, Walter J, Jung G, Hammes WP (2000) Characterization of reutericyclin produced by *Lactobacillus reuteri* LTH2584. Appl Environ Microbiol 66:4325–4333
118. Gänzle MG (2004) Reutericyclin: biological activity, mode of action, and potential applications. Appl Microbiol Biotechnol 64:326–332
119. Katina K, Sauri M, Alakomi HL, Mattila-Sandholm T (2002) Potential of lactic acid bacteria to inhibit rope spoilage in wheat sourdough bread. Lebensm Wiss Technol 35:38–45
120. Valerio F, De Bellis P, Lonigro SL, Visconti A, Lavermicocca P (2008) Use of *Lactobacillus plantarum* fermentation products in bread-making to prevent *Bacillus subtilis* ropy spoilage. Int J Food Microbiol 122:328–332
121. Bohn L, Meyer A, Rasmussen S (2008) Phytate: impact on environment and human nutrition. A challenge for molecular breeding. J Zhejiang Univ Sci B 9:165–191
122. De Angelis M, Gallo G, Corbo MR, McSweeney PLH, Faccia M, Giovine M, Gobbetti M (2003) Phytase activity in sourdough lactic acid bacteria: purification and characterization of a phytase from *Lactobacillus sanfranciscensis* CB1. Int J Food Microbiol 87:259–270
123. Lopez HW, Krespine V, Guy C, Messager A, Demigne C, Remesy C (2001) Prolonged fermentation of whole wheat sourdough reduces phytate level and increases soluble magnesium. J Agric Food Chem 49:2657–2662
124. Reale A, Mannina L, Tremonte P, Sobolev AP, Succi M, Sorrentino E, Coppola R (2004) Phytate degradation by lactic acid bacteria and yeasts during the wholemeal dough fermentation: a 31P NMR study. J Agric Food Chem 52:6300–6305
125. Osman MA (2004) Changes in sorghum enzyme inhibitors, phytic acid, tannins and in vitro protein digestibility occurring during Khamir (local bread) fermentation. Food Chem 88:129–134
126. Songre-Outtara LT, Mouquet-Rivier C, Icard-Verniere C, Rochette I, Diawara B, Guyot JP (2009) Potential of amylolytic lactic acid bacteria to replace the use of malt for partial starch hydrolysis to produce African fermented pearl millet gruel fortified with groundnut. Int J Food Microbiol 130:217–229

Chapter 11
Sourdough and Cereal Beverages

Jussi Loponen and Juhani Sibakov

11.1 Introduction

Worldwide, a number of fermented solid, semi-liquid, and liquid products exist that are produced by microbial incubations or fermentation utilising milk, meat, legumes, tubers, and cereals as raw materials. However, only with cereal substrates is it possible to produce fermented doughs, porridges, gruels as well as alcoholic and non-alcoholic beverages, and also non-dairy yogurt alternatives. This chapter will focus on traditional cereal-based non-alcoholic fermented beverages and intends to introduce products that differ in their manufacturing concepts. In principal, all porridges or gruels may be turned into "beverage mode" by adding more water to the recipes though sedimentation of solids may make this impractical. Nevertheless, processing technology and specialty ingredients nowadays allow stabilisation of these into homogenous and stable products. It is worth noting that most of the fermented cereal beverages are traditionally consumed in certain geographical regions and, thus, the majority can be considered as indigenous to specific groups of people.

11.2 Boza

Boza is a traditional cereal beverage made by fermentation of cooked, strained, and sugared cereal slurry. The term *boza* refers to millet (*Persian*) but, in practise, flour or semolina or cracked grains of wheat, maize, rice, or their mixtures are also used

J. Loponen (✉)
Fazer Group, Helsinki, Finland
e-mail: jussi.loponen@fazer.com

J. Sibakov
VTT Technical Research Centre of Finland, Espoo, Finland

M. Gobbetti and M. Gänzle (eds.), *Handbook on Sourdough Biotechnology*,
DOI 10.1007/978-1-4614-5425-0_11,

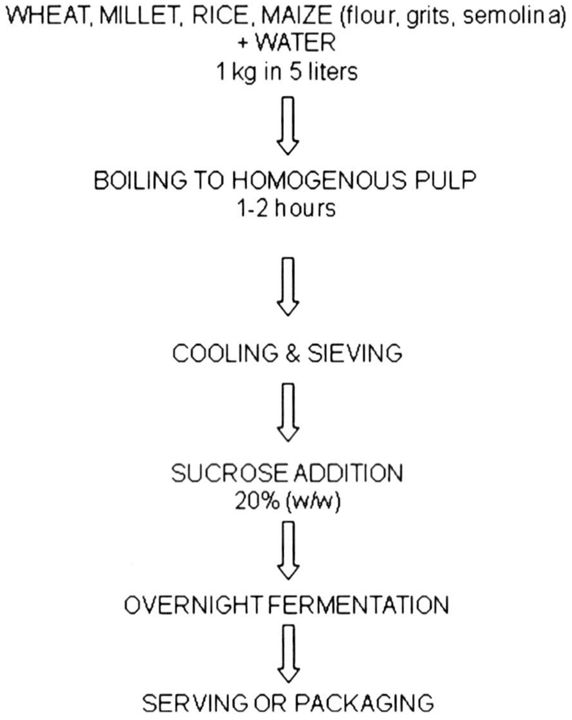

Fig. 11.1 Process chart for boza preparation

as raw materials. The end product is a fairly thick liquid with pale yellow appearance, and a mixed sweet and sour taste that is characteristic for boza. Starter cultures for boza production may originate from a previous boza fermentation, sourdough, pure cultures, or yoghurt. Arici and Daglioglu [1] review the characteristics of boza and its history more in detail. Turkey and Bulgaria are the heartlands of boza.

The process of boza fermentation is illustrated in Fig. 11.1 (based on [1]). Cooking of pre-processed (cleaned, milled) cereal raw materials into a slurry is followed by a cooling and straining stage, sucrose addition (up to 20% w/w), inoculation with a starter culture, overnight fermentation, and finally cooling and packaging (Fig. 11.1). Water is added during processing in order to adjust the consistency of the slurry. The fermented boza exhibits pseudoplastic rheological behaviour, i.e. its apparent viscosity decreases with increased shear rate [2, 3].

Hancioğlu and Karapinar [4] were among the first to monitor boza fermentations. During a 24 h fermentation, acidification took place, the pH dropped from 6.1 to 3.5, and final cell counts of lactic acid bacteria and yeasts were 5×10^8 and 8×10^6, respectively. Gotcheva et al. [5] observed that glucose accumulated and free amino nitrogen levels decreased during boza fermentation, of which the latter observation

Table 11.1 LAB and yeast strains isolated from boza fermentations. Literature data was collected using criteria that the total cell counts for LAB and yeasts in boza were at minimum levels of 10^7 and 10^5, respectively; lower cell densities were considered irrelevant

		Amylolytic	Bacteriocin	
LAB	*Weissella confusa*			[4]
	Lactobacillus fermentum			[4, 6, 7]
	Lactobacillus paracasei		×	[6, 8]
	Lactobacillus pentosus	×		[6, 8, 9]
	Lactobacillus rhamnosus		×	[6, 8]
	Pediococcus pentosaceus		×	[10]
	Lactobacillus sanfranciscensis			[4]
	Lactobacillus coryniformis			[4]
	Lactobacillus plantarum	×	×	[6–9]
	Lactobacillus coprophilus			[7]
	Lactobacillus brevis			[6, 7]
	Lactobacillus acidophilus			[7]
	Lactobacillus raffinolactis			[7]
	Leuconostoc mesenteroides		×	[4, 7, 11]
	Leuconostoc paramesenteroides			[4]
	Oenococcus oenos			[4]
Yeasts	*Saccharomyces uvarum*			[4]
	Saccharomyces cerevisiae			[4, 7]
	Geotrichum penicillatum			[7]
	Geotrichum candidum			[7]
	Candida tropicalis			[7]
	Candida glabrata			[7]

indicates that extensive protein breakdown did not occur during the fermentation process. Hayta et al. [2] showed that the soluble protein content increased during boza fermentation, which was probably induced by acidification. These values and tendencies are characteristic for boza fermentation. Indeed, the cooking stage prior to starter inoculation is an important feature of boza, because the heating practically inactivates the raw material associated enzymes and microbes, and consequently results in a semi-sterile base free from enzymes and microbes. This feature allows the usage and operation of starter cultures with specific activities. Of the lactic acid bacteria, especially heterofermentative lactobacilli and leuconostocs dominate boza fermentations (Table 11.1) whereas *Saccharomyces* species are the predominant yeasts. In addition members of the *Weissella*, *Lactococcus*, *Pediococcus*, *Geotrichum* and *Candida* families were identified (Table 11.1). Zorba et al. [12] used strains previously isolated from boza as starters for new fermentation in order to develop an optimal starter culture. A combination of *Leuconostoc mesenteroides* ssp. *mesenteroides*, *Weissella confusa* and *Saccharomyces cerevisiae* strains was suggested as a feasible starter culture for the controlled production of boza as the trio elicited the desired rheological and sensory properties of fully fermented boza [12].

In addition to improved sensory and rheological properties, the use of controlled starter cultures in boza fermentation aimed at incorporating bacteriocin-producing strains in boza fermentations; Todorov and Dicks [8, 10, 11] isolated a number of bacteriocin-producing strains from boza and also partially characterised many bacteriocins (Table 11.1). Bacteriocins are small proteinaceous metabolites that are excreted by the host strain during its stationary growth phase, and they exhibit lethal or static effects against bacteria closely related to the host, which may prolong shelf life and improve the safety of the products.

Boza, as a semi-sterile base, lacks the enzyme activity of cereal raw materials, which in sourdough fermentation, for instance, play a significant role. This sets requirements for starter strains to grow as well as offers some possibilities for food technologists to take advantage of specific catalytic activities of starter strains of even added exogenous catalysts. Petrova et al. [9] identified two amylolytic lactobacilli strains from Bulgarian boza. The strains, *Lactobacillus plantarum* Bom 816 and *L. pentosus* N3, used starch as a sole carbohydrate source, and the amylolytic activity of the cell-wall associated enzymes was optimal at pH 5.5 and 45 °C. No research on the synthesis of extracellular polysaccharides (EPS) exists though boza could serve as a potential matrix for this as it can be fermented with selected starter cultures (e.g. EPS-producing starter cultures) and its carbohydrate composition could be adjusted to meet the desired EPS-synthesis profiles. For instance, the synthesis of homopolysaccharides by glucansucrases or fructansucrases of LAB is boosted by high matrix sucrose content. In addition, future works could focus on investigations related to other bioactivities such as vitamin B profile and their changes during boza fermentation.

11.3 Togwa

Togwa is a Tanzanian beverage prepared from cereal porridge by liquefying the starch paste with malt followed by fermentation. The flour used to make the porridge is usually from maize, sorghum or finger millet, though cassava is a common ingredient as well. The addition of germinated sorghum or finger millet, with high amylolytic activity, induces saccharification of gelatinised starch and results in a sweetened slurry with reduced viscosity. Ripe togwa from a previous fermentation batch usually serves as a starter for the fermentation of the liquefied and saccharified slurry. Togwa is described as having an opaque, brownish appearance and a slightly floury mouthfeel [13]. A more detailed illustration of the manufacturing process for togwa is provided in Fig. 11.2 (Modified after [13, 14]).

Lactic acid bacteria and yeasts are the dominant microorganisms in togwa. Mugula et al. [14, 15] investigated some key fermentation parameters and identified dominant microorganisms from Tanzanian togwa samples. The final pH-values for togwa were between 3.1 and 3.5 and most of the LAB isolates were heterofermentative and roughly every third isolate produced dextran. Homofermentative *L. plantarum*, however, dominated fermentations and three of its isolates showed amylolytic activity

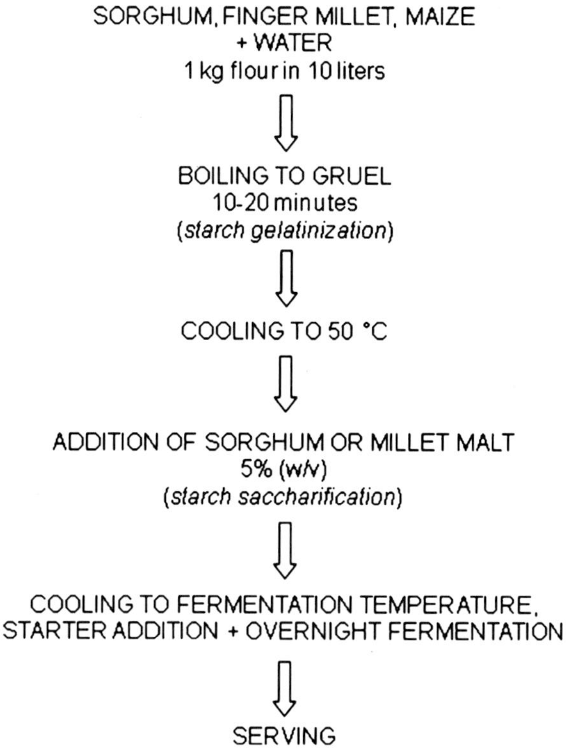

Fig. 11.2 Process chart for togwa preparation

as they hydrolysed starch [14]. The dominance of *L. plantarum* was probably because of its acid tolerance and its ability to effectively utilise different carbon sources. Obligative heterofermentative *Lactobacillus brevis, L. fermentum* and *W. confusa*, and homofermentative *Pediococcus pentosaceus* were also among dominant LAB strains in togwa [14]. *Issatchenkia orientalis* was the predominant yeast whereas *S. cerevisiae, Candida pelliculosa* and *C. tropicalis* were also among the most common yeast isolates (Table 11.2).

Hellström et al. [16] isolated a number of yeast strains from togwa fermentations, identified them using API and molecular techniques, and determined their phytase activity. Yeast strains with high phytase activity could serve as alternatives for mineral fortification or the addition of phytases in order to improve the mineral availability of fermented foods through elimination of the metal-chelating compound phytate. The lactic acid bacteria may also promote phytate elimination as the acidification increases the activity of endogenous cereal phytases [17].

In addition to the elimination of antinutritive compounds, the microflora of togwa may produce beneficial compounds such as B vitamins in food fermentations. Hjortmo et al. [18] investigated folate production in togwa fermented with previously isolated togwa-adapted yeasts [16]. Tetrahydrofolate and 5-methyl-tetrahydrofolate were the dominant forms. The concentration of the latter folate type substantially

Table 11.2 Dominant microbes isolated from togwa

	Organism	Relevant enzyme activities	Reference
LAB	*Weissella confusa*		[14]
	Lactobacillus fermentum		[14]
	Pediococcus pentosaceus		[14]
	Lactobacillus plantarum	Amylolytic	[14]
	Lactobacillus brevis		[14]
Yeasts	*Saccharomyces cerevisiae*		[14, 16]
	Issatchenkia orientalis (Candida krusei)	Phytase	[16]
	Pichia guilliermondii		[16]
	Kluyveromyces marxianus		[16]
	Hanseniaspora guilliermondii	Phytase	[16]
	Candida glabrata		[16]
	Pichia norvegensis		[16]
	Pichia burtonii		[16]
	Candida tropicalis		[14, 16]
	Candida pelliculosa (Pichia anomala)		[14, 16]

increased in a yeast-fermented maize base whereas the levels of tetrahydrofolate remained unchanged. The increase of total folate was highest with a strain of *Candida glabrata*, which resulted in a togwa with 23-fold folate content compared to the control incubation without yeast; the final folate concentration in the *C. glabrata* ferment was 700 ng per g dry matter [18].

Mugula et al. [19] investigated the proteolytic activity in togwa and verified that the added malt was the main source of proteolytic activity. The amino acids leucine, isoleucine, valine, ornithine, glutamic acid accumulated during togwa fermentations [19]. These amino acids are well known to influence flavour formation directly or indirectly as, for instance, Leu, Val, Ile serve as precursors for Strecker aldehydes that are typical amino-acid-derived flavour compounds in sourdoughs [20].

11.4 Mahewu

Mahewu is a traditional non-alcoholic spontaneously fermented beverage consumed by the Bantu people of Southern Africa, for example in Zimbabwe. Gruel prepared from maize or sorghum is cooled down and inoculated with sorghum malt (or wheat flour) and left to stand for 1–2 days. During the incubation period, saccharification and spontaneous fermentation take place. The dominant microflora of mahewu consists of lactic acid bacteria, mainly *Lactococcus lactis* subsp. *lactis* [21] as well as yeasts, although grain-associated fungi were also detected [22]. A major goal of using malt combined with fermentation is to reduce the amount of tannins, which are antinutritive polyphenols equipped with astringent sensory properties and therefore generally regarded as highly unfavourable compounds. Bvochora et al. [22] showed that the fermentation reduced the levels of tannins (specifically proanthocyanidins) by more than 50% during 36 h mahewu preparation. According to Oyewole [23] the pathogenic *Campylobacter*, *Escherichia coli* and *Shigella* bacteria could not survive in fermented mahewu products.

11.5 Bushera

Bushera is a fermented beverage traditionally consumed in Uganda; low-alcoholic bushera is produced using a fermentation period of 1 day whereas longer fermentation results in an alcoholic beverage that is not suitable for children [24]. Bushera is prepared from sorghum malt or millet malt. Its manufacture begins with the addition of malt flours of sorghum or millet to boiling water. During this boiling stage the starch gelatinises, with saccharification also likely taking place. After the boiling stage, the gruel is cooled down and fermentation is initiated by adding malt flour to the system. After 1 day of fermentation a beverage with low alcohol content is obtained.

11.6 Pozol

Pozol is a maize-based non-alcoholic beverage that originates from the Maya regions of South-Eastern Mexico and Guatemala (citations in [25]). The beverage is consumed by all family members, including infants, and it is a staple food of Mayans. The pozol beverage is made by suspending spontaneously fermented maize dough in water. Prior to fermentation, maize kernels are nixtamalised, i.e. cooked in lime (1%) water and the hulls are removed. After this stage a second boiling stage may be used or the nixtamalised dehulled kernels are directly washed and wet-milled [26]. The milled mass (called *masa* or *nixtamal*) is moulded into oval-shaped dough pieces, wrapped in banana leaves and left to stand at ambient temperature for one to several days. During this period of time, spontaneous fermentation takes place. At the start of incubation the pH of the dough is ~7 and during the first day it declines to below pH 5, while after 6 days the pH is below 4 [25, 27].

Lactic acid bacteria (mainly from the genera *Lactococcus, Lactobacillus, Leuconostoc* and *Weissella*) are numerically the dominant microbes (~10^9 cfu/g FW) whereas yeasts and fungi are also present in substantial numbers [26–28]. As in many other cereal fermentations amylolytic strains of LAB and yeasts as well as dextran-producing strains apparently have adapted and, thus, are significantly present in pozol [27]. Olivares-Illana et al. [29] characterised an inulosucrase active strain of *Leuconostoc citreum* isolated from pozol that synthesised inulin-type polymers. Of the fungus, all *Geotrichum* strains and nearly half of the yeast strains were able to utilise lactate as a sole carbon source, which demonstrates a good adaptation to sour conditions [27].

11.7 Chicha

Chicha is a fermented corn-based beverage consumed mainly in Southern Africa. The manufacturing technique is quite unique, because human saliva is used as a source of amylase enzyme, which converts the starch of corn into fermentable sugars. The primary microorganisms are yeasts, especially *S. cerevisiae*, bacteria from the species of *Lactobacillus*, *Leuconostoc* and *Acetobacter*, as well as moulds, such as *Aspergillus* [21].

11.8 Kishk

Kishk is a traditional fermented product, which is still consumed in Egypt, Syria and in many Arabic countries. It is made of dough, which contains salt and fermented skim milk. In addition, oat groats, oat flour, pre-cooked wheat groats (also called *burghol*) or wheat flour can be used as an ingredient of kishk. Most usually, the dough consists of a mixture of fermented milk and burghol. The dough is kneaded once a day for 6 days at 35 °C to hydrate and gelatinise the starch.

The fermentation of kishk is spontaneous and takes place during this period mainly by *L. plantarum*, *L. brevis*, *L. casei*, *Bacillus subtilis* and certain yeasts [21, 30].

After 6 days, the dough is moulded into small pellets, which are set into trays to dry in the sunshine for a week. The dried pellets are ground either traditionally by hand or in grain stores in industrial scale production. The beverage is produced by dissolving the ground pellets into hot water. Fermented kishk products have elevated levels of minerals and they are good sources of β-glucan and dietary fibre. Fermentation and drying also affects the shelf life of the product positively [30].

11.9 Kvass

Kvass is a fermented cereal beverage consumed in Eastern Europe. It is produced from rye malt, rye flour, stale rye bread and sucrose. In addition, barley malt or flour, as well as other cereals can be used as ingredients. A special type of kvass malt (fermented rye malt) is often used for the typical aroma characteristics of kvass. Kvass resembles Turkish boza with respect to the composition of the final product as well as the microflora [31, 32].

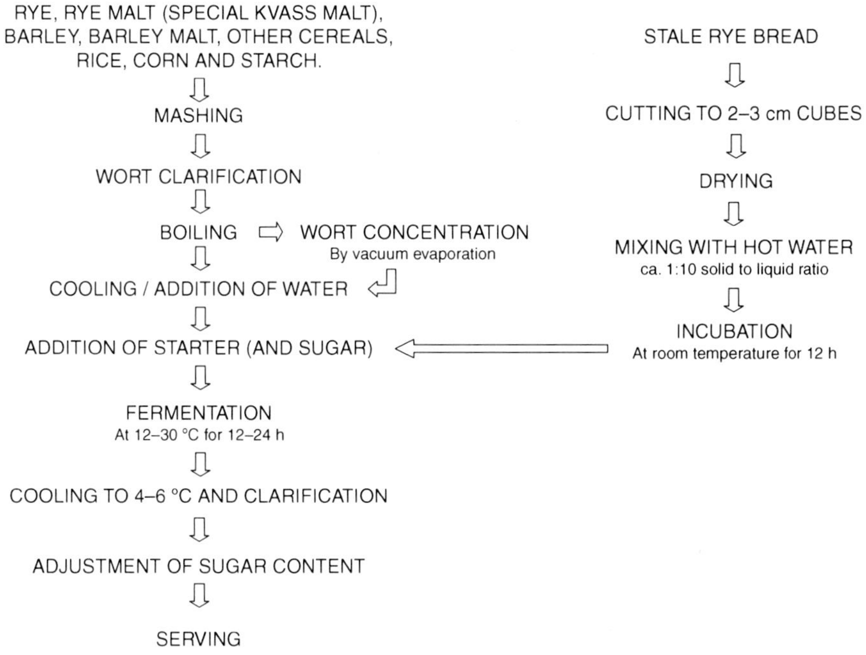

Fig. 11.3 Flow charts of two main kvass-making techniques (Modified according to [31, 32])

Baker's yeast (*S. cerevisiae*) is the predominant microorganism in kvass fermentations. In addition, lactic acid bacteria are added or already present as a latent infection in the malt or in the bakery/brewery environment [31–33]. Dlusskaya et al. [32] were among the first to characterise the microbes in kvass fermentations. They concluded that *L. casei*, *Lc. mesenteroides* and *Saccharomyces cerevisiae* were the dominating members of the kvass microflora.

Two main kvass-making techniques exist using either stale sourdough bread or malt as raw materials (Fig. 11.3). When fermentation originates from stale sourdough bread, all sugars necessary for yeast growth are derived from the bread-making process. In the second technique, gelatinised starch is cleaved by malt enzymes during mashing. In both techniques, the liquefied sugars are fermented by yeast and lactic acid bacteria at 12–30 °C for 12–24 h. During fermentation, no more than about 1% of the sugars are converted into lactic acid, CO_2 and ethanol [31, 32].

After fermentation, kvass is cooled down to 4–6 °C and clarified through filtration or centrifugation. Traditionally intensive clarification was not used and, thus, kvass contained high cell counts of viable yeasts and lactic acid bacteria, as no heating was used after fermentation. The sugar content might be adjusted by addition of another syrup after fermentation. As a rule of thumb, the quality of kvass can be determined through the amount of sugar added. In general, cheaper products contain a significantly higher amount of sugar than premium products, which are made with a bigger proportion of high-quality wort extract. The alcohol content of kvass varies and is influenced by the choice of ingredients, the mixture of microorganisms, the duration and temperature of fermentation, as well as by consumer preference. It can even be considered that kvass is spoiled if it contains more than 1% of ethanol [31, 32].

Before the nineteenth century, kvass production was limited only to private homes. Rye bread was soaked in water and spontaneously fermented with yeast and lactic acid bacteria. Several days of fermentation resulted in an alcohol concentration lower than 1.5 vol-%. Berries, fruits, herbs or honey can be added either before or after fermentation to adjust the flavour. Because of the high sugar and low alcohol concentration, and the absence of pasteurisation, kvass was only enjoyed for short periods of time and could not be stored [31].

In the nineteenth century, most kvass producers specialised in particular raw materials, leading to new flavours, such as apple, pear and peppermint kvass. At the same time the growing urbanisation and industrialisation made homemade kvass increasingly rare. In the 1960s, the success of the Coca-Cola Company inspired the Russian government to assign chemists in Moscow the task of developing an economical method for kvass production. In the early years of industrial production, the final product was filled and sold directly in large containers, which were commonly situated on trailers on street corners and in market places. Now the vast majority of kvass is sold in 1–3-l plastic bottles, which can be stored for 4–6 weeks [31].

11.10 Sourish Shchi

Sourish shchi is a Russian beverage with a low level of alcohol (below 2.5 vol-%). It was popular in Russia during the eighteenth and nineteenth centuries, but it is still in production. The origin of this beverage is most likely linked to the appearance of champagne in Russia as it required champagne bottles and a second fermentation. The six components usually involved in the preparation of sourish shchi (shchi is old Russian and means "six") were three kinds of malt: barley, rye, wheat; one kind of flour: wheat or rye; buckwheat and honey [34].

In the beginning of the process, malted and unmalted grains are mashed separately using hot water and then the two fractions are combined to produce a wort (Fig. 11.4). Fermentation starts spontaneously through the naturally occurring microbes found on grains, in the vessels and in the air. The wort is allowed to stand in a warm place for 12–24 h. Once the wort begins to ferment, there follows one or more inoculations with a "special sour" culture (from the sediment of the vessel from the previous fermentation). For example, Dankovtsev et al. [34] used *S. cerevisiae*, *S. uvarum* (*carlsbergensis*) and *S. minor* yeasts, as well as *L. brevis* as starter cultures. The majority of the spontaneous sour inoculum in the experiments of Dankovtsev et al. [34] consisted of wild and cultured yeast, such as the previously mentioned starter strains, as well as *Saccharomyces oviformis*, and strains from the genera *Torula* and *Candida*. The bacteria were identified as lactobacilli.

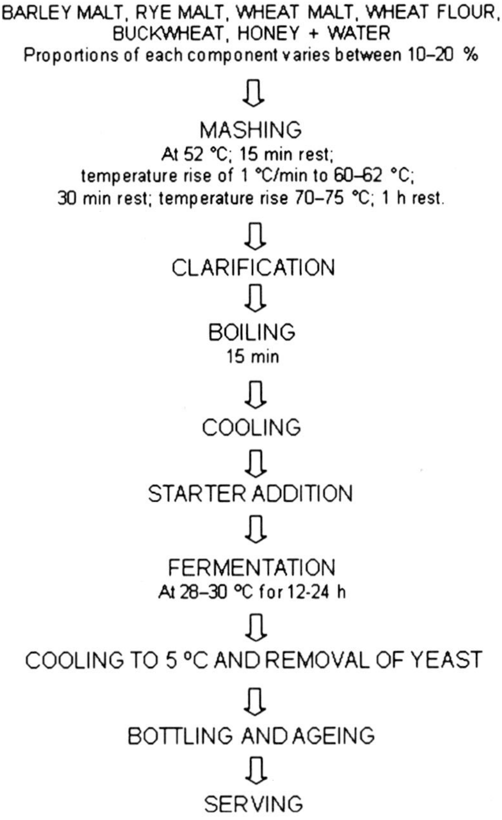

Fig. 11.4 Process chart for the preparation of sourish shchi

After the final inoculation, it is important to monitor the "culmination" of the fermentation, which is usually indicated by the thickness of the foam layer. When fermentation is almost complete, the wort is cooled down and the beverage bottled in champagne bottles. To ensure a sufficient level of carbon dioxide, sugar syrup or honey is added just before bottling. The bottles are allowed to end ferment at very low temperature. Sourish shchi differs from Russian kvass and other traditional beverages in its sparkling nature due to a high CO_2 concentration, the frothiness and its resemblance in taste to champagne [34].

11.11 Hulu-Mur

Hulu-mur is a Sudanese non-alcoholic beverage made of sorghum (*Sorghum bicolor*) and preferably using the cultivar Fetarita [35]. In the preparation of Hulu-mur, both malted and unmalted sorghum grains are used (Fig. 11.5). First unmalted sorghum dough is prepared and par-baked after which malted sorghum slurry and a starter is added. Sorghum sourdough (used for kisra bread making) is used as a

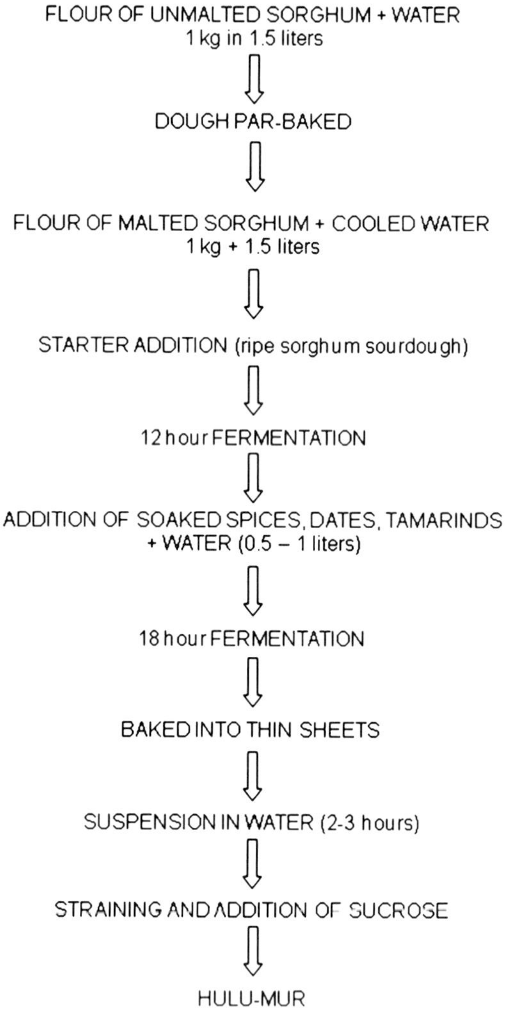

Fig. 11.5 Process chart for hulu-mur preparation

starter. During the first fermentation stage, acidification takes place and the pH drops from 5.9 to 3.7 [35]. After the first fermentation other ingredients (spices, dates, and tamarinds) are added with some water and the mixture is allowed to ferment a second time. Finally, the fully fermented dough is baked into small and thin sheets, and the dry sheets are suspended in water to prepare the hulu-mur drink. Relatively little research data on hulu-mur exist though Mahgoub et al. [36] investigated the micronutrient content and its changes in hulu-mur.

References

1. Arici M, Daglioglu O (2002) Boza: a lactic acid fermented cereal beverage as a traditional Turkish food. Food Rev Int 18:39–48
2. Hayta M, Alpaslan M, Köse E (2001) The effect of fermentation on viscosity and protein solubility of Boza, a traditional cereal-based fermented Turkish beverage. Eur Food Res Technol 213:335–337
3. Genç M, Zorba M, Ova G (2002) Determination of rheological properties of boza by using physical and sensory analysis. J Food Eng 52:95–98
4. Hancioğlu Ö, Karapinar M (1997) Microflora of Boza, a traditional fermented Turkish beverage. Int J Food Microbiol 35:271–274
5. Gotcheva V, Pandiella SS, Angelov A, Roshkova Z, Webb C (2001) Monitoring the fermentation of the traditional Bulgarian beverage boza. Int J Food Sci Technol 36:129–134
6. Botes A, Todorov SD, von Mollendorff JW, Botha A, Dicks LMT (2007) Identification of lactic acid bacteria and yeast from boza. Process Biochem 42:267–270
7. Gotcheva V, Pandiella SS, Angeloc A, Roshkova ZG, Webb C (2000) Microflora identification of the Bulgarian cereal-based fermented beverage boza. Process Biochem 36:127–130
8. Todorov SD, Dicks LMT (2006) Screening for bacteriocin-producing lactic acid bacteria from boza, a traditional cereal beverage from Bulgaria: comparison of the bacteriocins. Process Biochem 41:11–19
9. Petrova P, Emanuilova M, Petrov K (2010) Amylolytic *Lactobacillus* strains from Bulgarian fermented beverage boza. Z Naturforsch 65c:218–224
10. Todorov SD, Dicks LMT (2005) Pediocin ST18, an anti-listerial bacteriocin produced by *Pediococcus pentosaceus* ST18 isolated from boza, a traditional cereal beverage from Bulgaria. Process Biochem 40:365–370
11. Todorov SD, Dicks LMT (2004) Characterization of mesentericin ST99, a bacteriocin produced by *Leuconostoc mesenteroides* subsp. dextranicum ST99 isolated from boza. J Ind Microbiol Biotechnol 31:323–329
12. Zorba M, Hancioglu O, Genc M, Karapinar M, Ova G (2003) The use of starter cultures in the fermentation of boza, a traditional Turkish beverage. Process Biochem 38:1405–1411
13. Kitabatake N, Gimbi DM, Oi Y (2003) Traditional non-alcoholic beverage, Togwa, in East Africa, produced from maize flour and germinated finger millet. Int J Food Sci Nutr 54:447–455
14. Mugula JK, Nnko SA, Narvhus JA, Sørhaug T (2003) Microbiological and fermentation characteristics of togwa, a Tanzanian fermented food. Int J Food Microbiol 80:187–199
15. Mugula JK, Narvhus JA, Sørhaug T (2003) Use of starter cultures of lactic acid bacteria and yeasts in the preparation of togwa, a Tanzanian fermented food. Int J Food Microbiol 83:307–318
16. Hellström AM, Vázques-Juárez R, Svanberg U, Andlid TA (2010) Biodiversity and phytase capacity of yeasts isolated from Tanzanian togwa. Int J Food Microbiol 136:352–358
17. Reale A, Mannina L, Tremonte P, Sobolev AP, Succi M, Sorrentino E, Coppola R (2004) Phytate degradation by lactic acid bacteria and yeasts during the wholemeal dough fermentation: a 31P NMR study. J Agric Food Chem 52:6300–6305
18. Hjortmo SB, Hellström AM, Andlid TA (2008) Production of folates by yeasts in Tanzanian fermented togwa. FEMS Yeast Res 8:781–787
19. Mugula JK, Sørhaug T, Stepaniak L (2003) Proteolytic activities in togwa, a Tanzanian fermented food. Int J Food Microbiol 84:1–12
20. Hansen Å, Schieberle P (2005) Generation of aroma compounds during sourdough fermentation: applied and fundamental aspects. Trends Food Sci Technol 16:85–94
21. Blandino A, Al-Aseeri ME, Pandiella SS, Cantero D, Webb C (2003) Cereal-based fermented foods and beverages. Food Res Int 36:527–543

22. Bvochora JM, Reed JD, Read JS, Zvauya R (1999) Effect of fermentation processes on proanthocyanidins in sorghum during preparation of Mahewu, a non-alcoholic beverage. Process Biochem 35:21–25
23. Oyewole OB (1997) Lactic fermented foods in Africa and their benefits. Food Control 8:289–297
24. Muyanja CM, Narvhus JA, Treimo J, Langsrud T (2003) Isolation, characterisation and identification of lactic acid bacteria from bushera: a Ugandan traditional fermented beverage. Int J Food Microbiol 80:201–210
25. Wacher C, Canas A, Cook PE, Barzana E, Owens JD (1993) Sources of microorganisms in pozol, a traditional Mexican fermented maize dough. World J Microbiol Biotechnol 9:269–274
26. Wacher C, Cañas A, Barzana E, Lappe P, Ulloa M, Owens JD (2000) Microbiology of Indian and Mestizo pozol fermentations. Food Microbiol 17:251–256
27. Nuraida L, Wacher C, Owens JD (1995) Microbiology of pozol, a Mexican fermented maize dough. World J Microbiol Biotechnol 11:567–571
28. Escalante A, Wacher C, Farrés A (2001) Lactic acid bacterial diversity in the traditional Mexican fermented dough pozol as determined by 16S rDNA sequence analysis. Int J Food Microbiol 64:21–31
29. Olivares-Illana V, Wacher-Odarte C, Le Borgne S, López-Munguía A (2002) Characterization of a cell-associated inulosucrase from a novel source: a *Leuconostoc citreum* strain isolated from pozol, a fermented corn beverage of Mayan origin. J Ind Microbiol Biotechnol 28:112–117
30. Tamine AY, Muir DD, Khaskheli M, Barcley MNI (2000) Effect of processing conditions and raw materials on the properties of Kishk. 1. Composition and microbial qualities. Lebensm Wiss Technol 33:444–451
31. Bahns P, Michel R, Becker T, Zarnkow M (2010) Kvass – from homebrew to industrial production. Brauwelt Int 2:96–100
32. Dlusskaya E, Jänsch A, Schwab C, Gänzle MG (2008) Microbial and chemical analysis of a kvass fermentation. Eur Food Res Technol 27:261–266
33. Eisenberg A (1908) Process for the production of kvass, US Patent 938374
34. Dankovtsev AV, Vostrikov SV, Markina NS (2002) Fermentation studies on traditional Russian drink "Sourish Shchi". J Inst Brew 108:474–477
35. Agab MA (1985) Fermented food products 'Hulu Mur' drink made from sorghum bicolour. Food Microbiol 2:147–155
36. Mahgoub SEO, Ahmed BM, Ahmed MMO, El Agib ENAA (1999) Effect of traditional Sudanese processing of kisra bread and hulu-mur drink on their thiamine, riboflavin and mineral contents. Food Chem 67:129–133

Chapter 12
Perspectives

Michael Gänzle and Marco Gobbetti

12.1 Microbial Ecology of Sourdough

The characterization of microbiota of a large number of industrial and artisanal sourdoughs has greatly promoted our understanding of the microbial ecology of sourdoughs, and resulted in the description of more than a dozen new species ([1–3], Chap. 5). The origin of sourdough microbiota, however, remains unclear. Environmental contamination appears to play a role for yeasts found in sourdough [4]. However, the lactic microbiota of back-slopped sourdoughs are markedly different from the microbiota of spontaneously fermenting sourdoughs or the raw materials, excluding the raw materials as a significant source of most lactobacilli associated with sourdough. Moreover, back-slopped sourdoughs remain the only known source of several *Lactobacillus* species. Significant changes in the microbiota of sourdoughs were found during long-term propagation in artisanal bakeries [4]. In other cases, sourdough microbiota remained stable at the strain level over several decades of propagation. Genome sequencing of *Lactobacillus sanfranciscensis* recently provided an unprecedented snapshot of microbial evolution in the sourdough environment. Whole genome sequencing of two strains isolated from the same industrial sourdough before and after 18 years of continuous back-slopping revealed that the genome underwent only minimal changes during that time [5]. Human use of sourdough is thus unlikely to allow the evolution of specialized microbial species, or sourdough-adapted lineages of *Lactobacillus* species.

M. Gänzle (✉)
Department of Agricultural, Food and Nutritional Science, University of Alberta, Edmonton, Canada
mgaenzle@ualberta.ca

M. Gobbetti
Department of Soil, Plant and Food Science, University of Bari Aldo Moro, Bari, Italy

M. Gobbetti and M. Gänzle (eds.), *Handbook on Sourdough Biotechnology*,
DOI 10.1007/978-1-4614-5425-0_12, © Springer Science+Business Media New York 2013

Sourdough microbiota and the lactic microbiota of the intestine of humans and animals show remarkable overlap, indicating that intestinal lactobacilli are a potential source of sourdough-adapted strains [1]. The substrate availability of the upper intestine of mammals that consume cereal-based foods, the major site of colonization by lactobacilli in swine, poultry and rodents [6], is remarkably similar to sourdough. Sucrose and maltose are the major carbon sources and carbohydrate metabolism by sucrose phosphorylase or levansucrase and maltose phosphorylase contributes to the ecological fitness of lactobacilli. The intestinal origin of sourdough strains was recently demonstrated for *L. reuteri*, which is both a gut symbiont and a stable member of sourdough microbiota [1, 7]. Several *L. reuteri* lineages have evolved to become specific for their respective hosts, i.e. humans, swine, poultry or rodents [7]. Multi-locus sequence analysis and analysis of host-specific physiological and genetic traits assigned five *L. reuteri* isolates from back-slopped sourdoughs to rodent- or human-specific lineages. Comparative genome hybridization revealed that the sourdough isolate *L. reuteri* LTH2584 is genetically highly related to the rodent isolate *L. reuteri* 100-23. Taken together, these results demonstrate that sourdough isolates of *L. reuteri* that persisted in a back-slopped sourdough for over 50,000 generations initially originated from intestinal microbiota [8]. Likewise, *L. rossiae*, *L. pontis* and *L. fermentum* were described as stable members of intestinal and sourdough microbiota [9], and sourdough isolates of these species may also originate from the intestine of animals.

The overlap of intestinal and sourdough microbiota allows the development of probiotic cereal products employing probiotic lactobacilli as starter cultures. Vellie, oat bran fermented with probiotic lactobacilli, provides a conceptual template for such products [10]. Mesophilic lactobacilli are unlikely to originate from the intestinal tract of mammals. However, *L. sanfranciscensis* was isolated from the intestinal tract of fruit flies [11], which may provide a ubiquitous reservoir for the species.

Wheat and rye sourdoughs exhibit similar microbiota; their composition and activity depends mainly on the process conditions (see Chap. 5). Durum wheat sourdough, commonly used for bread baking in Southern Italy, was suggested to select for obligate heterofermentative lactic acid bacteria owing to the higher levels of maltose, sucrose and amino acids [12]. Moreover, cereal fermentations in Africa and South Asia predominantly use sorghum, millet, corn, rice, or teff as raw materials [13]. The dominant species of lactic acid bacteria in fermentations with these substrates only partially overlap with wheat and rye sourdoughs [14, 15]. The higher ambient temperature in tropical climates selects for thermophilic microbiota. Substrate-derived factors such as the carbohydrate availability and the presence of antimicrobial phenolic compounds also contribute to the establishment of divergent and substrate-specific microbiota. However, the specific effect of the raw materials on the microbial ecology of sourdough is currently not fully understood. The industrialization of food production in developing countries will result in a more standardized fermentation process and the production of starter cultures (see e.g. [16]), paralleling the development in the European baking industry in the past decades. This development will also lead to an improved understanding of the microbial ecology of cereal fermentations in tropical climates.

12.2 Sourdough and Product Quality

Current knowledge of the effect of carbohydrate metabolism, exopolysaccharide production, and amino acid conversion allows the targeted selection of starter cultures for improved bread quality (see Chap. 7). The availability of genome sequences for sourdough-adapted lactobacilli [7, 17, 18] not only furthers the understanding of the microbial ecology of sourdoughs, but also facilitates the elucidation of additional metabolic pathways converting amino acids, lipids, and phenolic compounds and their effect on bread quality. The elucidation of the effect of individual metabolic pathways on bread quality is supported by the development of tools related to comparative genomics, proteomic and transcriptomic analyses, and the development of novel tools for metabolic re-engineering of sourdough lactic acid bacteria [7, 8, 18]. A more comprehensive analysis of the effect of strain-specific metabolic traits on product quality in combination with a decrease of the expense associated with genomic analyses for starter cultures will facilitate *in silico* strain selection as a complement for the evaluation of cultures in application trials.

12.3 Sourdough and Nutrition

The comparison between baked goods made with baker's yeast or chemical leavening and those made with sourdoughs clearly evidences that sourdough bread is more digestible and has a higher bioavailability of minerals because substantial degradation of cereal components (e.g. proteins and phytate) occurs during fermentation [19]. This is already an undisputable reason for the preferred use of sourdough. However, public policy and industrial product development in developed countries are no longer predominantly related to the need to meet basic nutritional requirements by an increased bioavailability of macro- and micronutrients. In contrast, nutritional intervention or the formulation of functional food aims to prevent or to mitigate chronic diseases that are highly prevalent in developed countries. Chronic diseases that are substantially influenced by diet include inflammatory bowel disease, diabetes, celiac disease as well as cardiovascular diseases, obesity and the metabolic syndrome. Consequently, the formulation of functional food products has become an important component in industrial product development and product diversification. Examples include gluten-free products, products with reduced sodium content or an increased level of dietary fibre, and food enriched with antioxidative or bioactive compounds.

Wheat flour-based products such as bread, biscuits and breakfast cereals are characterized by relatively high values of glycaemic index and insulin index. Sourdough fermentation or the addition of soluble fibres represent the most promising tools to decrease starch digestibility and, consequently, the values of glycaemic index and insulin index. Various mechanisms are responsible for this effect but these remain to be elucidated in depth [20, 21]. Sourdough fermentation represents an indispensable biotechnology for making wholegrain bread, especially rye bread, and it may be used to modify fibre-rich cereal ingredients such as bran and germ, which improves the technological functionality of these components.

During the last decade, a number of studies [22–26] showed that fungal proteases or malt in combination with sourdough can result in complete hydrolysis of gluten during long-term fermentation. As recently shown by two in vivo challenges, wheat flour fermented with selected sourdough lactobacilli and fungal proteases is tolerated by celiac patients under remission. Sourdough fermentation also improves the sensory acceptability of naturally gluten-free products and may prevent the moderate cross-contamination by gluten.

Microbial metabolism during sourdough fermentation may also favour the synthesis, release, and/or the bioavailability of a number of functional compounds such as vitamins, phytochemicals, pre-biotic exopolysaccharides and bioactive peptides. For example, antioxidant peptides (e.g. lunasin) are liberated from cereal proteins during sourdough fermentation. These peptides may have promising preventive activities towards oxidative stresses that are associated with degenerative aging diseases (e.g. cancer and arteriosclerosis) [27].

In conclusion, the use of sourdough has evolved as an important tool in the development of functional baked goods (see Chaps. 10 and 11). Our knowledge on the intricate relationship between diet and chronic disease continues to increase. This knowledge allows one to take advantage of the formation or modification of bioactive compounds during sourdough fermentation to further expand the toolset for development of sourdough products with specific nutritional functionality.

12.4 Industrial and Artisanal Use of Sourdough

In European countries, 30–50% of the bread production includes the use of sourdough or sourdough products. In North America, sourdough is a small but rapidly growing segment of the market for baking improvers. The traditional use of sourdough as a leavening agent in artisanal bakeries retains its place in small or medium-sized, specialized bakeries. Particularly the production of regional specialities, for example Panettone, Pumpernickel or San Francisco Sourdough Bread, continues to rely on the traditional use of sourdough as a leavening agent. However, for a majority of industrial applications, the technological aim of sourdough use has shifted from its traditional use as a leavening agent to use as a dough acidifier or baking improver (see Chap. 1). This development is supported by the industry's efforts to develop "clean label" products. The use of sourdough allows bread production from the ingredients flour and water with optional addition of yeast and salt, and thus can replace numerous other ingredients to improve bread quality and shelf life. Moreover, in-house sourdough fermentation can achieve substantial cost savings. Ingredients, for example malt, emulsifiers, hydrocolloids, or enzymes, are replaced by the expertise and equipment needed for sourdough fermentation. However, few industrial bakeries master the automated and large-scale fermentation of sourdough. One of the limiting factors is the availability of equipment allowing fermentation control for use of sourdough as a leavening agent. Equipment for industrial sourdough fermentation is typically designed to carry out (semi-)automated batch fermentations according to the fermentation scheme of traditional procedures, and is thus incompatible with large-scale and continuous bread production [28]. Consequently, sourdough fermentation is increasingly carried out by specialized suppliers to the baking industry [29] . The use of sourdough in bakeries employs stabilized, usually dried, preparations that are shelf stable. This second line of sourdough products in addition to traditional fermentations allows for product innovation to match the specialized need of individual customers. Examples include ready-to-use, active sponge doughs, dried sourdough products enriched with exopolysaccharides or flavour compounds derived from the Maillard reaction, and starter cultures selected for specific metabolic traits for improved bread quality.

It is noteworthy that lyophilized starter cultures for direct inoculation of bread dough have not found widespread commercial use in baking applications, in contrast to the predominant use of starter cultures in meat and dairy fermentations. Freeze-dried cultures fail to develop the required metabolic activity in straight dough processes, and thus require revitalization in a pre-ferment or sponge dough prior to use. In-house propagation of sourdough with occasional restoration of the desired fermentation microbiota with cereal-based freeze-dried starter preparations is thus a preferred option for many bakeries.

The increasing use of sourdough as a baking improver also allows the inclusion of non-conventional organisms and raw materials. Continuous propagation of sourdough invariably selects for fermentation microbiota consisting of lactic acid bacteria and yeasts. The industrial production of baking improvers, however, can be started with other food-grade organisms that grow in cereal substrates and maintain dominance over one or a few stages of fermentations. Bifidobacteria [30], propionibacteria [31], fungi and acetic acid bacteria [32] all grow in cereal substrates and have been employed in experimental cereal fermentations. Moreover, traditional cereal fermentations employed in Africa, Asia, and Latin America for production of steamed bread, beverages, porridges, vinegar, or condiments provide a source of fermentation organisms that are highly adapted to cereal substrates. The metabolic potential of these organisms vastly differs from sourdough lactic acid bacteria and their use allows novel functionalities for baked products.

References

1. Vogel RF, Knorr R, Müller MRA, Steudel U, Gänzle MG, Ehrmann MA (1999) Non-dairy lactic fermentations: the cereal world. Antonie van Leeuwenhoek 76:403–411
2. Meroth CB, Walter J, Hertel C, Brandt MJ, Hammes WP (2003) Monitoring the bacterial population dynamics in sourdough fermentation processes by using PCR- denaturing gradient gel electrophoresis. Appl Environ Microbiol 69:475–482
3. De Vuyst L, Neysens P (2005) The sourdough microflora: biodiversity and metabolic interactions. Trends Food Sci Technol 16:43–56
4. Minervini F, Lattanzi A, De Angelis M, Di Cagno R, Gobbetti M (2012) Artisan bakery or laboratory propagated sourdoughs: influence on the diversity of lactic acid bacterium and yeast microbiotas. Appl Environ Microbiol. doi:10.1128/AEM.00572-12

5. Ehrmann MA, Behr J, Böcker G, Vogel RF (2011) The genome of *L. sanfranciscensis* after 18 years of continuous propagation. In: Abstract, presented at the 10th symposium on lactic acid bacteria, Egmond aan Zee. Accessed via www.lab10.org on 20 June 2012
6. Walter J (2008) Ecological role of lactobacilli in the gastrointestinal tract: implications for fundamental and biomedical research. Appl Environ Microbiol 74:4985–4996
7. Frese SA, Benson AK, Tannock GW, Loach DM, Kim J, Zhang M, Oh PL, Heng NC, Patil PB, Juge N, Mackenzie DA, Pearson PM, Lapidus A, Dalin E, Tice H, Goltsman E, Land M, Hauser L, Ivanova N, Kyrpides NC, Walter J (2011) The evolution of host specialization in the vertebrate gut symbiont *Lactobacillus reuteri*. PLoS Genet 7:e1001314
8. Su MSW, Oh PL, Walter J, Gänzle MG (2012) Phylogenetic, genetic, and physiological analysis of sourdough isolates of *Lactobacillus reuteri*: food fermenting strains are of intestinal origin. Appl Environ Microbiol 78:6777–6780.
9. Hammes WP, Hertel C (2006) The genera *Lactobacillus* and *Carnobacterium*. Prokaryotes 4:320–403
10. Salovaara HO (2006) Cereal-based alternatives to dairy snacks of yogurt-type. Paper presented at World Grain Summit: foods and beverages, San Francisco, 17–20 Sept 2006. http://www.aaccnet.org/meetings/Documents/Pre2009Abstracts/2006Abstracts/S-68.htm
11. Groenewald WH, Van Reenen CA, Todorov SD, Du Troit M, Witthuhn RC, Holzapfel WH, Dicks LMT (2006) Identification of lactic acid bacteria from vinegar flies based on phenothpic and genotypic characteristics. Am J Enol Vitic 57:519–525
12. Minervini F, Di Cagno R, Lattanzi A, De Angelis M, Antonielli L, Cardinali G, Cappelle S, Gobbetti M (2012) Lactic acid bacterium and yeast microbiotas of 19 sourdoughs used for traditional/typical Italian breads: interactions between ingredients and microbial species diversity. Appl Environ Microbiol 78:1251–1264
13. Nout MJ (2009) Rich nutrition from the poorest- Cereal fermentations in Africa and Asia. Food Microbiol 26:685–692
14. Vogelmann SA, Seitter M, Singer U, Brandt MJ, Hertel C (2009) Adaptability of lactic acid bacteria and yeast to sourdoughs prepared from cereals, pseudo-cereals and cassava and the use of competitive strains as starter cultures. Int J Food Microbiol 130:205–212
15. Sekwati-Monang B, Gänzle MG (2011) Microbiological and chemical characterisation of ting, a sorghum-based sourdough product from Botswana. Int J Food Microbiol 150:115–121
16. Keeratipibul S, Luangsakul N, Otsuka S, Hatano Y, Tanasupawat S (2010) Application of the Chinese Steamed bun starter dough (CSB-SD) in breadmaking. J Food Sci 75:596–604
17. Liu M, Nauta A, Francke C, Siezen RJ (2008) Comparative genomics of enzymes in flavor-forming pathways from amino acids in lactic acid bacteria. Appl Environ Microbiol 74:4590–4600
18. Vogel RF, Pavlovic M, Ehrmann MA, Wiezer A, Liesegang H, Offschanka S, Voget S, Angelov A, Böcker G, Liebl W (2011) Genomic analysis reveals *Lactobacillus sanfranciscensis* as stable element in traditional sourdoughs. Microbiol Cell Fact 10(Suppl 1):S6
19. Gobbetti M, De Angelis M, Corsetti A, Di Cagno R (2005) Biochemistry and physiology of sourdough lactic acid bacteria. Trends Food Sci Technol 16:57–69
20. De Angelis M, Damiano N, Rizzello CG, Cassone A, Di Cagno R, Gobbetti M (2009) Sourdough fermentation as a tool for the manufacture of low-glycemic index white wheat bread enriched in dietary fibre. Eur Food Res Technol 229:593–601
21. Maioli M, Pes GM, Sanna M, Cherchi S, Dettori M, Manca E, Farris GA (2008) Sourdough-leavened bread improves postprandial glucose and insulin plasma levels in subjects with impaired glucose tolerance. Acta Diabetologica 45:91–96
22. Di Cagno R, De Angelis M, Lavermicocca P, De Vincenzi M, Giovannini C, Faccia M, Gobbetti M (2002) Proteolysis by sourdough lactic acid bacteria: effects on wheat flour protein fractions and gliadin peptides involved in human cereal intolerance. Appl Environ Microbiol 68:623–633
23. Rizzello CG, De Angelis M, Di Cagno R, Camarca A, Silano M, Losito I, De Vincenzi M, De Bari MD, Palmisano F, Maurano F, Gianfrani C, Gobbetti M (2007) Highly efficient gluten degradation by lactobacilli and fungal proteases during food processing: new perspectives for celiac disease. Appl Environ Microbiol 73:4499–4507

24. Di Cagno R, Barbato M, Di Camillo C, Rizzello CG, De Angelis M, Giuliani G, De Vincenzi M, Gobbetti M, Cucchiara S (2010) Gluten-free sourdough wheat baked goods appear safe for young celiac patients: a pilot study. J Ped Gastroent Nutr 51:777–783
25. Greco L, Gobbetti M, Auricchio R, Di Mase R, Landolfo F, Paparo F, Di Cagno R, De Angelis M, Rizzello CG, Cassone A, Terrone G, Timpone L, D'Aniello M, Maglio M, Troncone R, Auricchio S (2011) Safety for patients with celiac disease of baked goods made of wheat flour hydrolyzed during food processing. Clin Gastroenterol Pathol 9:24–29
26. De Angelis M, Cassone A, Rizzello CG, Gagliardi F, Minervini F, Calasso M, Di Cagno R, Francavilla R, Gobbetti M (2010) Gluten-free pasta made of *Triticum turgidum* L. var. *durum*: mechanisms of epitopes hydrolysis by peptidases of sourdough lactobacilli. Appl Environ Microbiol 75:508–518
27. Coda R, Rizzello CG, Pinto D, Gobbetti M (2012) Selected lactic acid bacteria synthesize antioxidant peptides during sourdough fermentation of cereal flours M. Appl Environ Microbiol 4:1087–1096
28. Böcker G (2006) Grundsätze von Anlagen für Sauerteig. In: Brandt MJ, Gänzle MG (eds) Handbuch Sauerteig, 6th edn. Behr's Verlag, Hamburg, pp 329–352
29. Brandt MJ (2007) Sourdough products for convenient use in baking. Food Microbiol 24:161–164
30. Sanz-Penella JM, Laparra JM, Sanz Y, Haros M (2012) Assessment of iron bioavailability in whole wheat bread by addition of phytase-producing bifidobacteria. J Agric Food Chem 60:3190–3195
31. Kariluoto S, Edelmann M, Herranen M, Lampi AM, Shmelev A, Salovaara H, Korhola M, Piironen V (2010) Production of folate by bacteria isolated from oat bran. Int J Food Microbiol 143:41–47
32. Haruta S, Ueno S, Egawa I, Hashiguchi K, Fujii A, Nagano M, Ishii M, Igarashi Y (2006) Succession of bacterial and fungal communities during a traditional pot fermentation of rice vinegar assessed by PCR-mediated denaturing gradient gel electrophoresis. Int J Food Microbiol 109:79–87

Index

M. Gobbetti and M. Gänzle (eds.), *Handbook on Sourdough Biotechnology*,
DOI 10.1007/978-1-4614-5425-0, © Springer Science+Business Media New York 2013

C

D